A Functional Biology of Scyphozoa

A Functional Biology of Scyphozoa

Mary N. Arai

Professor
Department of Biological Sciences
Faculty of Science
University of Calgary
Calgary, Alberta, Canada
and
Senior Volunteer Investigator
Pacific Biological Station
Nanaimo, British Columbia, Canada

CHAPMAN & HALL
London · Weinheim · New York · Tokyo · Melbourne · Madras

Published by Chapman & Hall, 2–6 Boundary Row, London SE1 8HN

Chapman & Hall, 2–6 Boundary Row, London SE1 8HN, UK

Chapman & Hall GmbH, Pappelallee 3, 69469 Weinheim, Germany

Chapman & Hall USA, 115 Fifth Avenue, New York, NY 10001, USA

Chapman & Hall Japan, ITP-Japan, Kyowa Building, 3F, 2–2–1 Hirakawacho, Chiyoda-ku, Tokyo 102, Japan

Chapman & Hall Australia, 102 Dodds Street, South Melbourne, Victoria 3205, Australia

Chapman & Hall India, R. Seshadri, 32 Second Main Road, CIT East, Madras 600 035, India

First edition 1997

Typeset in Plantin 10/12 by Florencetype Ltd, Stoodleigh, Devon

ISBN 0 412 45110 7

A Catalogue record for this book is available from the British Library

Library of Congress Catalog Card Number: 96-84890

∞ Printed on permanent acid-free text paper, manufactured in accordance with ANSI/NISO Z39.48-1992 and ANSI/NISO Z39.48-1984 (Permanence of Paper).

To my family, past and present

Contents

Acknowledgements

This book could not have been written without the advice and help of many friends and colleagues whom I thank. The staff of the library at the Pacific Biological Station, G. Miller, P. Olson and M. Hawthornthwaite have been of invaluable assistance in obtaining the more obscure literature. Z. Kabata, A. Brinckmann-Voss and M. Reimer have assisted with translations of Russian and German papers. Literature searches and bibliographic work by E. Skinner were partially funded under Natural Sciences and Engineering Research Council grant A2007. Funds for library work and invaluable free time were provided by a Killam Resident Fellowship in the fall of 1993. M.J. Cavey and H.D. Arai have assisted with computer enhancement of illustrations. R.H. Brewer, D.R. Calder, P.F.S. Cornelius, J.H. Costello, D.G. Fautin, W.M. Hamner, P. Kremer, R.J. Larson, L.M. Passano, J.E. Purcell and J.N.C. Whyte read and commented on portions of the manuscript. R.J. LeBrasseur read the entire manuscript, providing stimulating comment and the irreverent perspective of a non-specialist. Finally I want to thank my family for their patience and encouragement while I devoted so much of my time to this project.

Preface

Scyphozoa have attracted the attention of many types of people. Naturalists watch their graceful locomotion. Fishermen may dread the swarms which can prevent fishing or eat larval fish. Bathers retreat from the water if they are stung. People from some Asiatic countries eat the medusae. Comparative physiologists examine them as possibly simple models for the functioning of various systems. This book integrates data from those and other investigations into a functional biology of scyphozoa. It will emphasize the wide range of adaptive responses possible in these morphologically relatively simple animals.

The book will concentrate on the research of the last 35 years, partly because there has been a rapid expansion of knowledge during that period, and partly because much of the previous work was summarized by books published between 1961 and 1970.

Bibliographies of papers on scyphozoa were included in Mayer (1910) and Kramp (1961). Taxonomic diagnoses are also included in those monographs, as well as in a monograph on the scyphomedusae of the USSR published by Naumov (Naumov, 1961). Most importantly, a generation of scyphozoan workers has used as its 'bible' the monograph by F.S. Russell (1970) *The Medusae of the British Isles*. In spite of its restrictive title, his book reviews most of the information on the biology of scyphozoa up to that date.

The expansion of knowledge since 1970 has not been even. It has been especially driven by the instances in which scyphozoa have impinged on human activities. We know more about the effects of cnidae on humans than on natural prey. There have been a number of studies on the effects of scyphozoa on fisheries, but we know very little about predation on scyphozoa. A great deal of new information on *Pelagia noctiluca* was generated because a 'bloom' in the Mediterranean Sea in the early 1980s affected tourism.

In other cases, however, the emphasis has indeed been on how a relatively simple animal is able to carry out its functions. The ways in which a simple nerve system transmits information have been examined, with particular reference to the properties of the bidirectional synapses in the nerve net. The ability of medusae to migrate horizontally using information from the sun has been established, although we do not yet understand the mechanisms. In this book I will pull together the diffuse literature, and give as balanced a view as possible of the biology of the group.

With the emphasis on functional biology, neither taxonomy nor morphology are extensively dealt with. However, Chapter 1 briefly introduces the design of each of the orders, and the Appendix lists by family those species that are mentioned in the text. Morphological structures are described in the context of their functions.

Terminology has been kept as simple as possible and is defined as it arises. Definitions are indicated in bold type in the index. Where greater detail on these subjects is desired the reader is referred to Russell (Russell, 1970), and to the review by Franc in the Traité de Zoologie (Franc, A., 1993).

Mary Needler Arai
Calgary
May 1996

1 *Design and relationships*

1.1 INTRODUCTION

The Scyphozoa constitute one of the four classes of living cnidaria. The members of the phylum Cnidaria are characterized by the possession of intrinsic cnidae: intracellular organelles consisting of a capsule and an attached hollow thread. Cnidarian animals consist of two epithelial body layers, the epidermis and gastrodermis, separated by a gelatinous connective tissue, the mesoglea. These three layers form a sac around the gastrovascular cavity or coelenteron which usually has a single opening, the mouth. Typically tentacles form a ring around the margin of an oral disc surrounding the mouth.

Cnidarians exhibit two adult body forms. One form, the medusa or jellyfish, is typically solitary, pelagic, with two saucer shapes of the three layers fused at the margins to form a bell with the mouth on the undersurface (subumbrella). The mesoglea is relatively thick. The other form, the polyp, is solitary or colonial, typically attached to a substrate with the mouth upwards. The mesoglea is relatively thin.

Other possible life history stages include a simple larva, the planula, and buds and cysts. The typical cnidarian life cycle includes a planula which develops into a polyp, which in turn asexually produces medusae which reproduce sexually (Figures 1.1, 6.1, 6.6). However, any of these stages can be reduced or absent, cysts may be included, and polyps may give rise asexually to more polyps (Figure 1.2) or may be the stage that reproduces sexually.

Fundamentally, scyphozoa are tetraradially symmetrical having many structures in multiples of four. Most medusae of the Scyphozoa

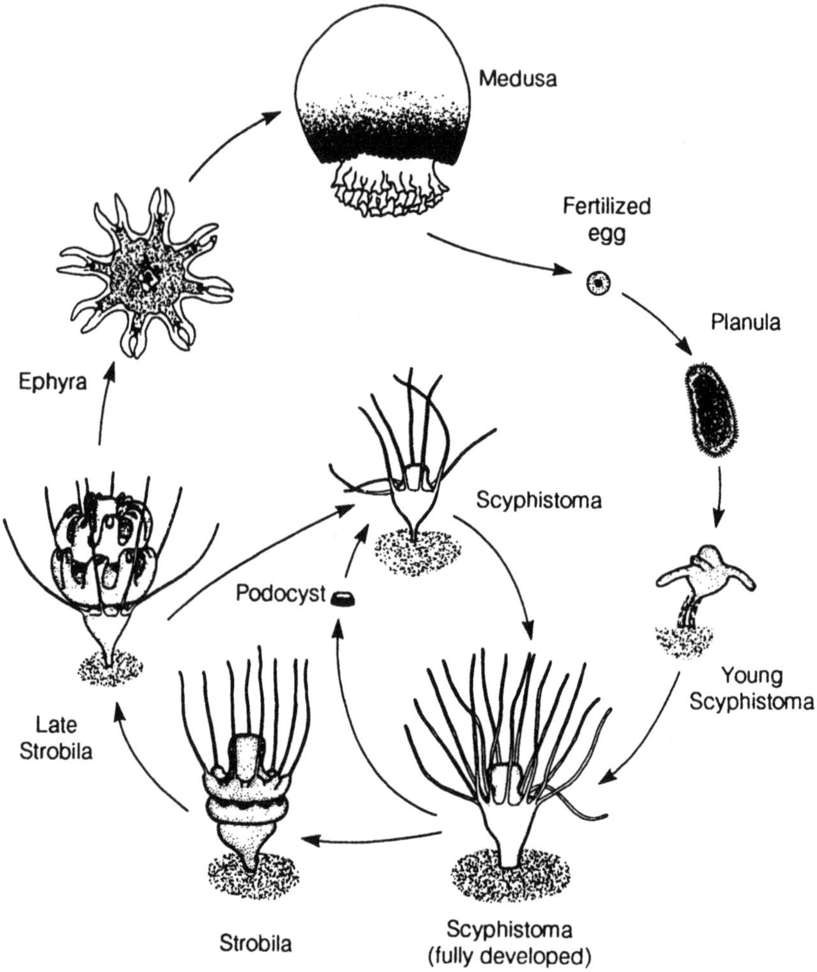

Figure 1.1 Life cycle of the rhizostome scyphozoan *Stomolophus meleagris*. The fertilized egg develops into a cilated planula larva which settles and forms a polyp, the scyphistoma. The scyphistoma can reproduce asexually either via a cyst, the podocyst, to form more scyphistomae, or by strobilation to form ephyrae which develop into medusae and reproduce sexually. (Source: Calder, 1982, with permission of D.R. Calder and *Biological Bulletin*.)

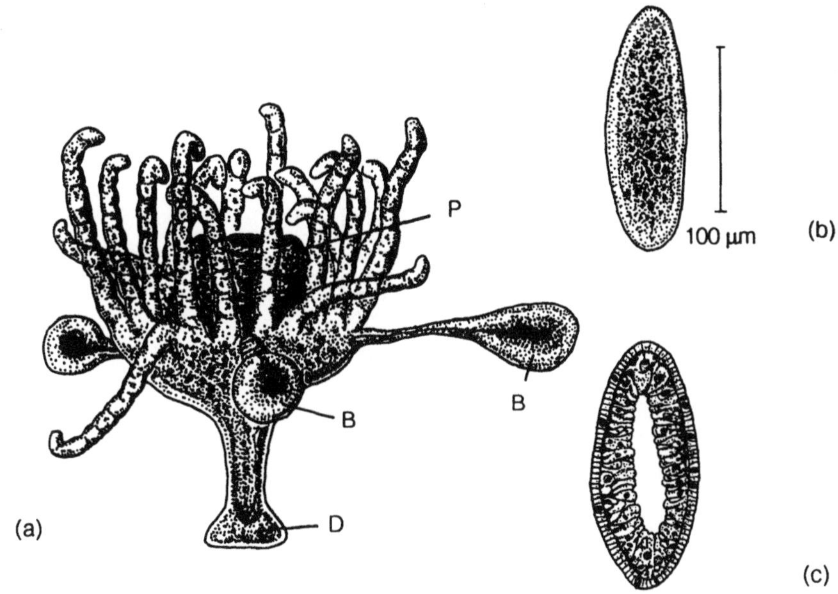

Figure 1.2 The interstitial scyphozoan *Stylocoronella riedli*. In asexual reproduction the polyp sheds buds which form unciliated planuloids and grow into new polyps. (a) Polyp; (b) free planuloid; (c) longitudinal section through a planuloid. B = bud; D = pedal disc; P = proboscis with mouth. (Redrawn from Salvini-Plawen, 1966, with permission of L. Salvini-Plawen.)

differ from those in the Cubozoa and Hydrozoa in lacking any shelf of tissue (velum or velarium) extending inward from the margin into the subumbrellar space. Scyphozoa lack a clearly defined pharynx leading in from the mouth such as is present in the Anthozoa. In Scyphozoa and Cubozoa there are gastric cirri in the stomach. The mesoglea maybe cellular and the gonads are gastrodermal in origin.

Scyphozoa are exclusively marine. Their medusae are found in pelagic habitats from the surface to very deep water. Their polyps are found attached to a wide variety of substrates. The only interstitial genus, *Stylocoronella*, includes two species of minute immature stauromedusan polyps, *S. riedli* and *S. variabilis* (see Salvini-Plawen, 1966, 1987; Kikinger and Salvini-Plawen, 1995) (Figure 1.2).

1.2 THE ORDERS: MORPHOLOGY AND LIFE CYCLES

In the Scyphozoa the medusoid stage typically predominates whereas the polypoid stage is very small. In many species the polyp is unknown,

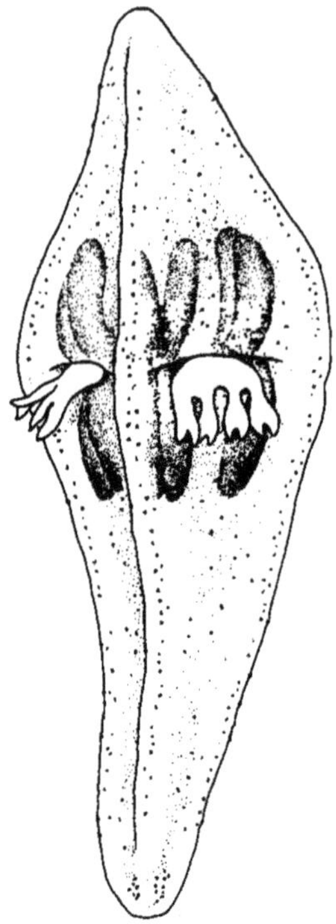

Figure 1.3 *Tetraplatia volitans*, Scyphozoa *incertae sedis*. Locomotory lappets arise from the equatorial groove, and gonads can be seen internally. Scale bar = 1 mm. (Source: Pagès, Gili and Bouillon, 1992, with permission of *Scientia Marina*.)

or has yet to be associated with the medusa in a complete life cycle. As will be seen in following chapters, much less is known of the physiology and ecology of the polyps than of the medusae.

There are four well recognized orders, described in the following sections. In addition at least one genus is less easily classified. *Tetraplatia* contains two species, *T. chuni* and *T. volitans* (see Ralph, 1959). These are very unusual pelagic medusae in which the subumbrella is convex so that the medusa is biconical with a partly exposed groove at its equator (Figure 1.3). Eight pairs of locomotory lappets arise from the groove, with eight statocysts in the clefts between the paired lappets but no tentacles. The presence of nematocysts and of epidermis and gastrodermis separated by mesoglea clearly place them in the Cnidaria, but there has been controversy for many years as to whether they have hydrozoan or scyphozoan affinities (Russell, 1970). Ralph (1960) placed them in a monogeneric family of coronate scyphomedusae, while Russell (1970) considered them as scyphozoa *incertae sedis*. They differ from all other scyphozoa in the lack of microbasic eurytele nematocysts (section 3.2.1).

1.2.1 Stauromedusae

The order includes small sessile or temporarily sessile polypoid medusae which attach to the substrate by an aboral adhesive disc on the exumbrella or an aboral stalk (Figure 1.4). The main body (calyx) has a central mouth on a short quadrangular manubrium, usually eight single primary tentacles, and eight clusters of hollow, capitate (knobbed) secondary tentacles which in most species are borne on eight arms. The four longitudinal septa of the gastrovascular space are each indented from the exterior by a deep funnel.

In sexual reproduction non-ciliated planulae larvae are formed which develop into polyps and then directly into the mature medusa. Settled planulae aggregates or polyps may also reproduce asexually by budding (Figure 1.2).

1.2.2 Coronatae

The order includes mostly bathypelagic to mesopelagic medusae each of which has a deep furrow (coronal groove) dividing the aboral surface (exumbrella) into a central disc and a peripheral zone (Figure 1.5). The peripheral zone has radial thickenings (pedalia), marginal lappets with interspersed sense organs, and solid marginal tentacles. There is a single mouth with simple lips, placed on a short manubrium. Radial septa fuse the subumbrellar wall of the gastrovascular cavity with the exumbrella between the pedalia to form peripheral pouches. Four crescent-shaped fusions form gastric septa partially separating the stomach from the peripheral pouches.

In most cases the life cycle of the medusa is unknown. Where known the polyps are solitary or colonial with firm periderm tubes of chitin (Figure 6.16). Juvenile medusae (ephyrae) are produced by transverse fission (strobilation) and develop into adult sexual medusae. Other coronate polyp species may lack medusae, reproducing sexually within the tube.

1.2.3 Semaeostomeae

The order includes large saucer-shaped adult medusae which lack the coronal grooves, pedalia and gastric septa of coronate medusae (Figures 1.6, 1.7). The margin is either divided into lappets or entire. Marginal sense organs with rhopalia (sensory clubs) arise from some

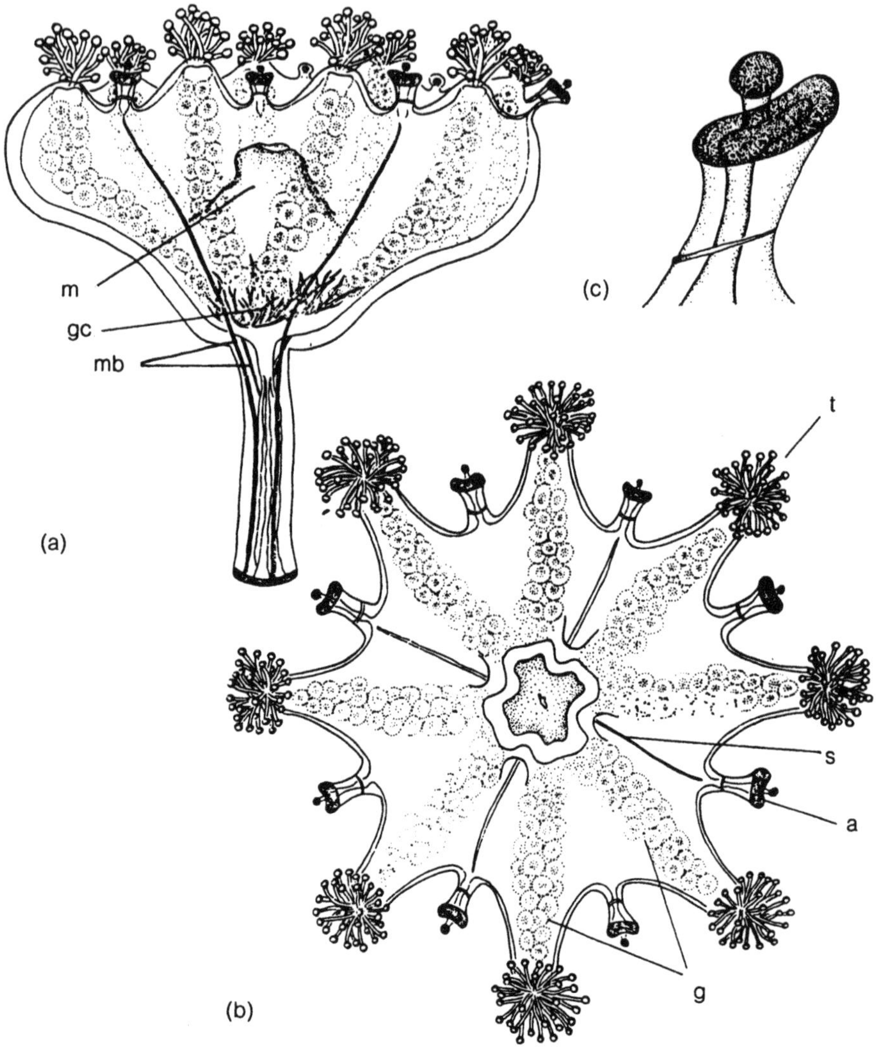

Figure 1.4 Stauromedusa *Haliclystus salpinx*. (a) Side view showing internal structures; (b) oral view; (c) anchor (primary tentacle). a = anchor; m = manubrium; g = gonads; gc = gastric cirri; mb = muscle bands; s = septum; t = secondary tentacles. (Redrawn from Berrill, 1962, with permission of National Research Council of Canada.)

or all of the niches between the lappets. Four oral arms with frilled or folded edges surround the single mouth. Numerous hollow tentacles are present in most species on the umbrellar margin. The peripheral zone around the stomach may contain radial pouches and/or a system of canals.

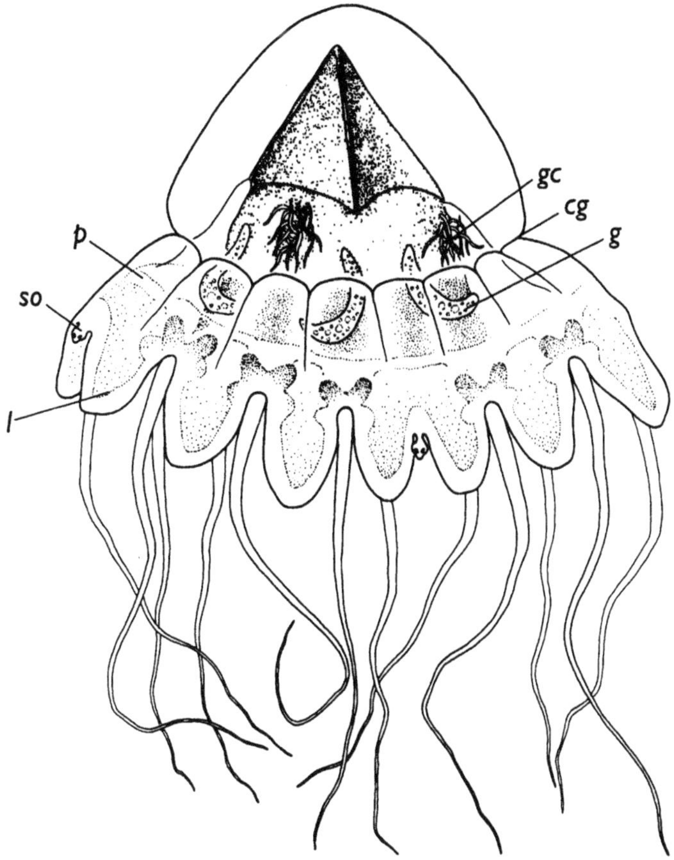

Figure 1.5 Young coronate scyphozoan *Periphylla periphylla*. cg = coronal groove; g = gonad; gc =gastric cirri; l = marginal lappet; p = pedalion; so = marginal sense organ. (Redrawn from Pagès, Gili and Bouillon, 1992, with permission of *Scientia Marina*.)

Planulae may develop into solitary non-tubed scyphistomae which strobilate to form ephyrae, or they may develop directly into the medusae (Figures 6.1, 6.15). The scyphistomae may also reproduce asexually.

1.2.4 Rhizostomeae

The order includes medusae which lack marginal tentacles and a central mouth, as well as lacking coronal grooves and pedalia (Figures 1.8, 1.9, 8.5). Four pairs of oral arms arise from the manubrium and fuse to form numerous mouth openings (ostia). The margin of the

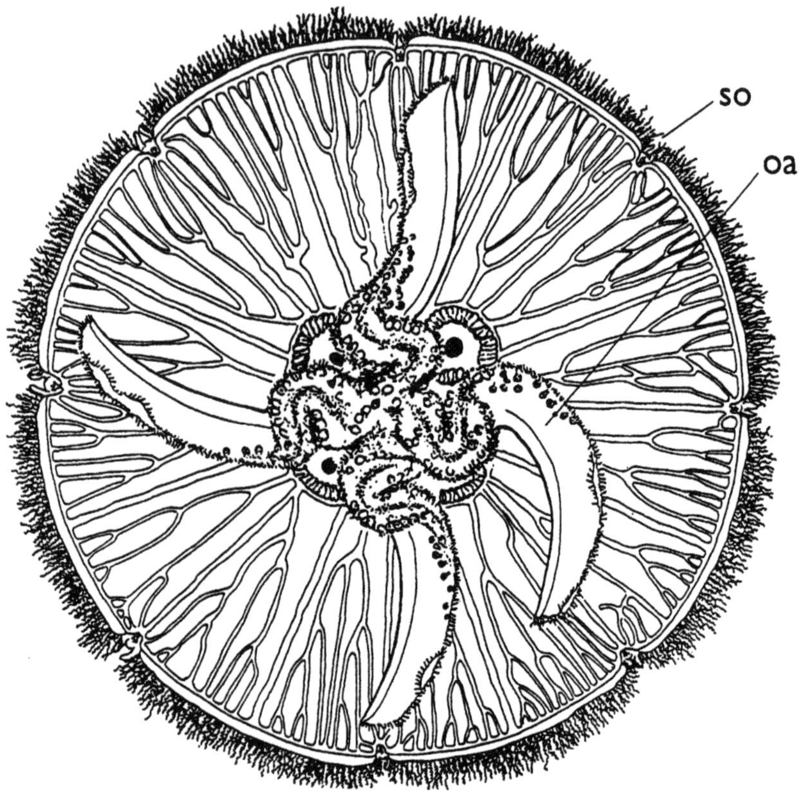

Figure 1.6 Semaeostome scyphozoan *Aurelia aurita.* Subumbrellar view of female specimen with brood pouches on the oral arms. oa= oral arm; so = marginal sense organ. (Redrawn from Russell, 1970, with permission of Cambridge University Press.)

umbrella is divided into eight or more lappets with eight or 16 marginal sense organs between them. The peripheral zone around the stomach contains a network of canals.

As in the Semaeostomeae, the solitary non-tubed scyphistoma may be present (Figures 1.1, 6.6) or lacking, may strobilate to form ephyrae, and may reproduce by budding.

1.3 RELATIONSHIPS AND ORIGINS OF THE CLASS AND ORDERS

The Phylum Cnidaria is clearly delineated by the presence or absence of cnidae. It is very unlikely that such a distinctive intracellular

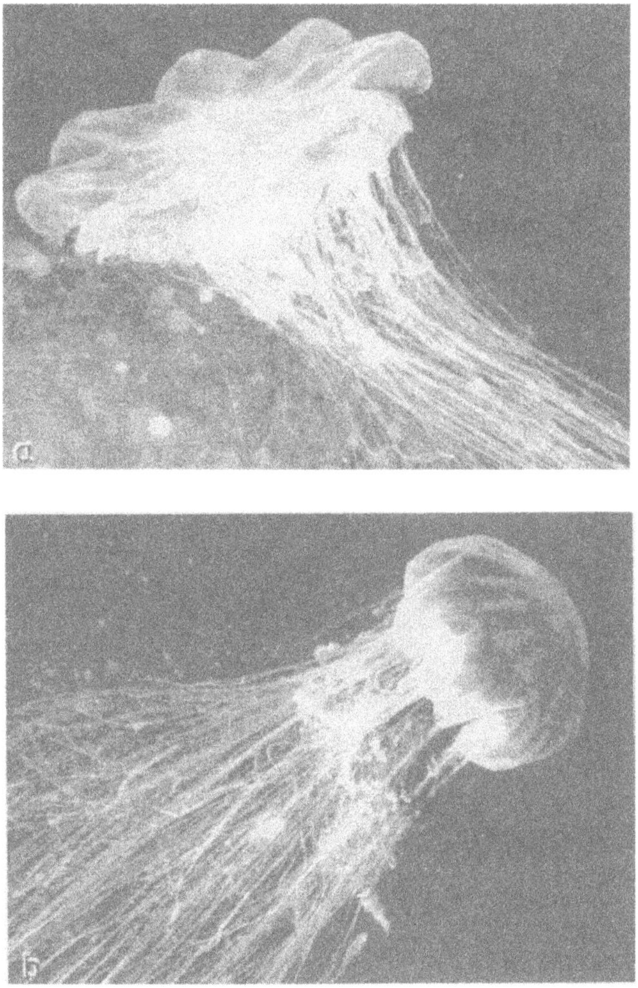

Figure 1.7 Semaeostome scyphozoan *Cyanea capillata*, swimming: (a) relaxed condition; (b) contracted condition. (Source: Gladfelter, 1972, with permission of Springer Verlag.)

organelle would have evolved more than once. However, the phylogenetic position of the Scyphozoa within the phylum has been a matter of speculation for many years.

Discussion about which is the primitive class, or which class most resembles the cnidarian stem, is intertwined with theories of the origin and diversification of the lower Metazoa. For example, if the original metazoan were a bilateral planula-like animal, then the Anthozoa, with

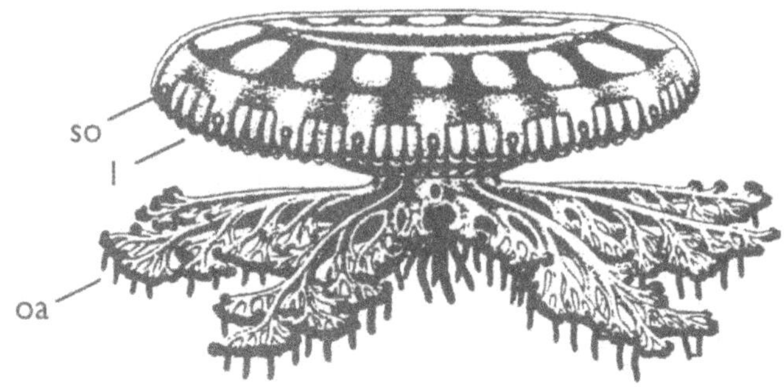

Figure 1.8 Rhizostome scyphozoan *Cassiopea xamachana*. l = lappet; oa = oral arm; so = marginal sense organ. Seventy percent natural size. (Redrawn from Mayer, 1910.)

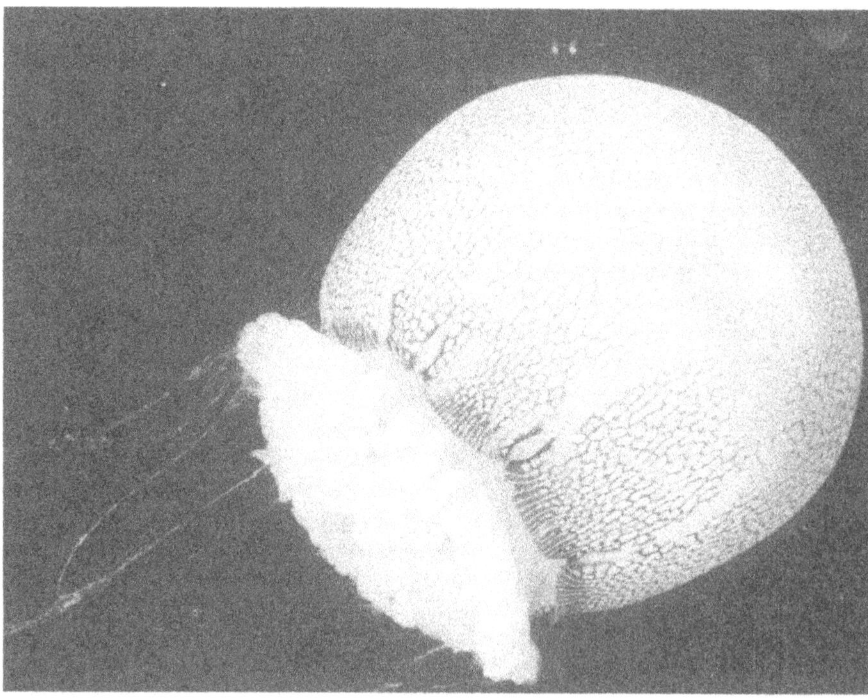

Figure 1.9 Rhizostome scyphozoan *Stomolophus meleagris*. Side view of specimen 6 cm in bell diameter. (Photograph courtesy of R.J. Larson.)

biradial members, would be favoured as the more primitive class. However, authors speculating that the first Metazoa were round (gastrula-like) organisms favour Scyphozoa, Cubozoa or Hydrozoa as the primitive class. Another question is whether the cnidarian stem was benthic and polypoid, or pelagic and medusoid. This question is plagued with problems of divergence and convergence where each form may have arisen or been lost more than once. For recent varied views on these subjects and reviews of the literature, see Bouillon (1981), Grasshoff (1984), Robson (1985), Stepanjants (1988), Willmer (1990) and Barnes and Harrison (1991).

There is little fossil evidence for these class-level relationships. The Pre-Cambrian Ediacarian fauna already included forms which have been tentatively assigned to Scyphozoa, Hydrozoa and Anthozoa (Scrutton, 1979; Glaessner, 1984; Wade, 1993). Often the forms with a rounded outline and radial symmetry have been interpreted as casts of scyphomedusae (Sun, 1986), but this interpretation has been questioned in many cases (Conway Morris, 1985, 1991; Jenkins, 1992). Experiments with modern medusae indicate that they would be unlikely to form casts and impressions similar to all the fossils (Norris, 1989; Bruton, 1991). Nevertheless it is possible that some scyphomedusae were present in the Pre-Cambrian.

Morphology and life cycles indicate a closer relationship between the classes Hydrozoa, Cubozoa and Scyphozoa than with the class Anthozoa (Werner, 1973; Salvini-Plawen, 1978; Petersen, 1979; Schuchert, 1993; Bridge *et al.*, 1995). The Anthozoa differ from the other classes in being polypoid, lacking any medusoid stage in their life cycle. They also differ in possessing a pharynx, an inturning of the body wall beneath the mouth to form an ectodermal gullet. As noted above, cnidae are diagnostic for the phylum Cnidaria. Hydrozoan, cubozoan and scyphozoan cnidae are more similar to each other than they are to cnidae of the Anthozoa, as is the structure of the enclosing cnidocytes (Mariscal, 1984; Bozhenova, Grebel'nyi and Stepanjants, 1988; Holstein and Hausmann, 1988). However, this division of the phylum into two branches does not indicate whether one branch or the other is primitive.

The closest class to the Scyphozoa is the Cubozoa, the members of which were formerly included in the Scyphozoa. The Cubozoa resemble the Scyphozoa in having life cycles with large and conspicuous medusae, in being tetraradiate, and in having gastrodermal gonads (Figure 1.10). The two groups have similar feeding and digestive structures (Larson, 1976) and nervous systems (Satterlie and Spencer, 1980). However, the cubomedusae are cuboidal with a tentacle or cluster of tentacles at each of the four corners of the bell,

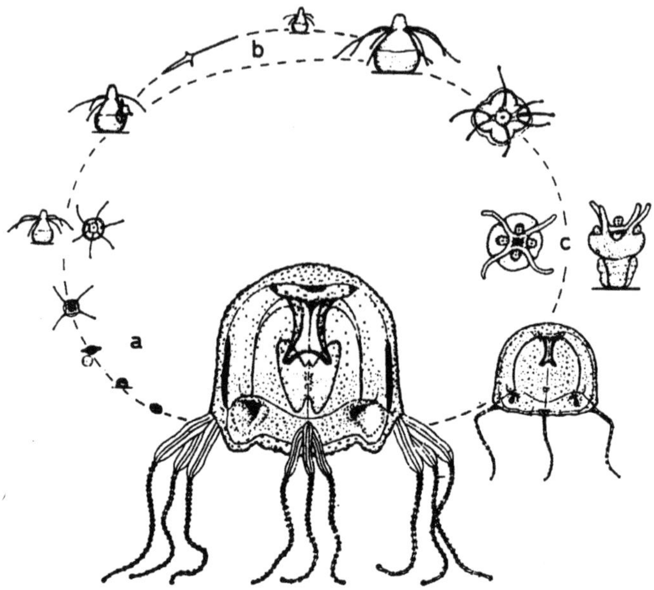

Figure 1.10 Life cycle of the cubomedusa *Tripedalia cystophora*. a = development of the planula into a young sessile primary polyp; b = asexual reproduction of the polyp by lateral budding of small secondary polyps; c = metamorphosis of the fully grown polyp into a single medusa, oral view to left and side view to right. (Source: Werner, 1973, with permission of Seto Marine Biological Laboratory.)

and there is a shelf-like projection (velarium) extending inward from the bell margin. The cnidae differ from scyphozoa in including microbasic mastigophores (Calder and Peters, 1975). The species may have a life cycle in which a single polyp metamorphoses totally into a single medusa, or the polyps may form buds in a way analogous to, and possibly homologous with, the Hydrozoa (Werner, 1975).

Some of these cnidarian relationships may be resolved by studies of molecular evolution, primarily by examining nucleic acids and proteins. New techniques for analysing molecular structure have led to a rapidly expanding literature. However, the old problems of detecting and taking account of convergent and parallel changes apply here as they did with analysis of anatomy and development (Willmer, 1990). Most studies so far published on molecular phylogeny include only one or two cnidaria, and do not add to our knowledge of relationships within the phylum. A few of these have included a scyphozoan as the representative cnidarian (Goldberg *et al.*, 1975a,b).

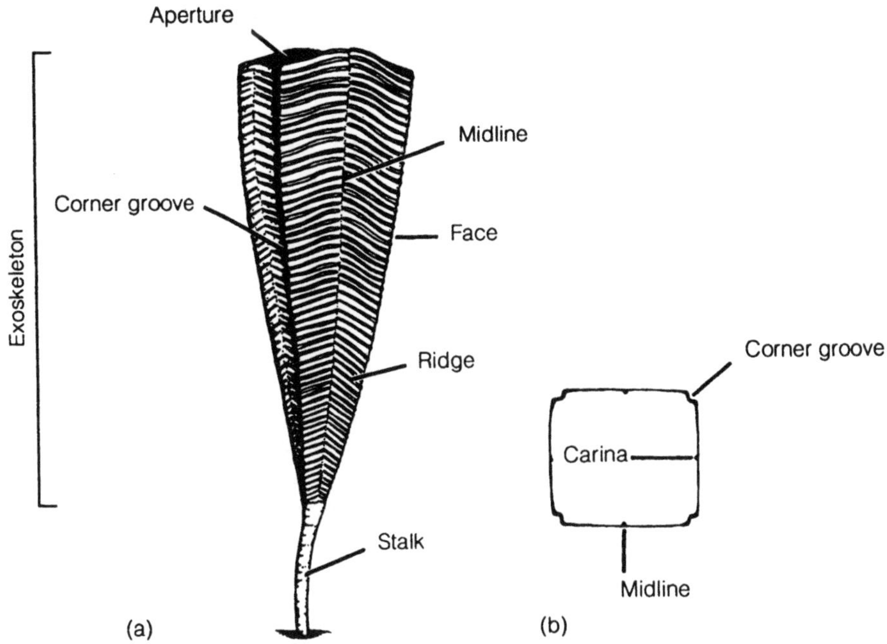

Figure 1.11 (a) Generalized external morphology of conulariid and (b) exoskeletal morphology shown in cross-section, based on *Paraconularia*. (Source: Babcock, 1991, with permission of Cambridge University Press.)

The most exciting work to date on molecular phylogeny of Cnidaria is that by Bridge *et al.* (1992) which examined the mitochondrial deoxyribonucleic acid (mtDNA) of 42 species. The mtDNA of most metazoan animals is circular in form. The anthozoan mtDNA is also circular but that of the Cubozoa, Scyphozoa and Hydrozoa is linear. This supports the divergence of the other classes from the Anthozoa, and the basal position of the Anthozoa within the phylum. Comparisons of the nucleotide sequences of ribonucleic acid (RNA) and of deoxyribonucleic acid (DNA) are so far available for fewer species, but also support a basal position for the Anthozoa (Hori *et al.*, 1982; Walker, W.F. and Doolittle, 1983; Hori and Osawa, 1987; Hori and Satow, 1991; Bridge *et al.*, 1995).

There is as yet little molecular data bearing on order-level relationships within the class Scyphozoa. Speculation from morphological and life cycle data has centered on extinct fossil groups. As noted above, scyphozoan medusae may have been present in the Pre-Cambrian. However, much of the speculation has centred on the later conulariids and bryoniids, both polypoid.

The conulariids had an elongate, four-sided pyramidal chitinophosphatic exoskeleton (Figure 1.11). They are present in the fossil record at least from the Ordovician to the late Triassic. Many authors, after Werner (1966), have regarded the conulariids as related to polypoid coronate scyphozoa. These living coronates, such as *Nausithoe racemosa* (Figure 6.16), also have chitinous tubes. However, this conclusion has been hotly debated. It is outside the scope of this book to review fully the evidence involved in this discussion. For two sides of the debate and a review of the literature, a comparison may be made between the views of two recent authors. Babcock (Feldman and Babcock, 1986; Babcock, 1991) holds the view that the conulariids were an independent, now extinct, phylum. Van Iten (1991a,b, 1992a,b) argues that they were cnidarian polyps, more closely related to the Scyphozoa than to the Anthozoa or Hydrozoa.

Bryonia, a fossil from the Upper Cambrian and Ordovician, is more generally accepted as having been a scyphozoan polyp (Glaessner, 1984; Bischoff, 1989). *Bryonia* and related forms have recently been placed in Bryoniida, an order of Scyphozoa which became extinct in the Permian (Bischoff, 1989). Bryoniids resemble coronates in possessing an elongate, conical tube with a circular to oval cross-sectional shape, and an apical attachment disc. Some members of the order are septate, with tetramerous symmetry.

The resemblance of the coronate polyp to the tubed conulariids and bryoniids does not necessarily mean that the coronates represent the most primitive order of modern scyphozoa as proposed by Werner (1973). A delicate cuticle surrounds the aboral end of the scyphistomae of the semaeostomes *Aurelia aurita, Chrysaora melanaster* and *Cyanea capillata* and the rhizostomes *Cassiopea andromeda, Rhopilema nomadica, R. verrilli* and *Stomolophus meleagris* (see Hyde, 1895; Gohar and Eisawy, 1961b; Chapman, D.M., 1966, 1968; Kakinuma, 1967; Widersten, 1969; Calder, 1973, 1982; Lotan, Ben-Hillel and Loya, 1992). This may be a vestigial tube indicating possible evolution of the Semaeostomeae or Rhizostomeae from a tubed polyp (Chapman, D.M., 1966; Widersten, 1969). The systematic position of the Stauromedusae (which entirely lack pelagic forms) also remains problematical (Uchida, 1973).

Structure of the sperm of the four orders supports the Stauromedusae as the most primitive order, whereas development of the oocytes is least specialized in the coronates (Hedwig and Schäfer, 1986; Eckelbarger and Larson, 1992, 1993) (section 6.2.2).

It is probable, based on developmental patterns common to the four orders, that the scyphozoan stem form was a tetraradiate polyp with four tentacles and four septa (Thiel, H., 1966; Uchida, 1969).

However, even this cannot be corroborated from the fossil record. The present evidence is insufficient to do more than speculate on phylogeny within the Scyphozoa. It is to be hoped that clarification will come from molecular data.

2 *Locomotion*

2.1 INTRODUCTION

Locomotion is necessary for coelenterates in order to reach food, to escape predation, to reach and select substrates, and to interact during reproduction (Mackie, 1974). The role of locomotion in contacting prey and in feeding will be discussed in sections 3.4 and 3.5.

In the Scyphozoa the most extensively examined type of locomotion is the swimming of the medusae. Swimming of most medusae is based on rhythmic contractions of the subumbrellar muscles which drive water out and move the animal by jet propulsion. Other modes of locomotion include gliding or somersaulting movements by the polyps and ciliary locomotion or creeping by the planulae. Flagellar locomotion used by the sperm will be discussed in section 6.2.2 on reproduction.

As in other animals, movement of scyphozoa requires support structures, contractile cells and nervous control. The present chapter will first describe the mesoglea, muscle, nerve and sense organs, and then how these are utilized for locomotion.

2.2 MESOGLEA

The mesoglea of cnidaria is an extracellular matrix between the ectodermal and endodermal epithelia. It consists of fibres embedded in a hydrated matrix, and may also contain cells. It varies greatly in extent and composition both from species to species and in different locations in the same species. In many medusae the mesoglea of the bell represents a high percentage of the volume of the animal.

The mesoglea may have a variety of functions, including maintenance of buoyancy and transparency, to be described later (sections 5.4.2 and 9.1.4). The main functions addressed here are the use of the mesoglea as a skeleton and for elastic recoil.

Cells are present in the mesoglea of some scyphozoa (Chapman, G., 1966). In some positions, such as the fishing tentacles of *Chrysaora quinquecirrha*, these are processes of muscle cells extending into the mesoglea to anchor the contractile epithelia (Burnett and Sutton, 1969; Perkins, Ramsey and Street, 1971). In symbiotic species, such as *Cassiopea xamachana*, zooxanthellae are present in mesogleal cells as described in section 4.5.1. However, cell populations are also present in some nonsymbiotic species such as *Aurelia aurita*, although absent in adult *Cyanea lamarcki* and *Chrysaora hysoscella* (see Chapman, G., 1953). It is not clear what function these cells serve, or how the cell population is originally derived in *A. aurita*. Once present in the polyp, there is high proliferative activity in the strobila and young ephyra sufficient for self-support and growth of this cell population (Napara and Chaga, 1992a,b). The cells are variable in inclusions and in shape, from smooth and rounded to shapes with filopodial projections (Chapman, D.M., 1974; Hentschel and Hündgen, 1980). Since they are not present in all species they are unlikely to be needed for formation of the typical mesogleal constituents.

It is not known how the mesoglea is formed. At least in species without mesogleal cells, it is probable that the fibrils are formed by self-assembly as is known in higher animals. Anderson and Schwab (1981) found membrane-bound packets of fine filamentous material in endodermal cells of *Cyanea capillata*. In one instance one of these packets was continuous with the mesoglea, and they speculated that the cells were secreting the material of the mesoglea.

2.2.1 Fibre composition

The fibres of the mesoglea vary from submicroscopical fibrils to large fibres crossing the main mesoglea layer of medusae umbrellas. The fishing tentacle of *Chrysaora quinquecirrha* has randomly oriented fibrils 6–7 nm in diameter (Burnett and Sutton, 1969). The scyphistomae have similar randomly oriented 8–9 nm fibrils which become oriented normal to the gastrodermal cell surfaces in constricting regions of the strobila (Bynum and Black, 1974). In medusae such as *Pelagia noctiluca* and *Lucernaria* sp., stout sinuous fibres with branched ends extend across the mesoglea between networks of smaller fibres beneath the

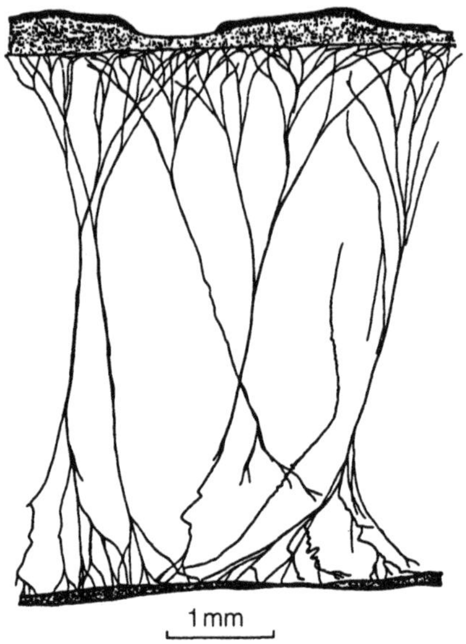

Figure 2.1 Radial section of the umbrella of *Pelagia noctiluca* to show the thick vertical fibres and their branches in the mesoglea. The exumbrellar tissue above the mesoglea is stippled. (Source: Chapman, 1959, with permission of G. Chapman and the Company of Biologists Ltd.)

ectodermal and endodermal epithelia (Figure 2.1) (Chapman, G., 1959; Elder and Owen, 1967).

The majority of the fibres consist of collagen-like protein. There are several types of evidence for this statement. The fibres split in such a way that the sum of the cross-sectional areas of the parts is equal to that of the parent fibre (Chapman, D.M., 1970b). The 50–66 nm banding characteristic of collagen has been observed in larger fibres, although it may be difficult to demonstrate in small fibrils (Chapman, G., 1959; Chapman, D.M., 1970b; Franc, S., 1988). The fibres are dissolved by the enzyme collagenase (Chapman, D.M., 1970b). The amino acid composition is similar to collagen (Bocquet *et al.*, 1972; Rigby and Hafey, 1972; Quensen, Black and Webb, 1981; Kimura, Miura and Park, 1983). Finally, when the chain structure of protein extracted from the mesoglea of *Stomolophus nomurai* was analysed it was found to be a heterotrimer, similar to vertebrate Type V collagen (Miura and Kimura, 1985).

Elastic fibres are also present which become wavy or helical when cut so that tension is released. These fibres differ chemically from elastin, which has not been demonstrated in scyphozoa. The large helical elastic fibres running across the mesoglea of medusae such as *Chrysaora quinquecirrha, Craterolophus convolvulus* and *Lucernaria* sp. stain with spirit blue and can be digested with elastase only following oxidation with reagents such as potassium permanganate (Elder and Owen, 1967; Elder, 1973; Chapman, D.M., 1974). In *Rhizostoma pulmo* collagenase-resistant fibres have staining properties similar to vertebrate oxytalan fibres (Bouillon and Coppois, 1977). This type of fibre may not have any observed banding as in *Pelagia noctiluca* (see Bouillon and Vandermeerssche, 1956; Chapman, D.M., 1974) or may have an indistinct beading of approximately 35 nm as in *Haliclystus auricula* (see Elder, 1966).

At interfaces between the mesoglea and surrounding epithelia an electron-dense band follows the surface contours of the cells, and is separated from them by an electron lucent region of uniform width (Burnett and Sutton, 1969; Chapman, D.M., 1970b). It is not known whether this band contains compounds such as laminin and fibronectin characteristic of the basement membranes beneath epithelia of higher animals (Pedersen, 1991).

2.2.2 Mechanics

In medusae, the recovery force expanding the bell, following contraction of the subumbrellar swimming muscles, depends on the mesoglea (Chapman, G., 1959; Gutmann, 1965, 1966). The mesoglea of the umbrella of seven species was examined by Gladfelter (1972, 1973). There is a subumbrellar layer below the gastrovascular cavity which serves as the base for the subumbrellar swimming muscle. Above the gastrovascular cavity, the thick exumbrellar mesoglea is involved in elastic recoil following the muscular contraction. The two layers are connected by radial anchoring ridges among the channels of the gastrovascular system (Figure 2.2).

The subumbrellar mesoglea is of a fibrous tough consistency that provides adequate anchorage for the muscle, but it is also thin and elastic enough to permit shortening of the muscle. As the muscle contracts it is thrown into folds.

The exumbrellar mesoglea of most species (not *Aurelia* or *Stomolophus*) is jointed with grooves on the subumbrellar side anchored to the exumbrellar surface by concentrations of the large elastic fibres (Figure 2.2). Between the joints the elastic fibres are more sparse and the collagenous mesoglea is relatively rigid. During contraction the muscles fold the mesoglea around the system of joints, stretching the elastic fibres. The elastic properties of this material in repetitive

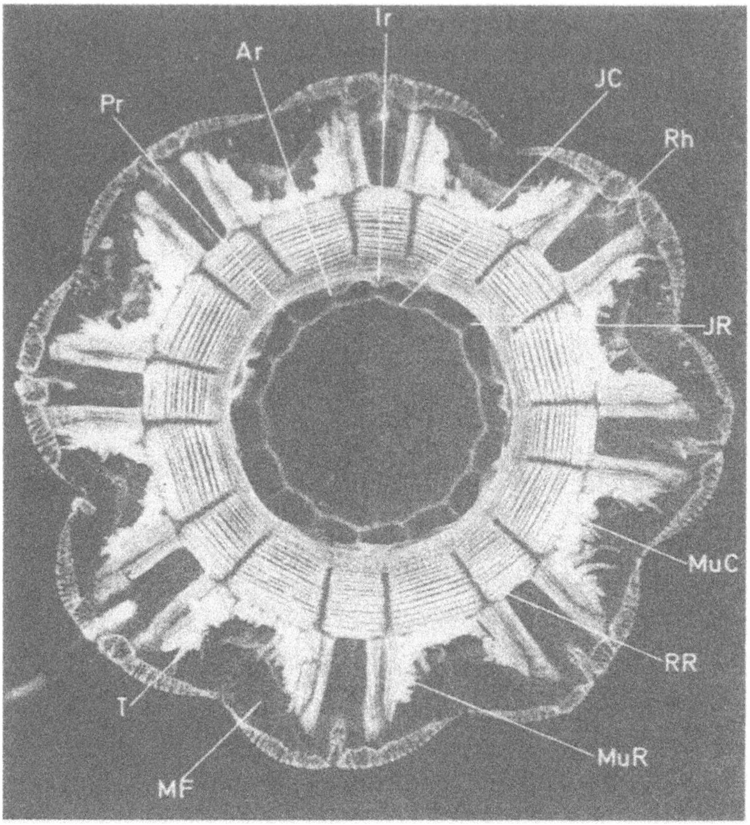

Figure 2.2 Subumbrellar surface of *Cyanea capillata* with oral arms and tentacles removed to reveal swimming muscles. The subumbrellar surface of the exumbrella with its mesogleal joints can be seen proximal to the ring of coronal muscle. Ar = adradius; Ir = interradius; JC = coronal mesogleal joint; JR = radial mesogleal joint; MF = marginal flap; MuC = coronal swimming muscle; MuR = radial swimming muscle; Pr = perradius; Rh = rhopalium; RR =radial anchoring ridge connecting subumbrellar and exumbrellar mesoglea; T = remnant of tentacle. (Source: Gladfelter, 1972, with permission of Springer-Verlag.)

stretch have not been measured. Alexander (1964) examined the stretch of *Cyanea capillata* mesoglea under a constant tensile stress. He showed that extension is rapid at first but declines until a very low value is reached after many hours.

Mesoglea is also present in the tentacles of medusae and in the columns and tentacles of polyps. There has been no investigation of the mechanical properties of mesoglea in the column of the polyp. By analogy with sea anemones, one function of the mesoglea is probably to resist stretch of the column wall. Tentacles of scyphistomae are supported by a solid core of cells.

The tentacles of medusae may be solid or hollow. Even the hollow tentacles need little mesoglea to resist stretch because of their small radius. (At a given interior pressure the tension on the wall of a cylinder is proportional to its radius.) On the contrary, in the fishing tentacles of *Chrysaora quinquecirrha* medusae, the mesogleal layer must be highly flexible to support folding of the ectodermal muscle as it shortens the tentacle to less than a thirtieth of its extended length. There is no circular muscle present in tentacles of either the polyps or medusae. Extension is at least partly due to recoil of mesoglea which has been strained during shortening of the longitudinal ectodermal muscle (Perkins, Ramsey and Street, 1971; Chapman, D.M., 1970b).

2.3 MUSCLE

Contraction of muscle cells depends on the relative movement of protein filaments which may be organized into intracellular myofibrils. The contractile myofibrils of scyphozoa are usually contained within epitheliomuscular cells. Each of these cells has an epithelial cell body and one or several basal processes containing myofibrils (Krasińska, 1914) (Figure 2.9). The basal processes may be directly attached to the cell body or be attached only by long thin connecting processes. Myofibrils may also run up through the cell body toward the outer surface of the cell. Occasionally, as in the septal muscles of the polyps, muscle cells lose their connection with the surface, i.e. they are not epithelial. They become more filiform with an elongated cell body lying adjacent to the myofibril.

2.3.1 General muscular anatomy

The swimming muscle of medusae such as *Cyanea capillata* is distributed in deep folds of the subumbrellar epithelium against corresponding folds of the mesoglea (Gladfelter, 1972; Anderson and Schwab, 1981). The proximal coronal muscle has circularly oriented fibrils and the distal radial muscle has radially oriented fibrils (Figure 2.2). Between each pair of radial muscle bands is a triangular septum of thinner peri-rhopalial tissue extending to the margin of the bell. The epidermal epitheliomuscular cells of this tissue form wide muscle processes with radial myofilaments (Figure 2.9). The peri-rhopalial tissue does not contribute to the swimming movements, although it contracts locally (Anderson and Schwab, 1981).

In other medusae the coronal muscle is always present but the position and extent of the radial muscle is very variable. It modifies the details of the swimming beat. In *Periphylla periphylla* strong radial deltoid muscles extend from the gastric region to the proximal margin of the coronal muscle. In the ephyrae of *Aurelia aurita* and *Pelagia noctiluca* the radial swimming muscles extend beyond the coronal muscle into the lappets (Matsuno and Hisamatsu, 1982; Rottini-Sandrini and Avian, 1983). The radial muscle of *Cassiopea ornata* has a pinnate arrangement (Thiel, M.E., 1976a). Radial swimming muscles are, however, poorly developed in *Chrysaora melanaster* and *P. noctiluca* (see Gladfelter, 1973). In the nonswimming Stauromedusae the coronal muscle is reduced to a narrow marginal band and radial subumbrellar muscles are present proximally (Gwilliam, 1960).

Elsewhere in medusae there are longitudinal oriented myofibrils in the epidermal epitheliomuscular cells of the tentacles (Perkins, Ramsey and Street, 1971; Westfall, 1973). Similar longitudinal muscle may also be present in manubria and oral arms.

In the Stauromedusae and polyps of other orders a strong longitudinal muscle cord runs down each septum of the calyx and into the stalk (Gwilliam, 1960; Widersten, 1966) (Figure 1.4). In the most extensively investigated scyphistoma, that of *Aurelia aurita*, there is also radial muscle in the oral disc and longitudinal muscle in the epidermis of the tentacles (Westfall, 1973; Chapman, D.M., 1965; Chia, Amerongen and Peteya, 1984). The extent and orientation of muscle in the gullet and column differ between Northeast Atlantic and Northeast Pacific populations. Circular gastrodermal muscles are absent, extension of the column or tentacles depending on the mesoglea.

2.3.2 Fine structure of muscles

When observed with an electron microscope, the myofibrils of swimming muscle of *Cyanea capillata* and *Cassiopea xamachana* medusae and of *Atorella* sp. and *Aurelia aurita* ephyrae are of the classically striated type with interdigitated thick (13–18 nm) and thin (5–7 nm) filaments arranged in longitudinally repeated units, the sarcomeres (Spangenberg, 1977; Blanquet and Riordan, 1981; Anderson and Schwab, 1981; Matsuno, 1981b; Matsuno and Hisamatsu, 1982; Matsuno, 1983). A number of structures in these sarcomeres are similar to those of other classically striated muscles (Figure 2.3). I-bands (thin filaments), A-bands with central H-zones (thick

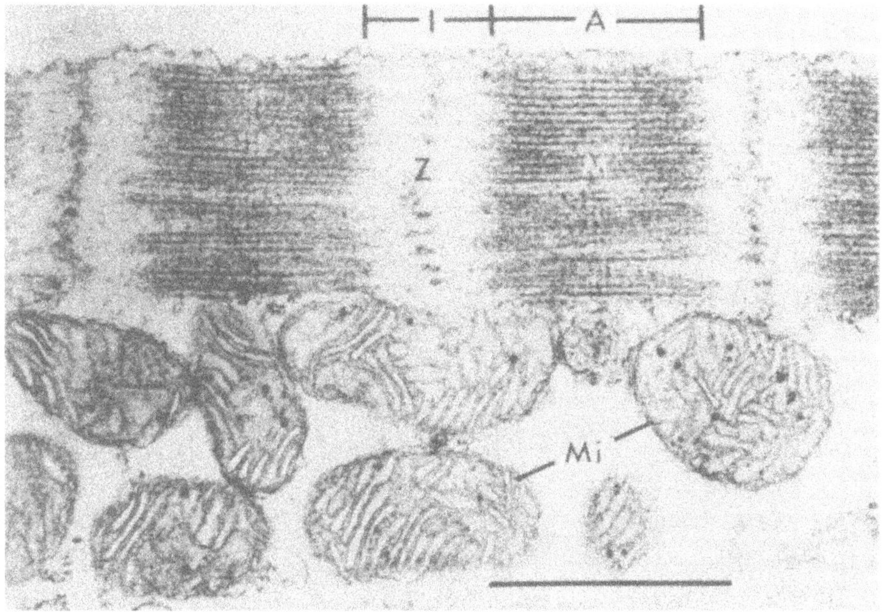

Figure 2.3 Longitudinal section through a swimming muscle fibre of *Cassiopea xamachana* showing a single myofibril and parallel rows of mitochondria. The striated myofibrils exhibit distinct I- and A-bands, Z-discs and M-lines in the H-zone. Mi = mitochondria. Scale bar = 1μm (Source: Blanquet and Riordan, 1981, with permission of R.S. Blanquet and American Microscopical Society.)

filaments) and M-lines can be identified although the relaxed sarcomeres are shorter (0.8–1.6 mm) than in higher animals. They are separated by Z-discs formed from accumulations of granules. In cross-sections of the outer A-band, where thick and thin filaments overlap, the thick filaments are arranged in a regular lattice with thin filaments arranged hexagonally around the thick ones. The morphological similarity to higher animals such as vertebrates and arthropods makes it likely that these scyphozoan muscles also contract with sliding of the thick and thin filaments past one another. That has not yet been demonstrated by comparison of contracted and relaxed tissue.

Classically striated muscle has so far only been identified in swimming muscle. In the epitheliomuscular cells of the polyps of *Atorella* sp., *Nausithoe punctata* and *Aurelia aurita* as well as the tentacular and peri-rhopalial muscles of *Chrysaora quinquecirrha* and *Cyanea capillata* medusae, the myofibrils may appear striated or smooth. However, the thick and thin filaments are not separated into sarcomeres by Z-discs,

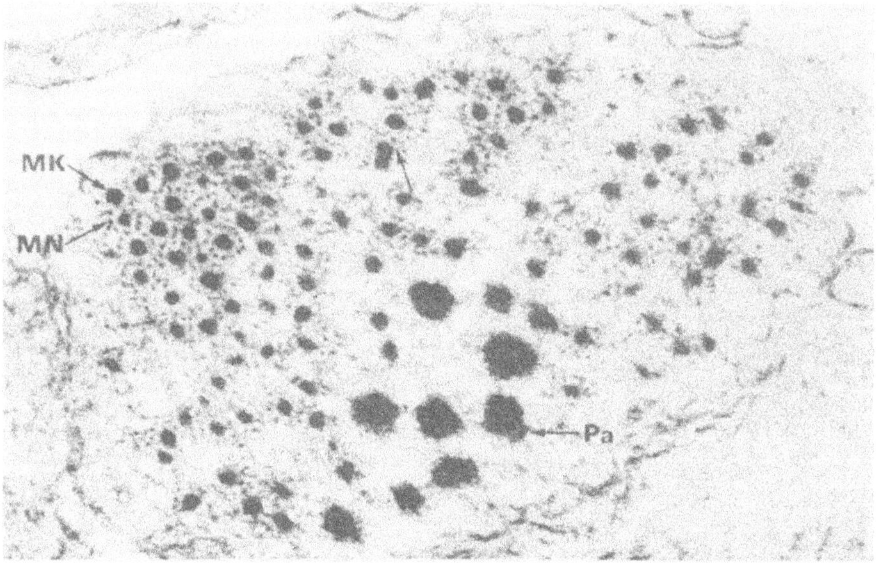

Figure 2.4 Cross-section of a muscle fibre in *Chrysaora quinquecirrha* fishing tentacle relaxed to 8.5 times contracted length. MK = thick myofilaments; MN = thin myofilaments; Pa = extra thick filaments. Possible bridges between thin and thick filaments are visible at the unlabelled arrow. × 82 000 (Source: Perkins *et al.*, 1971, with permission of Academic Press.)

even though the thick filaments may be in register leading to the striated appearance. In cross-sections the thin filaments are grouped around the thick ones (Figure 2.4) (Perkins, Ramsey and Street, 1971; Matsuno, 1981a, 1983; Anderson and Schwab, 1981; Chia, Amerongen and Peteya, 1984). The thin filaments are 4–11 nm in diameter and the thick filaments vary from 13 nm to 34 nm.

A third class of unusual extra-thick spindle or bar-shaped filaments may also be present in some of these muscles (Figure 2.4) (Perkins, Ramsey and Street, 1971; Matsuno, 1981a; Kawaguti and Yoshimoto, 1973). They have so far been observed in tentacles of *Chrysaora quinquecirrha* medusae and polyps of *Atorella* sp. and *Nausithoe punctata*. They show a periodic pattern at 100 nm and 13.5–15 nm intervals and vary in diameter from 20 to 90 nm. These filaments may be scattered among the thick and thin ones or concentrated in another area of the myofibril.

Perkins *et al.* (1971) examined *Chrysaora quinquecirrha* fishing tentacle muscle contracted, or relaxed to 7–20 times or to 30 times the contracted length. They found that the diameter of the thick filaments increased by about 40% during contraction (from an average diameter of 19 nm to an average diameter of 27.5 nm). The extra-thick filament type was present only in tentacles relaxed to 7–20 times the contracted length, i.e. not in fully contracted or fully relaxed muscle. Both of these observations are very interesting and should be verified. The possibility that thick filaments contract as well as slide past thin filaments in muscles of higher phyla is very controversial (Pollack, 1990). It is also unclear what the function may be of an extra filament type only present at intermediate muscle lengths.

It is not known what contractile proteins are present in scyphozoa. In muscle of higher animals myosin in the thick filaments interacts with actin in the thin filaments. A two-headed myosin has recently been extracted from sea anemones (Kanazawa *et al.*, 1993) which will interact with rabbit actin, i.e. myosin is clearly present in other coelenterates. However, actin has not yet been clearly demonstrated. Monospecific antibodies against vertebrate actin do not recognize 'actin' extracts obtained from *Aurelia aurita* or from the sea anemone *Actinia equina* (see de Couet, Mazander and Gröschel-Stewart, 1980; Thompson *et al.*, 1991). The pattern of banding in the extra-thick fibres suggests the presence of paramyosin (Perkins, Ramsey and Street, 1971; Kawaguti and Yoshimoto, 1973) but the protein has not yet been extracted.

Desmosomes observed connecting the ends of pairs of myoepithelial cells may transmit tension from the end of one cell to the next (Burnett and Sutton, 1969; Blanquet and Riordan, 1981; Anderson and Schwab, 1981). Desmosomes are thickened regions of the plasma membranes where adjacent cells are tightly attached. Lateral desmosomes have also been observed between adjacent striated muscle cells at the Z-lines of in-phase sarcomeres. These may transmit tension at angles to the axis of contraction. Perkins *et al.* (1971) found that the desmosomes of tentacle muscle are atypically labile, most common in contracted muscle but disappearing as the tentacles are extended.

Neuromuscular synapses have been observed on the connecting processes between the cell bodies of swimming and peri-rhopalial epitheliomuscular cells and the contractile processes (Anderson and Schwab, 1981). They may also be present on or near the contractile processes of epitheliomuscular cells of tentacles (Westfall, 1973). The junctions are asymmetrical chemical synapses with vesicles only on the neuronal side of the cleft. In *Atorella japonica* two types of vesicles are present in the same neuron; small (75 nm) clear vesicles near the presynaptic membrane and larger (120 nm) dense cored vesicles behind the line of clear vesicles (Matsuno and Kawaguti, 1991).

This may indicate the release of two transmitters at the same junction.

There is to date no evidence of structures conducting impulses from the cell surface into the contractile area of the epitheliomuscular cell. The sarcoplasmic reticulum in these cells may be absent or represented by a few small vesicles or by subsurface cisternae below the junctions (Westfall, 1973; Spangenberg, 1977; Blanquet and Riordan, 1981; Matsuno, 1981b; Anderson and Schwab, 1981; Matsuno and Hisamatsu, 1982).

2.3.3 Physiological properties of muscles

The force of contraction of a muscle is dependent on the cross-sectional area of the contractile material but weight moved is dependent on volume, therefore larger animals of a given shape tend to be more sluggish. In scyphozoa the larger species such as *Cyanea capillata* are slower moving, although the cross-sectional area of the swimming muscle is increased by folding of the muscle stratum against the mesoglea (Gladfelter, 1972). Myofibrils do not form multiple layers, but rather always remain in direct contact with the mesoglea.

Physiological properties of scyphozoan muscle are difficult to distinguish from those of the associated nerve and mesoglea. For example, the spectacular shortening of medusan fishing tentacles is due in part to the shortening of the myofibrils and in part to folding of the surrounding tissue (Perkins, Ramsey and Street, 1971). The only measurements of physical properties for the muscle of Cnidaria were made on the column muscle of *Pachycerianthus torreyi*, a ceriantharian anthozoan (Arai, 1965). A fixed reference length could not be defined due to extension of the mesoglea. Nevertheless, when the preparation of muscle and mesoglea was allowed to extend under a constant weight, the height of the twitch was shown to increase to a maximum and then decline, corresponding to the length–tension relationship of higher animals. Also as expected, the height of a twitch contraction and the shortening velocity both decrease with increased load.

Most experimentation on scyphozoan muscle has been done on the swimming muscles. These muscles contract following passage of an impulse in the motor nerve net (MNN) (described in section 2.5.3) which forms neuromuscular junctions with the myoepithelial cells. The impulses normally originate in the marginal centres, so removal of the marginal centres allows experimental stimulation of the nerve net at controlled frequencies. A single impulse causes a small contraction, which is of the same size at any strength of stimulus over threshold. Successive responses are larger, at moderate frequencies, up to a

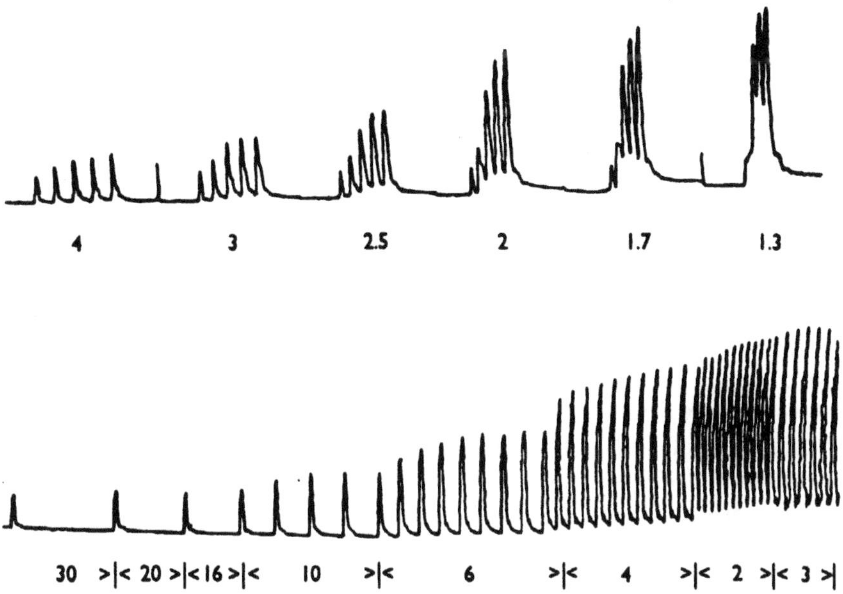

Figure 2.5 Facilitation in a strip preparation of *Rhopilema* sp. containing swimming muscle but without marginal bodies. Kymographic recording of muscle responses to electrical stimuli of constant strength, with interval between stimuli (in seconds) indicated. (Source: Bullock, 1943. Reprinted by permission of John Wiley & Sons, Inc.)

plateau height, i.e. there is facilitation. Figure 2.5 shows such a response for muscle of *Rhopilema* sp. (see Bullock, 1943). Similar facilitation has been observed in *Aurelia aurita* and *Cyanea capillata* (see Bullock, 1943; Pantin and Vianna Dias, 1952; Horridge, G.A., 1956a; Gwilliam, 1960). It is not known whether the facilitation is due to further recruitment of muscle fibres or to increased contraction of the originally stimulated fibres.

Another nerve net, the diffuse nerve net (DNN) (to be described in section 2.5.4) also has input to the swimming muscles which varies from species to species. In adult *Aurelia aurita* medusae only the MNN activity initiates swimming muscle contractions (Horridge, G.A., 1956a). However, in *Mastigias albipunctatus* there is a strong, rapid contraction following passage of an impulse through the MNN, and a somewhat slower, weaker contraction after passage of an impulse through the DNN (Passano, L.M., 1965). In *Cassiopea xamachana* an impulse in the DNN does not cause a contraction but it does facilitate (increase) the strength of a contraction caused by the MNN (Horridge, G.A., 1956a).

The neuromuscular delay (the time between passage of the impulse in the MNN and the contraction of the muscle) is very long. The delay also differs between different muscles, which allows sequential contraction of muscles in response to each impulse passing through the MNN. For example, in *Cassiopea xamachana* the radial muscle contracts 95 ms after the MNN pulse, followed by the coronal muscle 700 ms after the pulse (Passano, L.M., 1982).

The refractory period of the swimming muscle, during which the muscle cannot be restimulated, is also long. The absolute refractory period is approximately 0.7 seconds (Bullock, 1943). This prevents tetanic contractions resulting from the summing of the effects of successive stimuli and is necessary for the relaxation between repetitive swimming beats. The muscle contracts and relaxes at least partially before it can be restimulated. During frequent repetitive stimulation the refractory period may decrease, allowing somewhat greater summation but still not complete tetanus (Horridge, G.A., 1955).

Gwilliam (1960) recorded the contractions of stalk muscles of the stauromedusan *Haliclystus auricula*. His analysis was hindered by a considerable amount of spontaneous activity by the muscle. However, he was able to record large, smooth and prolonged contractions which were graded in contraction amplitude with increased frequency in the trains of stimuli. The fused contractions indicate a shorter refractory period than in the swimming muscles. These muscles are used to maintain the posture of the polyp, and so sustained contractions are desirable.

2.4 SENSORY RECEPTORS

Many of the receptors of scyphozoan medusae are concentrated in the marginal sense organs (typically multiples of four). Each of these organs includes a complex of sensory structures. There is always a club-like body, the rhopalium, with a terminal solid statocyst and an associated hood, which is sensitive to the position of the medusa. Information from each marginal sense organ is transferred to an associated marginal centre of the nervous system, and thence to the nerve nets. There may also be ocelli associated with the marginal sense organs which are sensitive to light. Functionally these responses of the marginal sense organs to light and gravity are best known. In addition the animals are also sensitive to other stimuli such as touch and various chemicals. Receptors for these modalities may be present in marginal patches of sensory epithelium, especially in pits, or may be outside the marginal sense organs such as on the tentacles. Similarly, although sense organs are not present in most stauromedusae or in the planulae and polyps of the other orders, these animals also respond to a variety of stimuli via more scattered sense cells.

2.4.1 Structure of the marginal sense organs

The marginal sense organs of *Aurelia aurita* have been most extensively investigated (Schäfer, 1878; Schewiakoff, 1889; Chapman in Russell, 1970; Chapman, D.M. and James, 1973). Each complex consists of a small hollow club, the rhopalium, with associated structures (Figure 2.6). The rhopalium projects from the umbrellar margin. It contains a small diverticulum of the gastrovascular cavity and a terminal statocyst formed of endodermal tissue filled with crystalline statoliths. It is situated in a niche between two adjacent lappets with an overlying extension of the umbrellar margin, the hood. The epithelium of the basal portion of the rhopalium contains patches of specialized sensory cells. The exumbrellar epithelium facing towards the hood is thickened into a touch plate with supporting and sensory cells (Pollmans and Hündgen, 1981). Also in the exumbrellar epithelium, peripheral to the touch plate and in the subumbrellar endoderm just proximal to the statocyst, are patches of pigmented cells forming ocelli. At the base of the rhopalium are two subumbrellar sensory pits, each lined with a

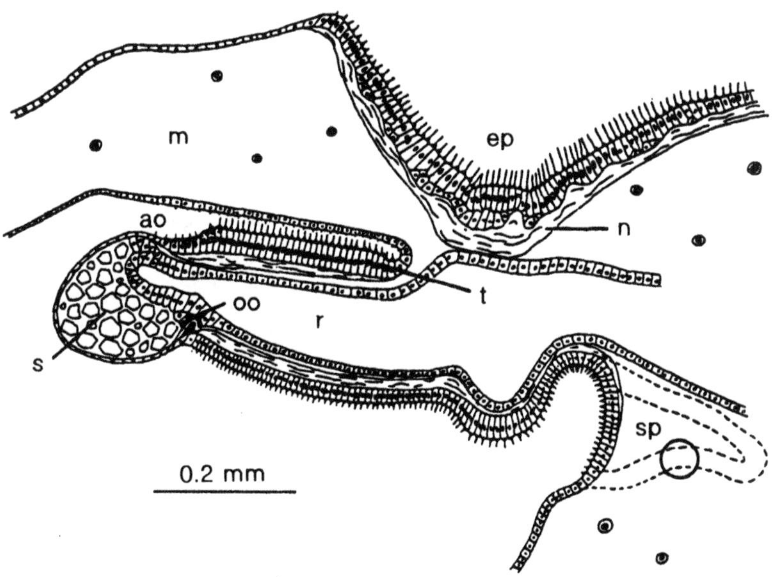

Figure 2.6 Radial section through a marginal sense organ of the medusa of *Aurelia aurita*. The subumbrellar pit is stippled in outline because each member of the pair is just to one side of the mid-line. ao = aboral ocellus; ep = exumbrellar sensory pit; m = mesoglea of hood; n = neurite layer; oo = oral ocellus; r = lumen of rhopalium; s = statocyst containing statoliths; sp = subumbrellar sensory pit; t = touch plate. (Source: Chapman and James, 1973, with permission of D.M. Chapman and Seto Marine Biological Laboratory.)

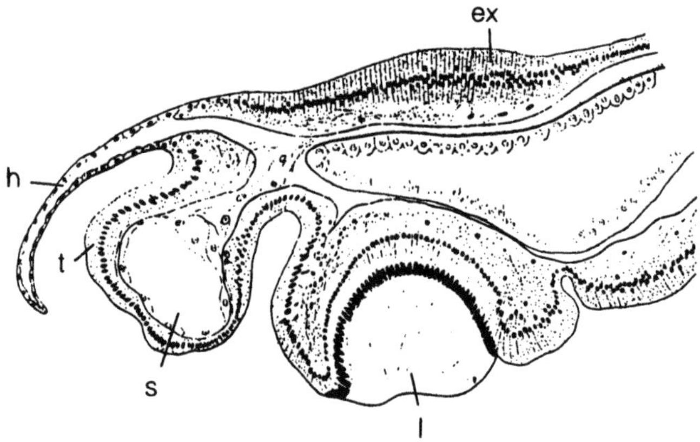

Figure 2.7 Radial section through a marginal sense organ of the medusa of *Paraphyllina intermedia*. ex = exumbrellar epithelium; h = hood; l = lens of ocellus; s = statocyst; t = touch plate. (Source: Maas, 1903.)

stratified epithelium including both columnar ciliated cells with basal axons and cells with intra-epithelial flagella (Chapman, D.M. and James, 1973; Pollmans and Hündgen, 1981). On the exumbrellar surface of the bell above the base of the rhopalium is another sensory pit, which is shallow, with radial folds (Maaden, 1939; Russell, 1970). The exumbrellar sensory epithelium overlies a thick layer of nervous and ganglion cells (Pollmans and Hündgen, 1981).

The marginal sense organs of other species of Semaeostomeae and Rhizostomeae may lack ocelli, but the general structure of the rhopalium and hood is very similar (Hesse, 1895; Bigelow, 1910; Bozler, 1926a; Wu, 1927; Russell, 1970; Titova, Vinnikov and Kharkejevich, 1979). In the Coronatae the hood is not formed by an extension of the umbrellar margin but is instead formed near the end of the rhopalium itself (Figure 2.7). The rhopalium may project from a basal cushion on the umbrellar margin, and may be cupped by a subumbrellar sensory bulb (Vanhöffen, 1900, 1902; Russell, 1970). The aberrant scyphozoan *Tetraplatia volitans* has saccular sense organs on the oral sides of the lappets each with a single statolith (Hand, 1955; Ralph, 1960). These have been considered as equivalent to the rhopalia of other scyphomedusae.

2.4.2 Photoreception

As indicated histologically by the presence of patches of pigmented cells, the marginal sense organs of a small proportion of species of

Semaeostomeae and Rhizostomeae contain ocelli. Each rhopalium of *Aurelia aurita* contains two ocelli: a flat ectodermal aboral one, and a smaller cup-shaped one on the oral side (Schewiakoff, 1889) (Figure 2.6). The smaller one is formed of a cup of pigmented endodermal cells surrounding a projection of cells from the ectoderm. Aboral ocelli are present on the rhopalia of *Cassiopea xamachana* (but not in *C. frondosa*) (see Bouillon and Nielsen, 1974). In the Stauromedusae, pigment spots are present on the oral surfaces of polyps of the genus *Stylocoronella* (see Salvini-Plawen, 1966, 1987). No lenses have been observed in these orders.

A few coronate species possess more complex ocelli. Lenses have been observed in *Nausithoe punctata* (see Hertwig and Hertwig, 1878). Maas (1903) described an ocellus on the oral side of a rhopalium of *Paraphyllina intermedia* which had a cup-shaped layer of pigment and a spherical lens (Figure 2.7). However *P. ransoni* does not possess ocelli (Russell, 1956). These coronate ocelli have not been investigated further.

Although pigmented cells are often associated with photoreception, clear proof of photoreception requires electrical recording of cellular responses to light. In other phyla pigmented cells may not be sensory, and the pigment may not be a photochemical, acting instead as a light barrier shielding the actual sensory cells. The presence or absence of pigmented ocelli in two species of the same genus such as *Cassiopea*, and the widespread sensitivity to light in species without pigmented ocelli, have led workers to question whether the pigmented cells of scyphozoa ocelli are light sensitive.

Electrical potentials have been recorded from the vicinity of the ocelli of *Aurelia aurita* stimulated by light, but have not been linked to individual cell types (Irisawa, Irisawa and Nishida, 1956; Yamashita, 1957). In ephyrae the development of the ability to respond to light-on or light-off stimuli temporally parallels the development of the oral ocelli (Yoshida and Yoshino, 1980). Using electron microscopy, the oral ocellus is seen to consist of a single cup-shaped layer of pigmented cells in the endoderm, surrounding a mass of ciliated cells in the ectoderm. It is considered that the latter are probably the sensory cells because they bear paired cilia which face the mesoglea rather than the exterior in the adult, and because they make synaptic connections with neurons (Yamasu and Yoshida, 1973; Pollmans and Hündgen, 1981). In *A. aurita* examined with freeze-fracture, these putative sensory cells of the oral ocellus show intramembranous particles similar to those of photoreceptors in higher animals (Takasu and Yoshida, 1984).

In the aboral ocellus of *A. aurita* (and in *Cassiopea xamachana*), only a single layer of ciliated and pigmented cells is present so that, if the structure is light sensitive, then the sensory cells are pigmented (Yamasu and Yoshida, 1973; Bouillon and Nielsen, 1974; Pollmans and Hündgen, 1981). There are low concentrations of intramembranous particles present in these cells (Takasu and Yoshida, 1984). Until direct recordings can be made from these ocelli, it must be concluded that the aboral 'ocellus' is unlikely to be photosensory.

2.4.3 Equilibrium reception

Equilibrium reception depends on the statocysts and sense cells (often organized into touch plates) of the rhopalia. The position of the rhopalial club is affected by gravity due to the heavy statocyst in its tip. This bends the sense cells toward or away from the hood as the animal tilts.

A statocyst is formed by endodermal lithocytes. The lithocytes secrete intracellular or extracellular crystalline or amorphous mineral deposits known as statoliths. In *Aurelia aurita, Chrysaora hysoscella* and *Cyanea capillata* the statoliths are composed of calcium sulphate dihydrate, i.e. gypsum (Spangenberg and Beck, 1968; Vinnikov *et al.*, 1981; Chapman, D.M., 1985). Gypsum is rare in biological systems, neverthless even in low sulphate sea water *A. aurita* does not utilize phosphate to form the more common calcium phosphate (Spangenberg, 1981).

Rhopalia and the associated statocysts of *A. aurita* are first formed during strobilation (Spangenberg, 1968b, 1991). Metamorphosis can be induced by thyroxine, so that statolith synthesis in the lithocytes can be studied at will in culture (Spangenberg, 1967). The statoliths are formed in calcifying vesicles and remain within intracellular vacuoles when completed (Spangenberg, 1976). Formation also occurs inside lithocytes of *Cyanea capillata* (see Vinnikov *et al.*, 1981).

The chemical reactions involved in the synthesis are not understood. As might be expected, normal formation of the calcium sulphate requires the presence of calcium and sulphate in the sea water bathing the medusa. However, the complex effects of other ions in the bathing medium on the mineralization system cannot yet be explained (Spangenberg, 1968b, 1979, 1981, 1986). In *Aurelia aurita*, while phosphate is not incorporated into the statoliths it enhances statolith synthesis, and acid phosphatase is present in the calcifying vesicles (Spangenberg, 1976, 1981). In *Cyanea capillata* the cytoplasm surrounding the calcifying vesicles during synthesis contains the enzyme carbonic anhydrase, which catalyses the reaction between carbon dioxide and water to produce carbonic acid (Aronova, Kharkeevich and Tsirulis, 1986).

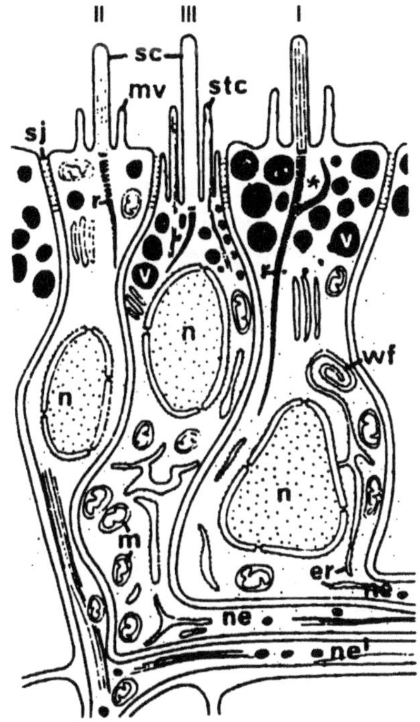

Figure 2.8 Receptor cell types I to III in the touch plate of the rhopalium of *Aurelia aurita*. er =endoplasmic reticulum; m = mitochondrion; mv = microvillus; n = nucleus; ne = neurites of receptor cells I and III; ne′ = neurite of receptor cell II; r = rootlet; stc = stereocilium; sj = septate junction; sc = sensory cilium; v = vacuole; wf = whorled myelin figure. (Source: Hündgen and Biela, 1982, with permission of M. Hundgen and Academic Press.)

The touch plates of *Aurelia aurita* are formed of ciliated epithelial cells (Chapman, D.M., 1974). Three types of sensory epithelia cells can be distinguished depending on whether microvilli or stereocilia surround the base of the cilium, and the presence or absence of a basal projection to the mesoglea (in addition to a basal neurite) (Hündgen and Biela, 1982) (Figure 2.8). The particular function of each of these cell types has not been identified. Similar sensory cells with motile cilia are present on the rhopalia of ephyrae prior to the development of the touch plate (Spangenberg, 1991). Ectodermal cells at the base of the statocyst of the coronate *Nausithoe punctata* correspond to the touch plate of *A. aurita* (see Horridge, G.A., 1969). Each cell extends a kinocilium toward the hood and bears an axon at its proximal end.

That the rhopalium acts as a gravity receptor can be shown by removing all but one of these sense organs. The contraction frequency of the medusa is then dependent on its vertical position relative to the remaining rhopalium. Early workers differed on which is the optimum position, (for review, see Passano, 1982). More recent workers agree that the contraction frequency is greatest when the remaining single rhopalium of *Aurelia aurita* or *Cyanea* sp. is uppermost (12 o'clock position) (Horridge, G.A., 1956a; Passano, L.M., 1982).

2.4.4 Other sensory responses

In addition to gravity and light, scyphomedusae show behavioural responses to other stimuli such as touch, various chemicals, pressure and temperature. It is not known whether all of these require specialized receptors. Temperature, for example, may act directly on the muscles concerned to affect the rates of contraction.

Sensitivity to mechanical and chemical stimuli is widely distributed. In *Aurelia aurita* polyps, mechanical stimuli on the tentacles (but not the column) causes a protective contractile response (Schwab, 1977a). This response shows habituation, i.e. a decrement in response following repeated stimuli (Johnson and Wuensch, 1994). As will be discussed in sections 3.4.2 and 3.5.3, chemical stimuli are important in attracting scyphozoa to prey and in controlling feeding behaviour in both polyps and medusae.

Observed receptors have not been correlated with mechano- or chemoreception. In medusae, ciliary currents, which might bring chemicals, are associated with the sensory pits of the marginal sense organs. Putative sensory cells may also be obtained by macerations of the subumbrellar epithelium (Krasińska, 1914). In both medusa and polyp there may be receptors associated with the cnidae (section 3.2.1). The only other sensory cells in the polyp, so far identified by electron microscopy, are present on the tentacles (Westfall, 1973; Chia, Amerongen and Peteya, 1984; Spangenberg, 1991). The surface of each of these cells has a cilium with a circle of microvilli and there is a single axon at the base of the cell. Earlier workers, using maceration or histological methods, also described similar cells in the epithelia of the calices of stauromedusae (Kassianow, 1901).

Although planulae are able to respond to their environment, particularly to complex clues for settling, the presence of sensory cells has not been firmly established. Widersten (1968) observed cells at the surface of the ectoderm of *Cyanea capillata* planulae which stained with methylene blue. Later workers, using electron microscopy, did not identify sensory or nervous cells in the ectoderm of planulae of *Haliclystus salpinx* or *Cassiopea xamachana* (see Otto, 1978; Martin, V.J. and Chia, 1982).

2.5 NERVOUS SYSTEM

The scyphozoan nervous system contains similar nerve cells (neurons) to those of higher animals. However, they are usually arranged in networks extending through the tissues between other cell types rather than in discrete nerves. For many years it was difficult to separate the neurons of these nerve nets from the surrounding tissue. Experimenters deduced the properties of the nerve from ingenious cutting experiments, and later by extracellular recording of electrical impulses. The results were reviewed by Passano (1982).

More recently a preparation has been developed which allows direct examination of the neurons from the peri-rhopalial area central to the marginal sense organs of *Cyanea capillata* medusae (Figure 2.9). The overlying myoepithelial cells can be removed and modern intracellular recording can be carried out on the neurons (Anderson and Schwab, 1984). This preparation has attracted the attention of neurophysiologists because of the significant discovery of unusual two-way transmission at the synapses between two neurons (for reviews, see Anderson and Spencer, 1989; Spencer, 1989). Caution must be used in generalizing from this preparation to other scyphozoan nerves.

2.5.1 Nervous system of medusae

The early experiments on nervous systems of scyphomedusae showed that at least three elements were involved in control of swimming. They include the marginal centres, associated with the marginal sense organs, which generate the swimming rhythm; the motor nerve net (MNN) innervating the swimming muscle; and a diffuse nerve net (DNN) bringing sensory information to the marginal centre (Passano, L.M., 1982) (Figure 2.19). It is less clear what other nervous elements may be present to control other functions.

The marginal centres, motor nerve net and diffuse nerve net will each be discussed in more detail in subsequent sections. The exact structures and locations of the marginal centre are not known (section 2.5.2). The term marginal centre refers to an endogenous pacemaker with input into the MNN, shown by cutting experiments to lie in the vicinity of the base of the rhopalium. The MNN (section 2.5.3) is a single functional unit, consisting of a network of relatively large fusiform bipolar neurons which extends in the subumbrellar ectoderm from the marginal centres over the swimming muscle. The DNN has been less exactly delineated. It was first defined as including any dispersed neurons that were not part of the MNN (Horridge, G.A., 1956b). It therefore included both motor and sensory cells of the exumbrella, tentacles and endoderm, as well as of the subumbrellar ectoderm including the manubrium. Recently authors have restricted the term to the primarily sensory through-conducting nerve net with input to the marginal centres, especially that present on the subumbrella (Passano, L.M., 1982; Anderson, Moosler and Grimmelikhuijzen, 1992). It will be used in this restricted sense in section 2.5.4.

If the term DNN is used in the restricted sense, it should then be possible to distinguish other nets either functionally or anatomically. The evidence for these nets is very fragmentary, but it is clear that various functions must be controlled separately from the through-conducting MNN and DNN. As a minimum these include:

1. local or regional contractions of the swimming muscles, including asymmetrical responses involved in turns and compensatory movements;
2. contractions of local elements of the animal such as individual tentacles, marginal lappets, or oral arms;
3. coordinated feeding responses;
4. control of horizontal and vertical migration;
5. control of cnidae discharge as described in section 3.2.3.

There may also be separation of sensory nets in the exumbrella, tentacles or endoderm from the subumbrellar DNN. Potentially the nervous system may be very complex.

There is good evidence for a motor network in each tentacle. Earlier workers considered that the contraction of the tentacles was controlled by the DNN. However, the response to pulses in the subumbrellar DNN is inconsistent, leading Passano (1982) to hypothesize facilitated junctions between the DNN and separate tentacle networks. Using immunocytochemical staining of tentacles of *Chrysaora hysocella*, *Cyanea capillata* and *Cyanea lamarcki* for RFamide-like peptides, Anderson, Moosler and Grimmelikhuijzen (1992) have shown a dense ectodermal nerve net which ends abruptly at the junction with the subumbrella (Figure 2.15). This net does not include sensory cells. In the two *Cyanea* species it includes a concentrated tract associated with a longitudinal muscle, indicating a motor function. Tapering processes from this net extend below the DNN to the base of the subumbrellar ectoderm. However, the interconnections between the two nets have not been identified histologically. Electrical activity of the marginal tentacles of *C. capillata* includes two types of discharges differing in amplitude and frequency, one of which may belong to this tentacular motor nerve net (Sviderskaya, Polyakova and Voskresensky, 1990).

Immunocytochemical staining indicated the presence of an exumbrellar nerve net in *Chrysaora hysocella* which is particularly dense near the margin (Anderson, Moosler and Grimmelikhuijzen, 1992). This net includes sense cells projecting to the ectodermal surface. A similar net has been described on the aboral surface of *Aurelia aurita* ephyrae using methyl blue staining (Horridge, G.A., 1956b). It is not known whether this net is continuous with the DNN.

A diffuse net (predominately of large bipolar neurons) in the endoderm lining the gastric cavity has been observed in *Rhizostoma pulmo, Cyanea capillata* and *Phacellophora camtschatica* by methyl blue staining (Bozler, 1927a; Passano, K.N. and Passano, L.M. 1971). A similar net is observed in *Chrysaora hysocella, Cyanea capillata* and *Cyanea lamarcki* following immunocytochemical staining for RFamide-like peptides (Anderson, Moosler and Grimmelikhuijzen, 1992). These are the largest neurons outside of the MNN. Primary sensory cells have been observed only on the gastric cirri, so the net is unlikely to be primarily sensory, but it has not yet been linked with its functions.

2.5.2 Marginal centres

The term marginal centre was first used by Passano (Passano, L.M., 1982) for the area which generates the rhythmic electrical potentials in the motor nerve net. It is a functional term. Passano had previously shown (contrary to some earlier workers) that the pacemaker output continued after the outer part of the rhopalium had been excised (Passano, L.M. 1973). Based on cutting experiments around the rhopalium, he deduced that the marginal centre lies close enough to the root of the rhopalium that it can be damaged when the immediate area of the rhopalium is removed, but it remains unimpaired when just the rhopalium is carefully removed.

The location of the marginal centre approximately corresponds to the aggregation of nerve cells at the base of the rhopalium which has been referred to as the marginal ganglion. However, other functions are also carried out by the ganglion. The area contains numerous sensory cells in the sensory pits (section 2.4.1). The ganglion of *Aurelia aurita* ephyrae, described by Horridge (1956b) from methylene-blue stained preparations, included both sensory cells and connections to the nerve nets. Subsequent electron microscopy established the absence of non-nervous glial cells, and presence of symmetrical synapses between the cells of the area, but did not further correlate cell type with function (Horridge, G.A. and Mackay, 1962; Horridge, G.A., Chapman and MacKay, 1962). The marginal centre therefore remains a 'black box' (Barnes, W.J.P. and Horridge, 1965; Passano, L.M., 1982).

In order to maintain the swimming rhythm of the medusa only one marginal centre need be present. This was shown independently by Eimer (1874, 1877) and Romanes (1876, 1877), who cut off the marginal sense organs until the swimming rhythm ceased when the final one was removed. Rhythmicity, including interactions between two or more centres, and between the centres and the MNN and DNN, has been investigated by many subsequent authors usually working on species of *Aurelia, Cassiopea* and *Cyanea* (see literature and discussion in Horridge, 1959; Lerner *et al.*, 1971; Passano, 1973, 1982; Murray, 1977; Voino-Yasenetskii *et al.*, 1979). The earliest workers monitored the contractions of the swimming muscle under various conditions. Subsequent authors recorded electrical potentials from the MNN and DNN with electrodes external to the tissues, and most recently it has been possible to record directly from the MNN.

The marginal centres generate action potentials in the MNN which are through-conducted to all parts of the net. Each single pulse causes a contraction of the swimming muscle. When pairs of pulses are released in *Cassiopea* and *Cyanea*, the second may fall within the refractory period before muscle recovery so that the muscle only responds to the first of each pair. Each centre is spontaneously active and potentially capable of maintaining the swimming rhythm, but the fastest centre will control the rate of the beats at any one time. Each potential from the fastest centre resets the endogenous rhythm of the other centres as it reaches them. If another centre increases its speed of generation, it will then set the rhythm.

The marginal centres receive various inputs that influence the rate of output. In addition to input from other marginal centres via the MNN, they are also influenced by information brought through the DNN, by information from the marginal sense organs, and probably by direct effects of factors such as temperature and light. The frequency of beating decreases with increased size of the medusa.

2.5.3 Structure and function of the motor nerve net

The presence of a nerve net controlling the subumbrellar swimming muscle was first suggested by cutting experiments. The term 'net' implies a diffuse, essentially two-dimensional assemblage of neurons.

Cutting experiments showed that impulses generated by marginal centres could take alternative pathways prior to excitation of the swimming muscle, implying the existence of a net rather than a tract of fibres.

The best-known conducting systems in coelenterates, such as the MNN, are through-conducting. In some other systems the wave of excitation initiated by a single stimulus stops before reaching the boundaries of the system. However, in through-conducting systems the responses mediated (in this case muscle contraction) are not graded in intensity with distance from the point of stimulation. The distance of excitation spread is independent of the total number and frequency of stimuli. This was shown in the 'entrapped wave preparation', a loop, including the subumbrellar swimming muscle and associated neurons, but with the oral structures and marginal centres of the medusa cut away. If a single wave of contraction was started it would circle the preparation at a fairly uniform rate for a few days, indicating that the system was continuously conducting. Results from further use of the preparation by Mayer (1906) and subsequent workers to examine the responses of nerve and muscle to ions, etc. were difficult to interpret due to the inability to separate muscle, nerve and surrounding tissue. Passano (1982) reviewed this extensive literature in the light of modern physiology.

Horridge (1953, 1954) first demonstrated directly that conduction depends on the presence of nerve fibres. It was known from histological staining that neurons, including a number of relatively large intercrossing bipolar cells, exist in the subumbrellar ectoderm of *Aurelia aurita* and *Rhizostoma pulmo* (see Schäfer, 1878; Bozler, 1927a,b). Using phase-contrast illumination, Horridge (and later Bergström, 1971) examined the conduction through narrow bridges of this tissue from *A. aurita*. He found that conduction depended on the presence of at least one of the large bipolar nerve fibres. He was also able to record conducted electrical impulses similar to those of higher animals. Subsequently examining this net over the muscles of *A. aurita* ephyrae, he described it as the giant fibre nerve net (GFNN), a misnomer as no scyphozoan neurons are very large (Horridge, G.A., 1956b). However, these are the largest neurons known in the Scyphozoa. More recently it has been renamed the motor nerve net (MNN) by Anderson and Schwab (1982).

The MNN has been examined most extensively in the preparation of the peri-rhopalial tissue epidermis of *Cyanea capillata* which was developed by Anderson and Schwab (1984). In this species a wedge of peri-rhopalial tissue without swimming muscle lies between the radial muscles bands and central to the marginal sense organs

Figure 2.9 Peri-rhopalial tissue epidermis of *Cyanea capillata*. The epidermis is composed of large, somewhat cuboidal, vacuolated epitheliomuscular cells. Part of a cell on the left has been cut to reveal its internal organization. Myofibrils are contained in the basal processes and also run up through the cell body toward the outer surface of the cell. Neurons of the motor nerve net (MNN) pass between the epitheliomuscular cells through gaps above the basal processes. (Source: Anderson and Schwab, 1981. Reprinted by permission of P.A.V. Anderson and John Wiley & Sons, Inc.)

(Figure 2.2). The relatively thin layers of epidermis, mesoglea, and gastrodermis separate part of the gastric canal system from the surrounding sea water. The epidermis (Figure 2.9) is composed of large, somewhat cuboidal, vacuolated, epitheliomuscular cells with the muscle tails forming a single layer of radial smooth muscle (Anderson and Schwab, 1981). The MNN neurons pass between the epitheliomuscular cells in gaps at the top of the muscle layer. The epitheliomuscular cells are connected by septate desmosomes so that they constitute a physical and diffusion barrier. They can be removed by osmotic shock or brief oxidation of the surface with sodium hypochlorite. As the basal processes of the epitheliomuscular cells are removed, the neurons settle on to the mesoglea. Provided a saline is present matching the ion content of the mesoglea, these neurons remain viable with functioning synapses. The neurons can be seen clearly on the acellular transparent mesoglea and intracellular recordings can be obtained (Figure 2.10).

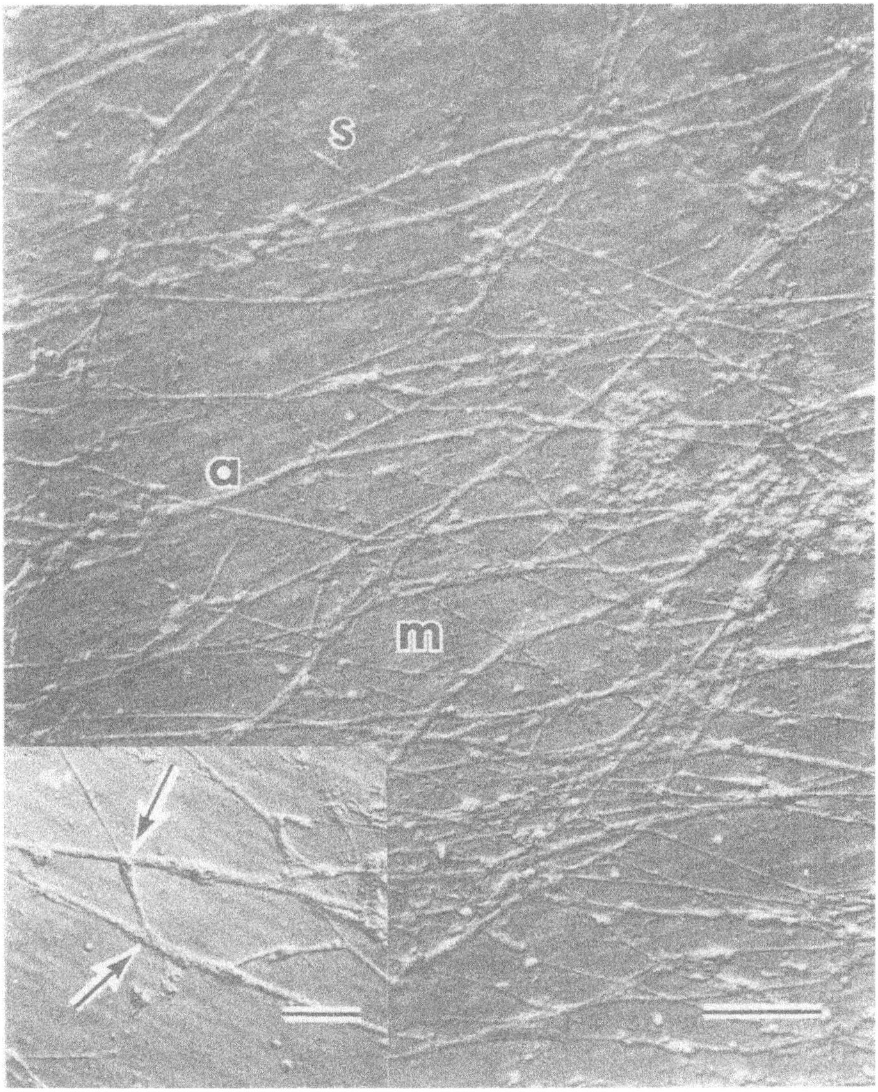

Figure 2.10 Peri-rhopalial preparation of the nerve net of *Cyanea capillata*. The epitheliomuscular cells that normally overlie these neurons have been removed, exposing the nerve net. The neurons now lie on the optically clear, acellular mesoglea. Two synapses (arrows) are shown at higher magnification in the inset. a = axon; m = mesoglea; s = somata (cell body). Scale bar on photograph = 100 μm; on insert = 50μm. (Source: Anderson and Spencer, 1989. Reprinted by permission of P.A.V. Anderson and John Wiley & Sons, Inc.)

In *Cyanea capillata* the MNN consists of criss-crossed millimetre-long bipolar neurons, with cell bodies and neurites (neuronal processes) 10–20 μm and 1–5μm in diameter, respectively (Anderson and Schwab, 1981) (Figure 2.10). The large unbranched neurites have often been referred to as axons although use of this term does not necessarily imply that conduction of the impulse is away from the cell bodies. The neurites contain unusual large vacuoles with cisternae-like inclusions (Anderson and Schwab, 1981). The neurites also contain many microtubules. With video analysis of microphotographs these microtubules can be observed to transport organelles along the neurite (Anderson *et al.*, 1986). Mitochondria move more slowly (0.99 μm/second) than smaller organelles.

The major role of the MNN is to transmit electrical impulses from the marginal centres and to stimulate the swimming muscle (section 2.3). Transmission of information along a neuron depends on maintenance of a resting membrane potential, stimulation, and then propagation of an action potential along the neurite. In long-distance transmission there must then be transmission across the synapse between each pair of neurons.

The resting membrane potential of the MNN neurons is –55 to –70 mV (Anderson and Schwab, 1983), i.e. the inside of the cells is negatively charged with respect to the outside. Such potentials are due to the diffusion of ions and are determined by the membrane's relative permeability to ions and the ionic differences across the membrane. In almost all known resting neurons, membrane pumps maintain a low sodium and high potassium concentration within the cell. The membrane at rest is relatively impermeable to sodium, i.e. no channels are open in the membrane through which this ion can move. Positive potassium ions diffuse outward through their channels down their concentration gradient (causing the interior of the cell to become negative), until at their equilibrium potential inward electrical forces balance the outward concentration gradient. The value of the equilibrium potential for any ion can be calculated from its concentration gradient across the membrane.

During the action potential, the potential of the stimulated membrane depolarizes (decreases toward 0) and may overshoot (reverse so the membrane becomes positive inside) for a brief period (Figure 2.11) (Anderson and Schwab, 1983). This depolarization causes depolarization in neighbouring portions of the neurite and triggers a similar sequence there, thus propagating along the neurite. The original portion of membrane then repolarizes, often showing a hyperpolarizing (greater than resting) afterpotential before returning to the resting level. In MNN neurons the depolarization is due

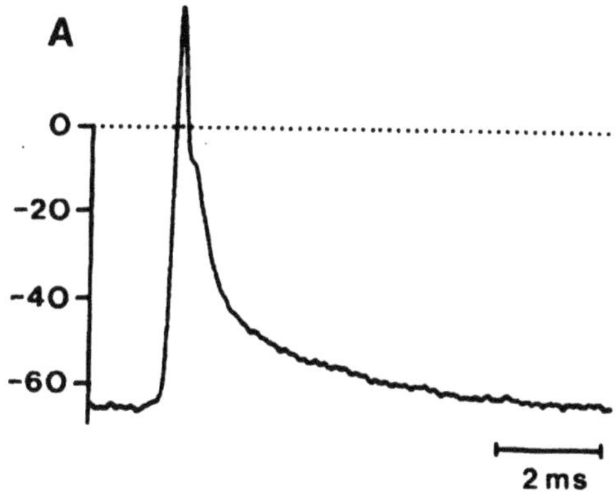

Figure 2.11 A single spontaneous action potential recorded from the axon of a motor neuron of a *Cyanea* sp. medusa. (Source: Anderson and Schwab, 1983, with permission of P.A.V. Anderson and the American Physiological Society.)

primarily to movement of sodium ions into the neuron through briefly open sodium channels (Anderson and Schwab, 1983; Anderson, 1987, 1989). The sodium channels are complex proteins with amino acid sequences similar to sodium channels of higher animals (Anderson, Holman and Greenberg, 1993). Repolarization depends on potassium efflux through two types of channels: one voltage-sensitive and one calcium-activated (Anderson and Schwab, 1983).

In the MNN, while the action potential changes last only 10–20 ms at any one point on the neurite, the membrane remains refractory (unable to carry another action potential) for a longer period. For an absolute refractory period of approximately 30 ms no stimulus can cause another action potential, then for a further relative refractory period of approximately 70 ms a stronger than normal stimulus is effective.

The MNN neurons are straight bipolar cells but they criss-cross one another randomly and extensively. At the intersections synapses form so that each neuron forms numerous synapses with different cells. The synapses can occur on any part of the cell, including the cell body. The two neurites at a synapse contain two directly opposing clumps of unusually large synaptic vesicles, indicating that chemical transmitters can be released by either neurite, i.e. that transmission is bidirectional (Anderson and Schwab, 1981) (Figure 2.12). Serial sections have revealed a complex structure in which the vesicles lie in a single layer against a region of membrane density of each terminal and are covered on the cytoplasmic side by a large, perforated cisternal sheet (Figure 2.13) (Anderson and Grünert, 1988).

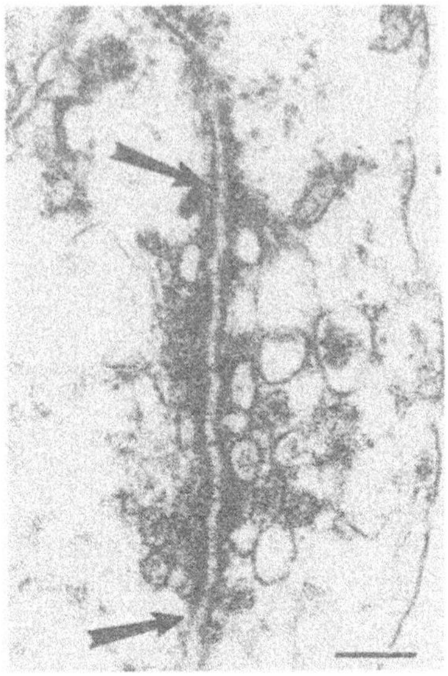

Figure 2.12 Synapse between two neurons of the motor nerve net of *Cyanea capillata*. The limits of the synapse are indicated with arrows. Note that vesicles occur on both sides of the synapse. Scale bar = 0.2 μm. (Source: Anderson and Schwab, 1981. Reprinted by permission of P.A.V. Anderson and John Wiley & Sons, Inc.)

It is possible to record from both the cells involved in a MNN synapse and to verify that the synapses transmit in either direction. When an action potential reaches the synaptic terminal of a presynaptic neuron it causes an excitatory post-synaptic potential (EPSP) in the post-synaptic cell. There is a 1 ms delay, presumed to be due to the release and diffusion of a chemical transmitter. The EPSP in turn causes an action potential in the post-synaptic cell (Figure 2.14) (Anderson, 1985; Anderson and Spencer, 1989). Depolarizations of 0 mV or more are required for transmitter release, so transmitter is released only by action potentials, not EPSPs, preventing continuous depolarization of the terminals.

2.5.4 Diffuse nerve net

That a second through-conducting network is present in the subumbrella of scyphozoa was first shown by Romanes (1877) in *Aurelia aurita*. This second net sometimes causes a wave of tentacular contraction which can be elicited separately from the swimming muscle contraction controlled by the MNN. Using a strip of muscle with only

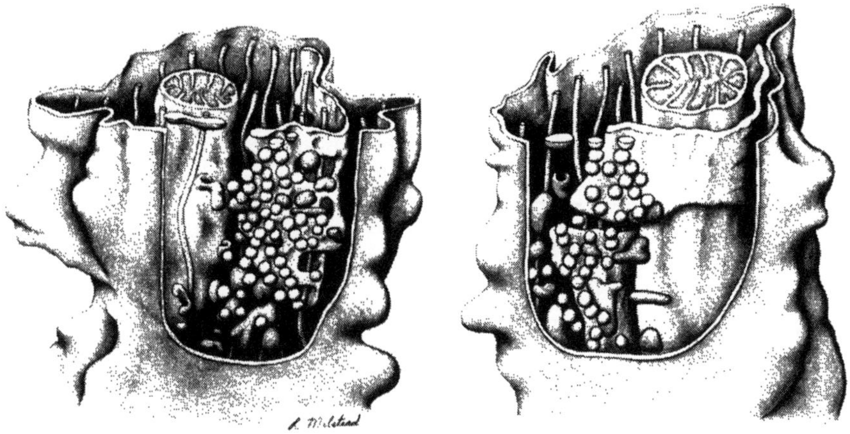

Figure 2.13 Bidirectional, excitatory chemical synapse from the jellyfish *Cyanea capillata*: drawing based on reconstruction from serial sections through one synapse. The synapse is viewed as if the two terminals were hinged and pulled open, with the synaptic membrane removed to reveal the interior of each terminal. Synaptic vesicles are drawn as light spheres, the bulbous cisternae are more irregular and darker, and both have a single, large elongate cisternal sheet covering their cytoplasmic side. A mitochondrion and several microtubules are also shown. Approximate magnification ×16 000. (Source: Anderson and Grünert, 1988. Reprinted by permission of P.A.V. Anderson and John Wiley & Sons, Inc.)

one marginal centre, Romanes found that when a wave of tentacular contraction reached a marginal centre it elicited a wave of muscular contraction after a delay of at least half a second. Horridge (1956a) was able to summarize several other instances of excitation that could cross the subumbrella without giving rise to a contraction wave en route. He referred all these responses to the diffuse nerve net.

When electrical recording from the tissue surface became possible it was found that, in addition to those from the MNN, potentials of different shape and amplitude could be recorded from a second subumbrellar net (Passano, L.M., 1965; Kokina, 1971). This slower but through-conducting net was able to elicit MNN impulses from the marginal centres (Passano, L.M., 1965, 1973, 1988). It varied from species to species in its other effects. It may have direct effects on the swimming muscles as well as the MNN (section 2.3.3).

These physiological properties have been ascribed to at least part of the diffuse network of multipolar neurons which may be stained on the subumbrella (section 2.5.1). So far direct recording from these neurons has not been achieved. They have not been identified in the

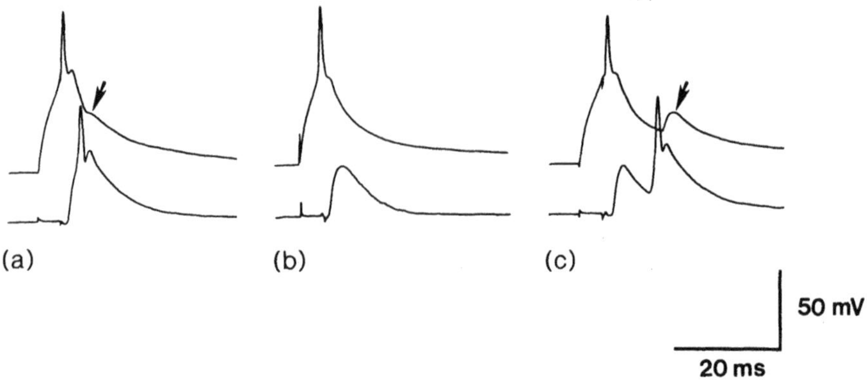

Figure 2.14 Intracellular recordings from pairs of motor nerve net neurons from *Cyanea capillata*, to show that the inter-neuronal synapses are bidirectional excitatory synapses. In all records stimuli are applied to the cell on the upper trace, i.e. the top trace is from the presynaptic cell and the lower trace is from the post-synaptic cell. (a) When an action potential is triggered in the post-synaptic cell with 1 ms delay after an action potential in the presynaptic cell, it in turn causes a notch (arrow) in the falling phase of the presynaptic action potential. (b) If the post-synaptic cell exhibits an excitatory postsynaptic potential (EPSP) which does not give rise to an action potential, then no response is found in the presynaptic cell. (c) If the action potential in the post-synaptic cell is delayed, it produces a 'return' EPSP in the presynaptic cell 1 ms later. (Source: Anderson (1985), with permission of P.A.V. Anderson and the American Physiological Society.)

peri-rhopalial preparation used for the MNN neurons, probably because they have been oxidized and removed with the epitheliomuscular cells (Anderson and Grünert, 1988). The correlation between the multipolar neurons and DNN function is therefore tentative (Passano, L.M., 1982; Anderson, Moosler and Grimmelikhuijzen, 1992).

2.5.5 Nervous system of polyps

Less is known about the nervous system in polyps than in medusae. Neurons have been identified in the epithelia of the tentacles, oral disc and muscle cord of the scyphistomae of *Aurelia aurita, Chrysaora quinquecirrha, Cassiopea andromeda* and *Cassiopea xamachana* (see Chapman, D.M., 1965; Korn, 1966; Loeb and Hayes, 1981; Chia, Amerongen and Peteya, 1984; Hofmann and Hellman, 1995). There is a concentration of neurons at the base of the tentacles. Ciliated sensory cells are also present in the tentacles (Westfall, 1973; Chia, Amerongen and Peteya, 1984). Neurosecretory cells are present during strobilation and budding (section 6.4). The limited nervous system supports a limited behavioural repertoire of local movement plus feeding and a protective spasm involving tentacles and column (Chapman, D.M., 1965; Schwab, 1977a).

In the polyp of the coronate *Atorella japonica* a more dense nerve plexus forms a ring on the upper column below the tentacles (Matsuno and Kawaguti, 1991).

The exumbrellar surface of the stauromedusa *Haliclystus auricula* is relatively insensitive to mechanical stimuli but the tentacles and oral surfaces are very sensitive (Gwilliam, 1960). The conducting system is diffuse, possibly due to a single nerve net. In the tentacles, axons are associated with basal myofilaments of the epitheliomuscular cells (Westfall, 1973). The basal disc has large nerve bundles which may be involved in attachment or detachment (Singla, 1976). Locomotion in these animals involves somersaulting (section 2.6.3).

2.5.6 Transmitters

In higher animals transmission between neurons is due to movement of chemical transmitters across a synaptic cleft between the cells, or to electrical coupling via gap junctions. In the latter case, cell interiors are directly linked by aqueous channels through gap junctional particles, allowing movement of compounds (including experimental dyes) between the two cells. Gap junctions have been identified in hydrozoa where they form the basis for epithelial conduction. To date no dye coupling or other evidence for gap junctions has been found in scyphozoa (Anderson and Schwab, 1981; Mackie, Anderson and Singla, 1984; Anderson, 1985).

As noted above, vesicles have been observed at scyphozoan synapses, indicating that transmitters may be released. A number of workers have investigated the chemical nature of these transmitters. Chemicals may be added to the media bathing whole specimens or preparations, and behavioural or physiological responses may be monitored. Alternatively the presence of particular chemicals may be investigated by extraction, or mapped histologically with specific fluorescent or immunocytochemical compounds. Either method must be followed by examination of the physiological effects of the putative transmitters on particular post-synaptic cells in order to verify their role as neurotransmitters.

Based on what is known in higher animals, neurotransmitter candidates include acetylcholine, catecholamines including dopamine, norepinephrine and epinephrine, other amines including serotonin (5-hydroxytryptamine, 5-HT), peptides including those similar to Phe-Met-Arg-Phe-amide (FMRFamide), and amino acids including GABA (gamma-aminobutyric acid) and taurine (Martin, S.M. and Spencer, 1983). As detailed below, none of these have been shown unequivocally to act as transmitters in scyphozoa.

Acetylcholine and its agonists and antagonists have no effect on contraction rate when applied to contracting segments of *Cyanea* sp. (see Horridge, G.A., 1959). Acetylcholinesterase, the enzyme which inactivates acetylcholine, cannot be demonstrated histochemically in the swimming motor neurons of *Cyanea* (see Scemes, 1989). Acetylcholine is therefore not a neurotransmitter in the MNN. However, the sense cells of the rhopalium stain for acetylcholinesterase (Aronova *et al.*, 1979), so acetylcholine may be active in other neurons.

The amino acid L-dopa and its catecholamine derivative dopamine have been extracted from the tissues of scyphomedusae but the amounts vary greatly in different species (Carlberg and Rosengren, 1985). The indolamine serotonin was also extracted from the tentacles of *Cyanea lamarcki*, but not from the closely related *C. capillata*. Serotonin can be reliably identified with immunocytochemical methods, and it was only found in the mucus-producing ectodermal gland cells of *C. lamarcki*, not in the neurons (Elofsson and Carlberg, 1989). Tryptamine accelerated the contraction rate of *Cyanea* sp., but was less effective on (Horridge, G.A., 1959) or inhibited (Schwab, 1977b) *Aurelia aurita*. Biogenic amines are clearly present in scyphozoa, but they have not been shown to function at the level of an individual synapse.

The FMRFamide-like peptides are putative neuromodulators. Immunocytochemistry has shown the presence of peptides including the Arg-Phe-amide (RFamide) sequence in a network in the subumbrella of *Pelagia* sp. ephyrae (Grimmelikhuijzen, Graff and Spencer, 1988). The antiserum stains much of the cells so that the extent of the staining network can be judged. In *Chrysaora hysocella* medusae the antiserum revealed nerve nets in the ectoderm of the subumbrella and exumbrella, of both faces of the oral lobes, of the tentacles, especially at the base (Figure 2.15), and in the endoderm lining the subumbrellar and exumbrellar surfaces of the gastric cavity (Anderson, Moosler and Grimmelikhuijzen, 1992). Staining was not associated with either the bipolar MNN cells or the rhopalia. In *Cyanea capillata* and *Cyanea lamarcki* there were also small nerve nets associated with clusters of cnidocytes in the tentacles. In *C. capillata* there is staining of the marginal rhopalia (Carlberg *et al.*, 1995). Three RFamide peptides have been isolated from this species (Grimmelikhuijzen and Westfall, 1995). Immunoreactive neurons have been found in several developmental stages of *Cassiopea* spp. (Hofmann and Hellman, 1995). While these compounds are certainly present in the neurons, they

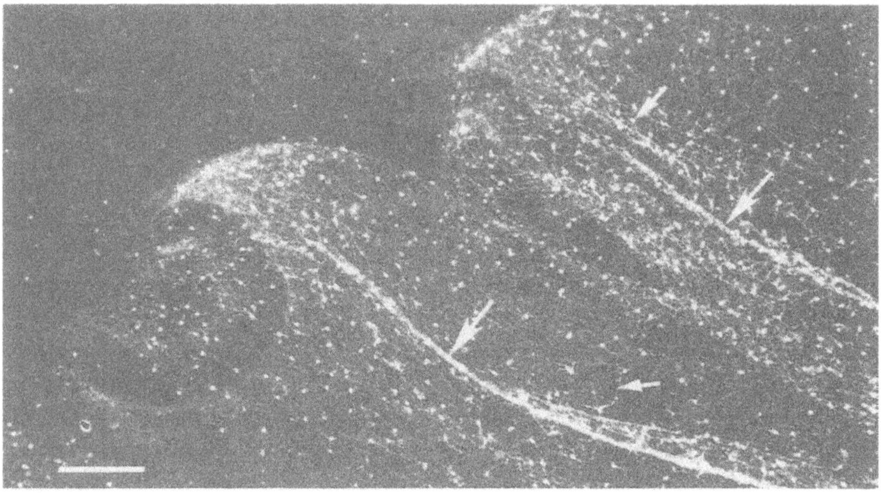

Figure 2.15 Micrograph of a whole mount of *Cyanea capillata* showing the base of two tentacles stained with RF-amide antiserum. The superficial tentacular nerve net (small arrows) and tentacular nerve tract (large arrows) are evident on both tentacles. Scale bar: 250 μm. (Source: Anderson, Moosler and Grimmelikhuijzen, 1992, with permission of P.A.V. Anderson and Springer-Verlag.)

have not yet been shown to function as transmitters. The wide distribution within each cell suggests instead diffuse release of the peptide and a modulatory or other role (Mackie, 1990).

The other putative neurotransmitters are amino acids. Gamma-aminobutyric acid (GABA) inhibits electrical activity of the marginal tentacles of *Cyanea capillata* (see Sviderskaya, Polyakova and Voskresensky, 1990). Antisera against the sulphonated amino acid taurine stain the peri-rhopalial portion of the MNN nerve net of *Cyanea capillata* (see Carlberg *et al.*, 1995). Double-labelling experiments demonstrated that some endodermal neurons were both taurine-immunoreactive and FMRFamide-immunoreactive, indicating that neurons may be utilizing multiple neurotransmitters or neuromodulators.

2.6 LOCOMOTION

Previous sections have considered the functions of individual tissues used in locomotion: mesoglea in section 2.2, muscle in section 2.3, sensory receptors in section 2.4 and nerve in section 2.5. The more complex integrated behaviour of locomotion will be covered in the present section.

2.6.1 Physical dynamics of swimming

A few medusae move by peristalsis. One example is the deep-water semaeostome medusa *Deepstaria reticulum*. This medusa has a voluminous thin-walled umbrella with a coronal muscle near the bell margin and a diffuse muscle extending over much of the subumbrella (Larson, Madin and Harbison, 1988). Peristaltic contraction waves may pass up or down the umbrella moving the medusa slowly along (Figure 3.10). The coronal muscle is used to purse the umbrella shut during feeding (section 3.5.1). However, swimming of most scyphomedusae depends on contraction of the coronal (and if present the radial) muscles to produce a jet of water, and the elastic recoil of the mesoglea to restore the resting shape.

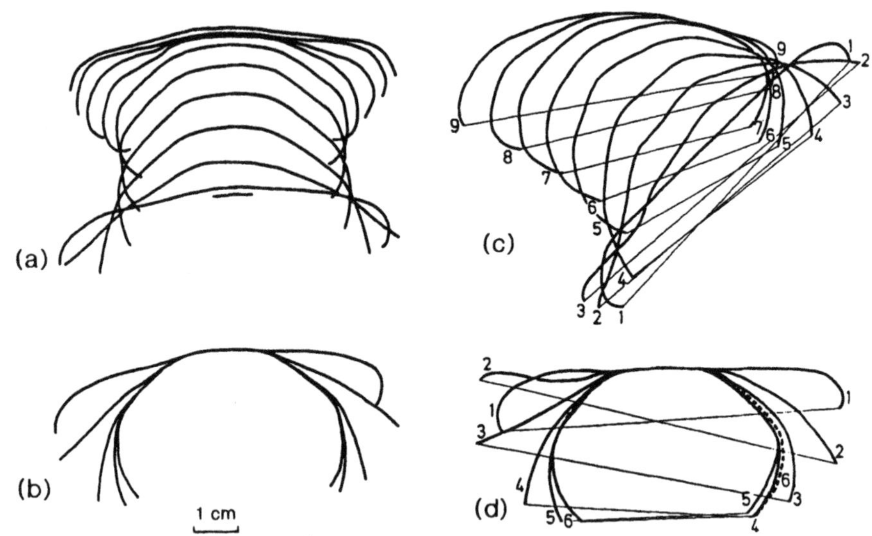

Figure 2.16 Sequential exumbrellar outlines of *Cyanea capillata* during straight swimming and turning, traced from cinematographic sequences; time interval 2/9 s. (a) Straight swimming showing actual change of position of umbrella (animals filmed against a fixed grid); (b) straight swimming, outlines from a swimming sequence superimposed; (c) turning, showing actual change in position; (d) turning, superimposed. (Source: Gladfelter, 1972, with permission of Springer-Verlag.)

The swimming stroke of *Cyanea capillata* has been most extensively analysed (Gladfelter, 1972) (Figures 1.7, 2.16). The coronal muscle draws the umbrella peripheral to the central disc inward and downward around a mesogleal joint. The radial muscle then causes a further flexion of the peripheral portion of the umbrella around radial mesogleal joints. There is an initial backward thrust of the umbrella on the water. There is also production of an outward jet of water with each contraction, and then inward currents as the bell recoils (Figure 3.6 shows similar currents in *Aurelia aurita*). These inward currents may be utilized for prey capture (section 3.4.1). There is sufficient drag on the medusa that by the end of recovery virtually all forward progress ceases, so that the animal must accelerate again during the next beat.

In turning there is an initial strong contraction on the side toward which the turn will be made (Figure 2.16). When the other swimming muscles also contract, the originally active side continues to be more bent due to its 'headstart', and the asymmetrical contraction pivots the medusa around the quadrant of the initial contraction.

The details of muscular distribution and of jointing in the mesoglea may differ in other semaeostome scyphomedusae, but the principles on which swimming is based are very similar. In *Aurelia aurita* there is a brief period of negative velocity, or backwards motion, during the refilling of the subumbrellar cavity (Costello and Colin, 1994). *Chrysaora melanaster* and *Pelagia noctiluca* lack radial muscles. Nevertheless, as in *Cyanea*, progression depends on a backward thrust on the water followed by extrusion of a jet of water from the subumbrellar cavity (Gladfelter, 1973). A young *P. noctiluca* reached a maximum velocity of about 4 cm/s during the swimming cycle, but averaged a progression of only 2 cm/s due to the rapid deceleration during the recovery phase of the cycle. The sequence in turning is also similar to that in *Cyanea capillata*.

The rhizostome medusa *Stomolophus meleagris* differs in having a more globular umbrella, no tentacles and a short oral-arm cylinder (Larson, 1987a). The circular swimming muscles cover about 80% of the subumbrellar surface. Progression depends primarily on production of an outward jet of water, without the backward thrust of the umbrella on the water seen in the more saucer-shaped *Cyanea capillata*. With greatly reduced drag compared with *C. capillata*, there is little acceleration or deceleration between pulsations. This last point does not apply to all rhizostomes, as many filter water through massive sieve-like oral arms which would increase drag. *S. meleagris* medusae swim for sustained periods at speeds up to 15 cm/s (Larson, 1987a; Shanks and Graham, 1987).

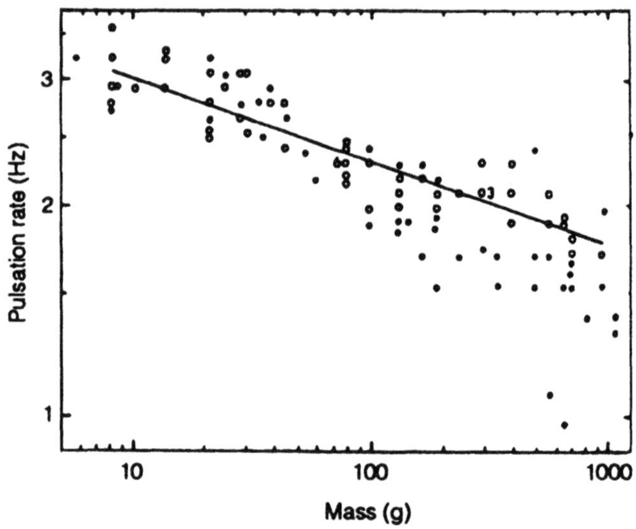

Figure 2.17 Mass vs pulsation rate of *Stomolophus meleagris*. Open circles = medusae in pools; solid circles = medusae in respiration chambers. (Source: Larson, 1987a, with permission of R.J. Larson and National Research Council Canada.)

The energetic costs of swimming and the constraints on size, shape and swimming behaviour have been examined for theoretical model medusae swimming by jet propulsion (Daniel, 1983). A swimming medusa requires energy to produce the jet for thrust and to deform its bell. Daniel found that the acceleration reaction is the largest instantaneous force measured during the contraction cycle. However, drag is the dominant average force to be overcome during continuous swimming. Prolate (cigar-shaped) medusae would maximize efficiency and minimize the cost of locomotion. The more flattened oblate medusae may be inefficient for locomotion but better able to generate feeding currents (section 3.4.1).

As will be discussed in section 5.2.2, approximately 50% of the oxygen consumption of *Pelagia noctiluca* and *Stomolophus meleagris* is related to the energy needs of locomotion (Davenport and Trueman, 1985; Larson, 1987a). If it is assumed that the metabolic substrate is primarily protein, the net metabolic costs of transport per unit mass and distance can be estimated from respiration rates and swimming speeds. For *S. meleagris* these metabolic costs range from 2 J/kg/m for a 5 g medusa to 1 J/kg/m for a 1 kg medusa (Larson, 1987a). Values for other medusae will vary with various factors, especially drag.

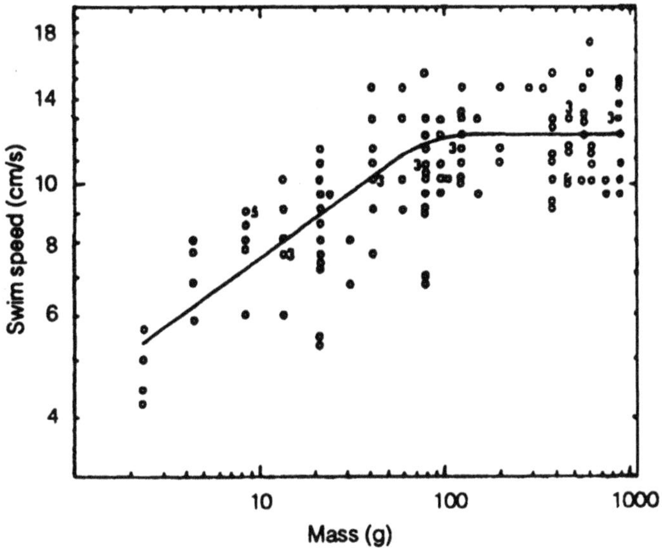

Figure 2.18 Mass vs swimming speed of *Stomolophus meleagris* (data from medusae in pools of diameter 2 or 3 m). (Source: Larson, 1987a, with permission of R.J. Larson and National Research Council of Canada.)

Nevertheless, these estimates suggest that the cost of transport for medusae is low compared with crustacea, and is similar to that of fish.

Pulsation rates decrease with increasing size or age. For example, for *Stomolophus meleagris* held in 3 m pools, pulsation rates ranged from 3.6 to 1.7 pulsations per second over a mass range of 10–1000 g (Larson, 1987a) (Figure 2.17). Similar decreases in pulsation rate with size have been observed for *Cassiopea andromeda, Cassiopea xamachana, Chrysaora quinquecirrha, Cyanea capillata, Drymonema dalmatinum, Phacellophora camtschatica and Pseudorhiza haeckeli* (see Mayer, 1906; Fancett and Jenkins, 1988; Gohar and Eisawy, 1961; Gatz, Kennedy and Mihursky, 1973; Larson, 1987c; Strand and Hamner, 1988).

However, the effects of size on locomotion are complex. The changes in pulsation rates are not directly correlated with changes in velocity, since there is also variability in physical parameters with size. For *Stomolophus meleagris* swimming speed increased from about 5 cm/s at 2 g to 12 cm/s at 70 g, but above 70 g remained nearly constant (Figure 2.18). A number of factors may be involved. Larger animals can exert greater force. However, as noted in section 2.3.3, the ratio between force of contraction of muscles (dependent on cross-sectional area) and mass moved (dependent on volume) decreases as size increases. Resistance to acceleration is directly proportional to the mass of the animal. Drag increases with increasing speed. Daniel (1983) incorporated some of these factors into his model and predicted that there would be an optimum size for locomotion of medusae of a particular shape.

2.6.2 Nervous control of swimming

Figure 2.19 summarizes what is known about the nervous control of swimming as described in section 2.5. The presence of the through-conducting motor nerve net, innervating the swimming muscle and carrying action potentials rhythmically generated by the marginal centres, is well documented. So is input to the centres from the marginal sense organs and the diffuse nerve net, and interaction between the centres through the motor nerve net. However, this does not explain all the phenomena of swimming.

One question is how simultaneous contraction is achieved in medusae which may be a metre in diameter. The motor nerve net can conduct in all directions from whichever marginal centre is leading and ultimately stimulate all the swimming muscle. However, that requires time for conduction. Most researchers have used relatively

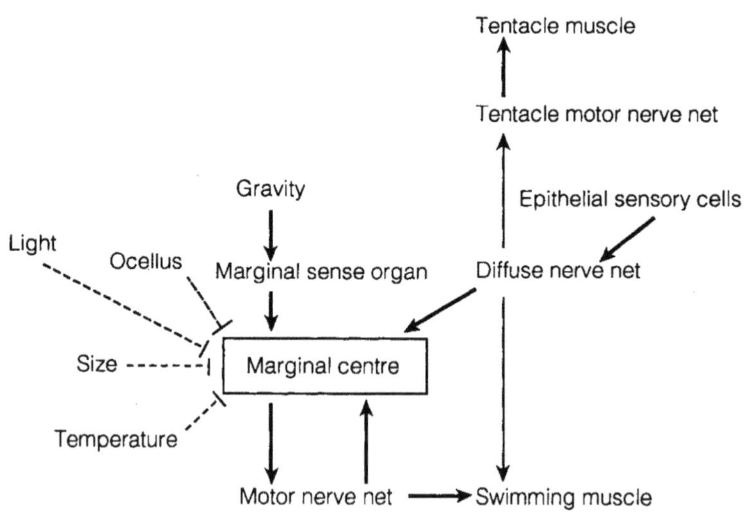

Figure 2.19 Behavioural control mechanisms of semaeostome medusae. Broad lines represent connections found in all investigated medusae; thinner lines represent connections present in some investigated medusae; dashed lines represent tentative connections. Arrows show excitatory action; bars show inhibitory action.

small animals where a pulse from one leading centre can reach and reset the other centres before they fire. (They may then fire during the refractory period of the nerve.) In larger medusae it is possible that another mechanism for synchrony is present. In *Cassiopea xamachana,* motor nerve net pulses reaching a marginal centre differ in form from those outgoing from the centre (Passano, L.M., 1965). It is interesting to speculate that this may allow differential neuromuscular delay between near and far muscles as has been documented in hydrozoa (Spencer, 1982).

Turning of the medusa requires an initial localized contraction on one side of the umbrella, followed in the same cycle by simultaneous contraction of all the swimming muscles (section 2.6.1). However, it is not clear how the localized contraction is initiated. The motor nerve net is through-conducting, and there is no histological evidence of gap junctions between muscle cells, so it is presumably due to another nerve net (for discussion, see Passano, L.M., 1982). Subsequent contraction could then be controlled by the through-conducting motor nerve net. The turning reaction can be elicited as a 'righting reaction', dependent on information from the gravity receptors of the marginal sense organs. It can also be elicited by more complex stimuli allowing directional migration and recovery from turbulence (sections 8.3 and 8.4). This suggests that there may be marginal centres integrating this information and controlling the localized nerve net, as well as the marginal centres controlling the MNN.

It is not known how the marginal centres are affected by some factors that influence the rate of pulsation. The effects of size were mentioned in the previous section. Light is known to affect the rate of pulsation of some medusae where ocelli are not known. For example, light decreases the pulsation rate of *Pelagia noctiluca* (see Axiak, 1984). It is possible that there are direct effects on the neurons of the marginal centre.

Temperature also affects the rate of pulsation. It is not known whether this is a direct effect on the marginal centres, or whether there are receptors present. In most cases temperature increase causes an increase in pulse rate over the normal temperature range, but the response falls off at higher temperatures. Responses of this type have been described for *Aurelia aurita, Cassiopea andromeda, Cassiopea xamachana, Chrysaora quinquecirrha* and *Pelagia noctiluca* (see Mayer, 1914a; Thill, 1937; Gohar and Eisawy, 1961a; Mangum, Oakes and Shick, 1972; Gatz, Kennedy and Mihursky, 1973; Dillon, 1977; Rottini-Sandrini, 1982; Heeger and Möller, 1987; Malej, 1989a; Avian, Rottini-Sandrini and Stravisi, 1991). Acclimation to temperature will be discussed in section 8.2.1.

2.6.3 Locomotion of polyps

Stauromedusae such as *Haliclystus salpinx, Kishinouyea corbini* and *Lucernaria quadricornis* can move about by somersaulting (Berrill, M., 1962; Larson, 1980). This process requires reversible adhesion of the basal disc as well as either the primary tentacles (anchors) between the arms or the secondary tentacles on the arms. The arms of *K. corbini* bear secondary tentacles with adhesive tips and also an adhesive pad formed by the fusion of several secondary tentacles. One or more arm tips adhere to the substrate, the basal disc releases and the medusa flips by contracting the coronal and radial muscles (Larson, 1980). The basal disc of *Haliclystus stejnegeri* contains cells with dense-cored rods of secretory material, which passes out of the cells on to the substrate through finger-like processes at the apex of each cell (Singla, 1976). There are also contractile supporting cells, with microfilaments and associated axons, that may be involved in detachment of the disc.

Scyphistomae are attached to the substratum by the pedal disc or by pedal stolons. The pedal discs of *Cyanea capillata* and *Aurelia aurita* contain desmocytes, cells which form 'rivets' of protein tonofibrillae binding the mesoglea through the epidermis to the substrate (Widersten, 1966; Chapman, D.M., 1969). In spite of this attachment mechanism, scyphistomae of *A. aurita* may glide slowly along the substratum with the pedal disc and stalk preceding the clumped tentacles (Spangenberg, 1964). The pedal stolon of *Chrysaora hysocella, Chrysaora quinquecirrha* and *A. aurita* is an elongated tendril extending from the stalk region of the polyp (Chuin, 1930; Gilchrist, 1937; Cargo and Rabenold, 1980; Schmahl, 1985a). It may attach, contract and pull the polyp towards its point of attachment.

2.6.4 Locomotion of planulae

The planktonic planulae of semaeostome scyphozoa such as *Aurelia aurita, Cassiopea xamachana, Cyanea capillata,* and *Cyanea lamarcki,* have fairly uniformly distributed ciliation (Widersten, 1968; Martin, V.J. and Chia, 1982). The planulae of *Chrysaora quinquecirrha* are at first round or oval. Within two to three hours they become pyriform and begin to move through the water with the broad end directed anteriorly (Littleford, 1939). As such planulae swim they rotate around the longitudinal axis (Figure 6.11). *A. aurita* planulae can attain relatively high speeds (Konstantinova, 1966; Berger, Lukanin and Khlebovich, 1970; Khlebovich, 1973); at approximately 200 μm length, they can swim at 420 μ/s (Konstantinova, 1966). This represents over 100 times their body length each minute.

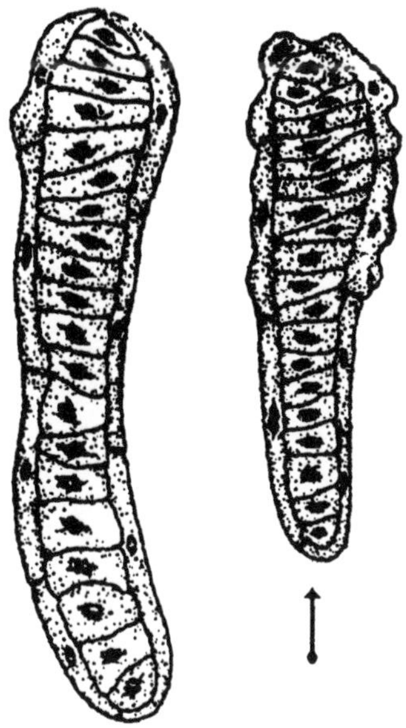

Figure 2.20 Creeping locomotion of planula of *Manania distincta*. (a) Extended planula; (b) anterior portion contracted, drawing the posterior part forward. (Source: Hanaoka, 1934.)

Viscosity, rather than inertia, is the predominant force acting on the larva (Chia, Buckland-Nicks and Young, 1984). The relationship between inertia and viscosity is expressed by the Reynolds number (Re), which may be approximated as follows: Re = (density of sea water) × (length of larva) × (mean swimming velocity)/(dynamic viscosity of sea water). As the size or speed of the larva decreases, the effect of viscosity increases. Larvae as small as planulae do not coast or glide. Streamlining, used to minimize friction in an inertial glide, becomes unimportant. Efficient movement depends only on a configuration allowing efficient operation of the cilia as they push against the viscous fluid.

The planulae of the Stauromedusae are not planktonic. Planulae such as those of *Haliclystus salpinx, H. stejnegeri* and *Manania distincta* lack ciliation on the ectodermal surface (Hanaoka, 1934; Otto, 1976). They creep about the substrate (Figure 2.20). The planulae have a constant number of endodermal cells which can elongate and retract as they move. Microfilaments encircle these endodermal cells (Otto, 1978). A sticky substance is secreted to attach the planula to the substrate temporarily and allow it to creep along.

3 *Feeding*

3.1 INTRODUCTION

Coelenterates use a variety of sources of nutrition including animal prey, dissolved organic matter and substances derived from symbiotic algae (Sebens, 1987). This chapter will discuss the acquisition of prey. The next chapter will discuss the digestion and assimilation of the prey, as well as the other sources of nutrition.

Cnidaria, as implied by the name, possess unique intracellular organelles, the cnidae. Because of their importance in feeding, this chapter will first discuss their functioning and then more general aspects of feeding behaviour.

3.2 CNIDAE

Cnidae are complex intracellular secretory products characteristic of the phylum Cnidaria. Each cnida consists of a microscopic capsule containing a coiled hollow thread-like tubule. When it is stimulated the tubule is discharged, everting or turning inside out much like the finger of a glove, while remaining attached to the capsule.

Many cnidae have tubules that are able to penetrate human skin. They may be barbed and may contain various toxins which cause painful stings. Unlike certain species of cubomedusae, scyphomedusae are unlikely to kill humans outright. Although Sir Arthur Conan Doyle attributed a death to *Cyanea capillata* in his story *The Adventure of the Lion's Mane* (Doyle, 1930), *C. capillata* is in fact not a life-threatening species unless there is hypersensitization through successive contacts. However, the perceived threat of scyphozoan stings may drive people away from beach resorts. As a result, there has been more research done on the toxins and their effects on humans than on the basic mechanisms of function of the cnidae and their use by the medusae for feeding and defence.

3.2.1 Structure and classification

The capsule of a cnida consists of two or more layers (Figure 3.1), and, based on the amino acid content, it is composed of collagenous material (Stone, Burnett and Goldner, 1970). Unlike other collagen, however, at least the inner walls of discharged cnidae may be dissolved by dithioerythritol indicating the presence of disulphide bonds (Mariscal and Lenhoff, 1969; Mariscal, 1971). The presence of high concentrations of sulphur has been confirmed by X-ray microanalysis (Tardent *et al.*, 1990).

Each capsule is sealed by a single trapdoor-like operculum (Figure 3.1). Although it may be important in understanding control of discharge, little is known of its structure, except that it may have a laminar appearance (Westfall, 1966; Sutton and Burnett, 1969; Burnett, 1971).

The tubule is a cylindrical structure that is continuous with the apex of the capsule. All of the cnidae of scyphozoa are nematocysts, characterized by a tubule lacking accessory hollow tubules or longitudinal folds along its length. In many everted nematocysts there is a basally enlarged region of the tubule: the shaft. When everted both the shaft and the distal tubule may bear external spines. When inverted the tubule forms tripointed cross-folds (Sutton and Burnett, 1969). The shaft usually remains fairly straight while the remainder of the tubule is coiled into the capsule.

A complex classification of nematocysts, based largely on the structure of the tubule, has been developed for the whole phylum (Weill, 1930, 1934). A glossary of terms applicable to scyphozoa is given in Table 3.1. Unfortunately many of the details may be difficult to see without a scanning electron microscope. Slight differences in the tubule diameter or in spine size, or even the presence of small spines may not be obvious with a light microscope. The atrichous (non-spined) nematocysts described by early workers are, in most cases, seen to be armed with small spines when examined with a scanning electron microscope.

All scyphozoa so far examined, other than *Tetraplatia volitans*, contain heterotrichous microbasic euryteles, i.e. nematocysts with a

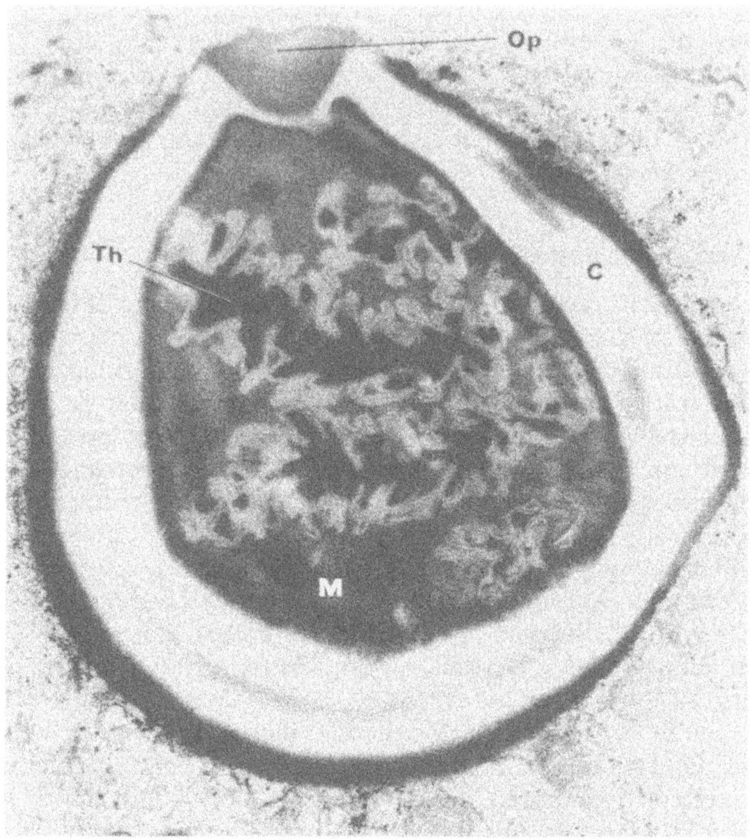

Figure 3.1 Section of an undischarged cnida of *Chrysaora quinquecirrha*. The nematocyst is a haploneme without a well-defined shaft so the tubule is completely coiled within the capsule with transverse folds. C = capsule; M = matrix; Op = operculum; Th = thread (tubule). ×18 000. (Source: Sutton and Burnett, 1969, with permission of J.W. Burnett and Academic Press.)

tubule with a well-defined short shaft which is dilated distally and bears spines of unequal size (Calder, 1983) (Figure 3.2). However, haplonemes (nematocysts with tubules without well-defined shafts) are also usually present (Figures 3.1, 3.3). The nomenclature of the types of haplonemes is still debated and their occurrence among representatives of the class is poorly known (Wang and Xu, 1990; Avian, Del Negro and Rottini-Sandrini, 1991; Östman, 1991). In both euryteles and haplonemes the tubule has a terminal opening. Recently there have been preliminary descriptions of a new type of cnida from *Pelagia noctiluca* in which the distal part of the tubule has a pointed dart with

Table 3.1 Glossary of terms applied to scyphozoan cnidae

Term	*Definition*
a-haploneme	hapoloneme with capsule pyriform, tubule short, regularly coiled inside capsule
A-haploneme	haploneme with capsule ovate, tubule very long, irregularly coiled inside capsule
alpha-haploneme	haploneme with capsule ellipsoidal to reniform, thread short, regularly coiled inside capsule
Anisorhiza	tubule of uneven diameter
Astomocnide	tubule closed at the tip
Atrichous	tubule without spines
Barb	barb-shaped spine or small secondary extension
Basitrichous	tubule with spines at base
Birhopaloid	shaft with one distal and one more proximal dilation
Cnidoblast	cell containing a developing cnida
Cnidocil	modified cilium of cnidocyte
Cnidocyte	cell containing a mature cnida
Cnidome	ensemble of cnidae types present in a species or other taxonomic unit
Eurytele	with shaft dilated distally
Haploneme	with tubule without a well-defined shaft
Heteroneme	with tubule with a well-defined shaft
Heterotrichous	spines of shaft, or of tubule, of unequal size
Holotrichous	tubule with well-developed spines along whole length, arranged in three rows
Homotrichous	spines all of approximately equal size
Isorhiza	with tubule of approximately the same diameter throughout (in practice for at least the distal half)
Merotrichous	with spines medially along the tubule
Microbasic	with shaft short, less than three times capsule length
Nematoblast	cell containing a developing nematocyst
Nematocyst	cnida with tubule lacking accessory hollow tubules or longitudinal folds (all scyphozoan cnidae are nematocysts)
Nematocyte	cell containing a mature nematocyst
Operculum	trapdoor-like structure sealing the apical opening of the undischarged capsule at the junction of the inverted tubule and the capsule wall
O-haploneme	haploneme with capsule sub-spherical
Polyspira	with capsule linguiform, tubule moderately long, irregularly coiled inside capsule
Rhopaloid	with shaft of unequal diameter
Shaft	basally enlarged portion of the tubule
Spine	armature decorating the surface of an everted tubule, usually barb-shaped
Stomocnide	tubule with terminal opening
Tubule	the portion of the cnida that everts during discharge
Thread	see Tubule (obsolete term)

Sources: modified from Weill (1934), Calder (1974a), Watson and Wood (1988), Bozhenova (1988), Wang and Xu (1990) and Östman (1991).

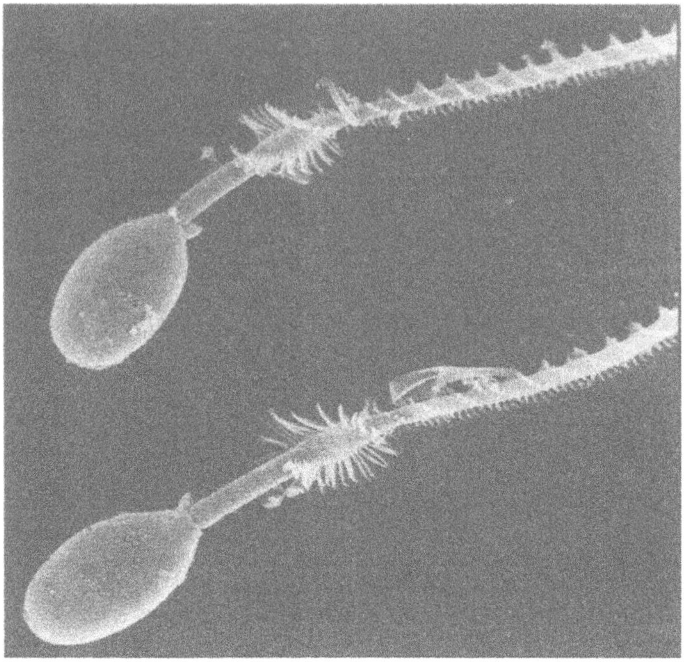

Figure 3.2 Scanning electron micrograph of two everted cnidae of *Cyanea capillata*. The heterotrichous microbasic euryteles have short shafts (which are dilated distally) with larger spines than those on the remaining portion of the tubule. This is a common nematocyst type in scyphozoa. ×2000. (Courtesy of C. Östman.)

closed apex (Avian, Del Negro and Rottini-Sandrini, 1991; Avian, Rottini-Sandrini and Bratina, 1991).

Mature nematocysts are contained within nematocytes. The nematocysts are oriented within the nematocyte with the operculum toward the apical surface of the cell. Scanning electron microscopy of the surface shows that nematocysts in *Chrysaora quinquecirrha* and *Cassiopea xamachana* discharge through a flagellum stereociliary complex (Blanquet and Wetzel, 1975; Mariscal and Bigger, 1976). In *C. xamachana* the central flagellum is present on the nematocyte and the surrounding stereocilia on three to five neighbouring cells. This complex has been less examined than the corresponding very intricate cnidocil apparatus of the Hydrozoa. The latter consists of a long cnidocil (a highly modified cilium), an outer ring of stereocilia, an inner ring of short microvilli and a complex system of rods and fibrils, the fibrillar collar, surrounding the nematocyst and the base of the

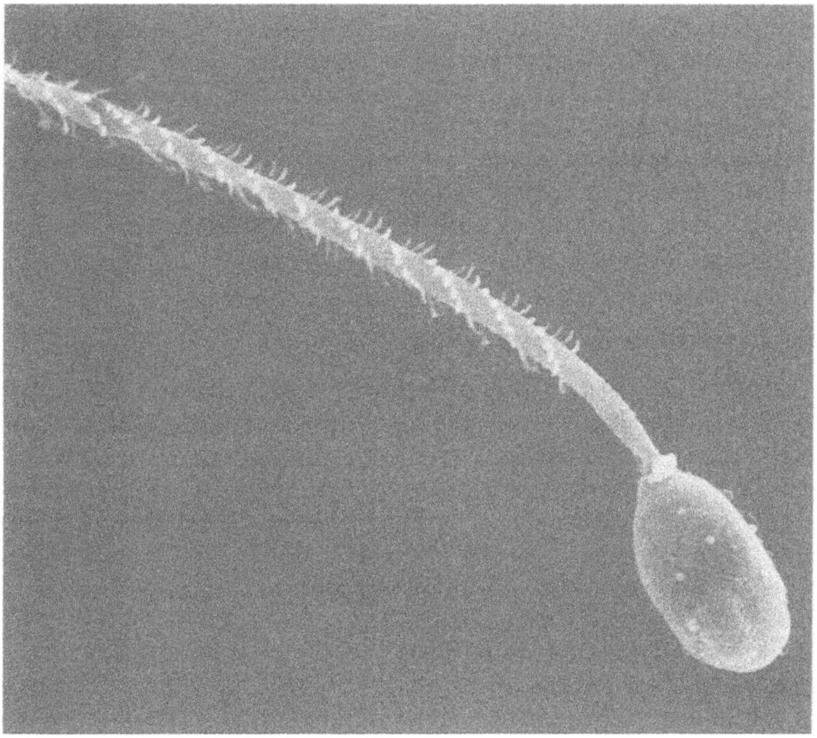

Figure 3.3 Scanning electron micrograph of a cnida of *Cyanea capillata*. This isorhiza haploneme nematocyst has a tubule of approximately equal diameter without a well defined shaft. An opercular flap can be seen at the base of the everted tubule. ×3300. (Courtesy of C. Östman.)

cnidocil (Holstein and Hausmann, 1988). The inner microvilli and possible elements of the fibrillar collar have also been observed in *Aurelia aurita* (see Chapman, D.M., 1974; Westfall, 1966; Heeger and Möller, 1987). It is probable that the flagellum stereociliary complex will prove to be morphologically similar to the cnidocil apparatus.

Nematocysts appear first in the planula and are present throughout the remainder of the life cycle. They are present over much of the body but are especially concentrated on the tentacles, near the mouth, and internally on gastric cirri. The complement of types may differ from one stage to another in the life cycle; for example the polyspiras of *Aurelia aurita* are present only in the polyp and occasionally in newly released ephyrae (Calder, 1983).

In spite of the difficulties in determining details and terminology at the electron microscope level, it is generally accepted that at the light level the cnidomes (the ensembles of nematocyst types and sizes) are characteristic of particular species at particular life-cycle stages. This has allowed use of nematocysts for taxonomic and systematic purposes (Papenfuss, 1936; Calder, 1971, 1972, 1977, 1983; Widersten, 1973). Nematocysts present in the gut contents have also been used to examine the diet of turtles which are predators of medusae (Den Hartog, 1980; Den Hartog and van Nierop, 1984; van Nierop and den Hartog, 1984).

3.2.2 Formation and migration

Nematocysts are formed within **nematoblasts**, cells which will mature into the nematocytes. During the development of the cnida, the tubule is synthesized in the cytoplasm outside the capsule. The developing tubule is associated with a well developed Golgi apparatus (Burnett, 1971). In the Hydrozoa and Anthozoa it has been shown that the tubule then inverts into the capsule, ready for eversion during discharge (Watson, 1988). The forces involved in this process are not understood.

Formation of the mature nematocyst also involves migration of the nematoblast from its site of origin to its final position in the animal. In hydrozoans, nematoblasts move as individual amoeboid cells for considerable distances (Campbell, 1988). Migration has not been analysed in scyphozoans. This may be because the migration is usually only for very short distances in the more thoroughly investigated planulae and pelagic medusae of this class, whereas it is more extensive in the polyps and stauromedusae. In the planulae of *Cassiopea xamachana* the migration is only from the base of the epidermis to the free surface of the ectoderm (Martin, V.J. and Chia, 1982). Krasiñska (1914) found nematoblasts in all tissues of *Pelagia noctiluca* medusae where nematocytes were found and concluded that extensive migrations did not occur. On the other hand, Komai (1935) found pockets of developing nematoblasts in the septal mesogloea of *Stephanoscyphus* sp. polyps. He believed the nematocytes then migrate to their definitive sites. Similarly the nematocytes of *Haliclystus octoradiatus* and *Lucernariopsis campanulata* are formed in reservoirs in the septa of the calyx and migrate into the tentacles (Weill, 1925, 1935).

Once discharged, nematocysts are not reusable. As a result, nematocyst formation constitutes a considerable energy cost to the animal. To date, there are no measurements of rates of nematocyst formation by scyphozoa. Polyspira nematocysts degenerate during strobilation in *Aurelia aurita* (Spangenberg, 1965b). This may also constitute a loss or the degeneration products may be recycled.

3.2.3 Discharge

Discharge of the nematocyst includes eversion of the tubule and often release of the capsule from the nematocyte. Based largely on data from the Hydrozoa, there have been a number of theories of how eversion occurs. Most theories focus on an increase of internal pressure due to osmotic uptake of water, on release of previously stored tension, or on combinations of the two (Tardent, 1988; Hidaka, 1993; Watson and Mire-Thibodeaux, 1994). Discharge of the nematocysts of *Pelagia noctiluca* is associated with a swelling and then a decrease in capsule size (Salleo *et al.*, 1986; Salleo, La Spada and Denaro, 1991). The swelling supports the theory that osmotic pressure is at least in part responsible for the discharge. However, discharge of isolated nematocysts using enzymes such as trypsin which cannot penetrate the capsule (Salleo, La Spada and Alfa, 1983) indicates involvement of the outer capsule wall, possibly in release of tension.

It is not clear how the ionic contents of the nematocyst might be involved in the discharge. The matrix of nematocysts contains high concentrations of cations such as potassium, magnesium and calcium (Mariscal, 1988; Tardent *et al.*, 1990) and of poly(gamma-glutamic acid) polyanions (Weber, 1991). Some of these may have other functions such as the activation of toxins. They are apparently bound during the resting state since the capsular wall is freely permeable to small molecules such as methylene blue (Salleo, La Spada and Alfa, 1983; Salleo, 1984). Free calcium, (revealed by the light emission of aequorin in the surrounding medium) is released from the nematocysts of *Pelagia noctiluca* prior to discharge (Salleo *et al.*, 1988; Salleo, La Spada and Denaro, 1991). This early release may indicate a derivation from the capsule wall as tension is released, rather than from capsular fluid which would continue to be ejected during the discharge.

Discharge can be induced in isolated nematocysts by non-physiological agents such as trypsin (Salleo, La Spada and Alfa, 1983), various ions (Salleo *et al.*, 1984a,b) and the calcium ion sequestering agents sodium citrate and sodium EDTA (Kem and Östman, 1991). However, discharge normally occurs from within the nematocyte.

Activation of discharge by mechanical and chemical stimuli requires reception of the stimuli at the surface of the epithelial cells, which then triggers the actual discharge mechanism. Both Ca^{2+} channels and stretch-activated channels in the cell membranes are involved. Blockage of either type of channel selectively with lanthanum or gadolinium chlorides inhibits discharge of haplonemes of the oral arms of *Pelagia noctiluca* (Salleo, La Spada and Barbera, 1994). The connection between these events at the surface of the cells and discharge is not understood. The apical portions of the nematocyte and surrounding cells are extremely complex (section 3.2.1). Electrical stimulation can cause potential changes in isolated cnidocytes, but these potential changes do not correlate well with the rate of discharge (Anderson and McKay, 1987; McKay and Anderson, 1988).

In some cases the nematocyst/nematocyte complex acts as an independent effector. This has been most clearly demonstrated in *Nausithoe punctata* eggs. There are scattered small isolated cnidocytes on the surface of the exterior mucus coat of the eggs. Euryteles in the cnidocytes evaginate following mechanical stimulation or contact with predators (Carré and Carré, 1980). In other cases the probable receptor for stimuli is the flagellum stereociliary complex involving interaction between the nematocyte and the surrounding cells. There may also be nervous input. Neuronematocyte synapses like those present in the Hydrozoa and Anthozoa have not yet been demonstrated in the Scyphozoa (Westfall, 1987). However, staining of *Cyanea capillata* and *C. lamarcki* tentacles with an antiserum against the anthozoan neuropeptide Antho-RFamide indicates small discrete nerve nets associated with clusters of cnidae (Anderson, Moosler and Grimmelikhuijzen, 1992).

3.2.4 Toxins

During discharge many nematocysts inject venom, including inert fluids, salts and toxins, i.e. materials having a known negative influence on biological systems. Injury may occur directly by action of the toxins or indirectly by involvement of immune reactions. Little is known about the effects of the toxins on other invertebrates or fish which might be the normal targets, but there is an extensive literature on the effects on humans and other mammals.

Most toxins are proteinaceous molecules, many of which target plasma membranes (Hessinger, 1988; Walker, M.J.A., 1988). An example is rhizolysin from nematocysts of *Rhizostoma pulmo* which is a high molecular weight protein with haemolytic activity on rat erythrocytes (Cariello *et al.*, 1988). Haemolysins have also been found in nematocysts of *Cyanea capillata* and *Chrysaora quinquecirrha* (see Long and Burnett, 1989; Long-Rowe and Burnett, 1994). A phospholipase has been isolated from *Rhopilema nomadica* tentacles (Lotan *et al.*, 1995). Another toxin of *C. quinquecirrha* is cytotoxic because it creates monovalent cation selective channels in lipid membranes and hence depolarizes the membranes of muscle and nerve (Cobbs *et al.*, 1983; Dubois, Tanguy and Burnett, 1983). A toxin from *Aurelia* sp.

nematocysts probably has similar activity (Kihara *et al.*, 1988). *C. quinquecirrha* venom also contains other enzymes including a collagenase, both an alkaline and an acid protease, an endonuclease and possibly a separate factor increasing calcium influx (Lal *et al.*, 1981; Neeman, Calton and Burnett, 1981; Calton and Burnett, 1982a,b; Lin, W.W., Lee and Burnett, 1988). In addition a small lipid mediator of leukocyte chemotaxis, leukotriene B4, has recently been found (Czarnetzki, Thiele and Rosenbach, 1990).

Most toxins have not been correlated with particular nematocyst types due to the difficulty in differential extraction. The large number of toxins would be expected to lead to a complex suite of reactions, varying with the tissue into which the venom is injected and with the types of nematocysts discharged.

The delivery of the toxin probably varies with nematocyst type. At least 85% of the nematocysts present in the fishing tentacles of *Rhopilema nomadica* are isorhiza haplonemes (Avian *et al.*, 1995). Immunocytochemistry reveals phospholipase toxin located in folds along the outer surface of the inverted, undischarged tubule (Lotan *et al.*, 1995). As the tubule everts the toxin comes to lie inside the discharged tubule, concentrated against the bases and in the lumina of the hollow spines. It is probable that the high hydrostatic pressure within the discharging capsule causes toxin to be discharged through the system of spines.

Contact of humans with scyphozoa may result in a wide variety of clinical responses ranging from no detectable effect to local skin reactions and severe pain, muscle weakness and cramps, and cardiac, respiratory and renal malfunction. There may be recurrent symptoms associated with detectable antibodies. Reactions following contact of various portions of the skin with various species of scyphomedusae differ greatly in severity. Nematocysts from some species are normally harmless because they are unable to penetrate the skin of the hands or other extremities. They may nevertheless cause reactions if applied to the eyes or lips. Further discussion of clinical data such as this is not pertinent to a book on the biology of the Scyphozoa *per se*. The extensive literature can be accessed using recent reviews by Burnett *et al.* (1986, 1987), and Burnett (1991a,b).

There is no very effective treatment for scyphozoan stings of people. A wide variety of topical agents, including previously recommended vinegar and baking soda, have been used unsuccessfully to deactivate unfired tentacle fragments of *Chrysaora quinquecirrha* before removal from the victim's skin (Burnett, 1991a). Recently Heeger *et al.* (1992) have shown that some sun lotions decrease the discharge of *Cyanea capillata* nematocysts, although the necessary constituents of the lotions were not isolated. There is also no good topical method for controlling pain because the painful sensations appear rapidly, deep in the skin. Systemic analgesics are eventually beneficial for pain relief, but require too long a time to act to be effective against the immediate dermal pain.

3.2.5 Functions

The main functions ascribed to nematocysts have been prey capture and protection from predators. Possible protection from predators will be discussed in section 9.1.

The functions of nematocysts during feeding of *Aurelia aurita* on herring larvae have been examined by Heeger and Moller (1987). Heterotrichous microbasic euryteles and isorhiza haplonemes with numerous very small spines on the tubule were observed with electron microscopy. Both types were present on the exumbrella and in nematocyst batteries on the tentacles of the medusa. Following contact with a herring larva the nematocysts were discharged. The tubules of the euryteles penetrated almost completely into the prey. Probably the basal spines fix the euryteles to the prey, while venom is injected from the tubule into deep tissue layers of the prey causing the observed paralysis of the larva. The tubules of the haplonemes only penetrated for a third of their length. Possibly their function is to entangle prey organisms.

Nematocysts may also deliver digestive enzymes further into the tissues of already paralysed prey than would be possible with surface application. Microbasic euryteles have been found in the gastric cirri of *Rhopilema verrilli* and *Deepstaria reticulum* (see Calder, 1972; Larson, Madin and Harbison, 1988). They may deliver digestive enzymes such as proteases, or simply attach the prey so that the enzymes released by the cirri are more effective.

3.3 TYPES OF PREY

3.3.1 Prey in diets of scyphomedusae

The diets of a number of species of scyphomedusae have been examined, since the pioneering work of Lebour (1922, 1923). Research of this type has been of interest because the medusae may eat larvae of commercially important fish. Table 3.2 summarizes data on the stomach contents of various field-caught scyphomedusae as percentage of prey numbers. Less quantitative information or data presented as a

Table 3.2 Stomach contents of field-caught scyphomedusae, as percentage of prey numbers

Species	*Prey*	%	*Source*
Aurelia aurita	copepods	45	Mironov, 1967
(40 specimens,	tintinnids	30	
28–160 mm)	veligers	11	
	Oikopleura	5	
	cladocera	3	
	Noctiluca	2	
	chaetognaths	2	
Aurelia aurita	veligers	56	Kerstan, 1977
(379 specimens,	copepods	32	
80–260 mm)	barnacle larvae	7	
	cladocera	4	
Aurelia aurita			Möller, 1980b
(961 specimens,	copepods	56	
11–20 mm)	herring	30	
	cladocera	13	
	hydromedusae	1	
(1200 specimens,	crustacea	63	
36–50 mm)	herring	34	
Aurelia aurita	copepods	77	Hamner, Gilmer and
(20 specimens, large)	veligers and trochophores	22	Hamner, 1982
Aurelia aurita	copepods	48	Matsakis and Conover,
(189 specimens,	hydromedusae	34	1991
10–150 mm,	eggs	12	
85 empty)	diatoms and ciliates	< 6	
Aurelia aurita			Olesen, Frandsen and
(55 specimens,	copepods	100	Riisgard, 1994
28–34 mm)			
(17 specimens,	rotifer	93	
2.5 mm)	tintinnids	7	
Chrysaora quinquecirrha			Purcell, 1992
(150 specimens,	copepods	55	
> 18 mm)			
(240 specimens,	copepods	71	
> 18 mm)			
Chrysaora quinquecirrha	copepods/cladocera	72	Purcell *et al.*, 1994
(80 specimens,	fish eggs	21	
> 31 mm)	fish larvae	1	

Table 3.2 (continued)

Species	*Prey*	%	*Source*
Chrysaora quinquecirrha	protozoa	61	Haven and Morales-
< 6 mm	rotifers		Almo
		23	in Purcell, 1992
Cyanea capillata	fish larvae	44	Plotnikova, 1961
(103 specimens,	ctenophores	28	
40–700 mm,	hydromedusae	28	
72 empty)			
Cyanea capillata	larvacea	31	Fancett, 1988
	cladocera	29	
	fish eggs/larvae	14	
	copepods	11	
	hydromedusae	9	
	ascidia	3	
Drymonema dalmatinum	medusae	100	Larson, 1987c
(13 specimens,			
5 empty)			
Pelagia noctiluca	fish eggs	43	Larson, 1987d
(50 specimens,	copepods	29	
10–40 mm,	cumacea	14	
2 empty)	chaetognaths	14	
Pelagia noctiluca			Giorgi *et al.*, 1991
(51 specimens,	copepods	67	
19 empty)	cladocera	11	
	chaetognaths	10	
	gastropods	3	
	euphausiids	2	
	fish larvae	1	
	mysids	1	
(38 specimens,	copepods	44	
9 empty)	decapods	39	
	cladocera	7	
	fish eggs/larvae	3	
	chaetognaths	2	
	amphipods	1	
Periphylla periphylla	copepods	100	Fosså, 1992
(39 specimens,			
15 empty)			
Phacellophora camtschatica	fish larvae	27	Purcell, 1990
(6 specimens)	larvacea	27	
	gelatinous zooplankton	24	
	meroplankton	15	
	copepods	7	

Table 3.2 (continued)

Species	*Prey*	%	*Source*
Pseudorhiza haeckeli	fish eggs/larvae	41	Fancett, 1988
	copepods	33	
	larvacea	8	
	decapod larvae	5	
	cladocera	4	
Stomolophus meleagris (165 specimens, 21–83 mm)	veligers	71	Larson, 1991
	copepods	16	
	tintinnids	9	
	larvacea	3	

percentage of the medusae containing a prey item is also available for *Cyanea* sp. (see Brewer, 1989), *Pelagia noctiluca* (see Zavodnik, 1991) and *Phacellophora camtschatica* (see Strand and Hamner, 1988). For earlier work on diets see also the tables in Alvariño (1985). Without comparative digestion rates for the prey, these data are only of qualitative significance. Nevertheless some generalizations on diets may be made.

Scyphomedusae are primarily carnivores. They do not utilize macrophytes. Although some phytoplankton may be ingested, the amount is not significant in comparison with zooplankton capture. For example, although Mironov (1967) found 20 species of phytoplankton in the stomachs of *Aurelia aurita* they represented less than 1% of the weight of the food.

Many species of Semaeostomeae and Rhizostomeae use a wide selection of zooplankton when it is available (Mills, 1995). For example, the stomach contents of *Aurelia aurita* medusae from a variety of locations have been examined (Orton, 1922; Southward, 1955; Hüsing, 1956; Mikhailov, 1962; Loginova and Perzova, 1967; Mironov, 1967; Kerstan, 1977; Möller, 1980b; Hamner, Gilmer and Hamner, 1982; Matsakis and Conover, 1991). The diet may include diatoms, protozoa, other medusae, ctenophores, polychaete larvae, nematodes, rotifers, larvae of lamellibranch and gastropod molluscs, chaetognaths, various arthropod larvae, copepods, cladocera, appendicularians and fish larvae. Although numerically less important than copepods and other small arthropods, larger animals such as fish larvae and chaetognaths are a significant proportion of the diet.

Less is known about the diet of coronate medusae. Larson (1979) has summarized the data indicating also a broad range of prey including gastropod veligers, copepods, shrimp, chaetognaths and fish.

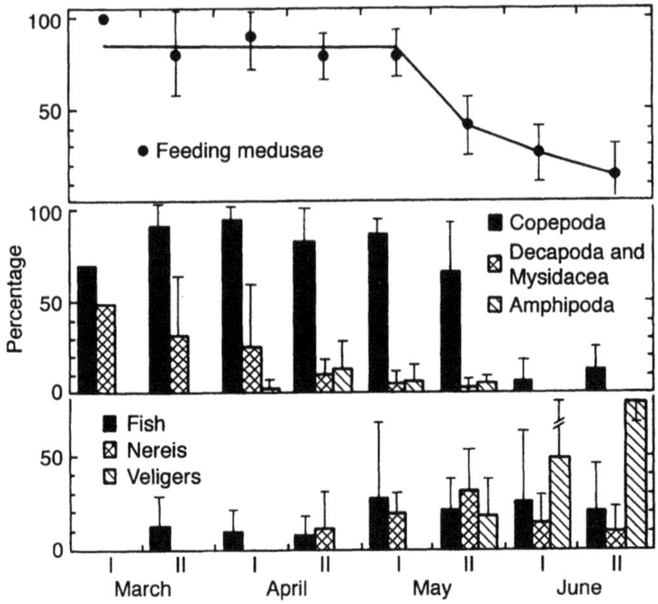

Figure 3.4 The percentage (±SD) of feeding *Cyanea* medusae (1980–1986) in the Niantic River estuary, and the average percentage (±SD) which contained the indicated taxon in their gastrovascular cavity in the half-month, showing annual succession of prey items. Note decrease in numbers of feeding medusae with onset of reproduction in May and subsequent deterioration. (Source: Brewer, 1989, with permission of R.H. Brewer and *Biological Bulletin*.)

Opportunistic predators may show a yearly succession of items appearing in the diet corresponding to the peak populations of particular prey species. Such seasonal variation in diet has been demonstrated for *Aurelia aurita* (see Loginova and Perzova, 1967; Kerstan, 1977), *Cyanea* sp. (see Brewer, 1989) (Figure 3.4) and *Pelagia noctiluca* (see Giorgi *et al.*, 1991). Careful comparison of the diet with the available prey populations at the same site and date shows that some selection also occurs, as will be discussed in section 3.6.1.

3.3.2 Prey of polyps

Much less is known about the feeding of the scyphozoan polyps than of the medusae.

Stauromedusae are able to catch crustacea and other animals present on the same substrate as the polyp (Hirano, 1986b). *Lucernaria quadricornis* feeds primarily on amphipods and small gastropods (Berrill, M.,

1962). *Manania gwilliami* may contain copepods and amphipods in the gastric cavity (Larson and Fautin, 1989).

The semaeostome and rhizostome polyps eat a variety of pelagic organisms. For example, polyps of *Aurelia aurita* have been fed *Artemia*, copepods, decapod larvae, larval molluscs and fish larvae in the laboratory (Lebour, 1923; Cargo, 1974, 1984; Groat, Thomas and Schurr, 1980; Spangenberg, 1965a; Gröndahl, 1988b). They will also eat planulae or other polyps of both *Cyanea capillata* and their own species (El-Duweini, 1945; Gröndahl, 1988b). It is not known how many of these items are utilized in the field. Scyphistomae of *Chrysaora quinquecirrha* will ingest 69% of oyster veligers that contact the tentacles, and digest 48% of those ingested (Purcell *et al.*, 1991). They may also be predators of juvenile mysids and their *Artemia* food if allowed to contaminate mysid cultures (Hutton *et al.*, 1986).

3.4 CONTACT WITH PREY

3.4.1 Medusae encounter probabilities

Medusae can remain still as 'ambush' predators, or swim through the water as 'cruising' predators. The comparative advantages of these two strategies for planktonic animals has been examined by the mathematical model of Gerritsen and Strickler (Gerritsen and Srickler, 1977; Gerritsen, 1980). The model assumes that:

1. the animals are points in a homogeneous three-dimensional space;
2. the animals move at random and are randomly distributed;
3. the predator has an encounter radius given by its sensory system.

The number of encounters will then depend on the population densities, speeds of the two species and the encounter radius of the predator. The first two assumptions are not strictly true for coelenterate predation, and the effects of turbulence are not considered. Nevertheless the model provides a useful background for discussion. It predicts two optimal strategies:

1. cruising predators which prey upon slower moving prey;
2. ambush predators which prey upon faster moving prey.

An extension of the model predicts that if movement is not random, a swimming predator will maximize encounters with prey by swimming at right angles to prey movement (Gerritsen, 1980).

In the case of opportunistic predators, such as medusae, with prey of various speeds, it may be most advantageous to swim at an intermediate speed while searching, and possibly to vary speed following contact with the prey. As expected, subsurface *Pelagia noctiluca* swim slowly and constantly at an optimum speed (Madin, 1988; Malej, 1989a). The medusa acts as a cruising predator for slowly moving prey, but as an ambush predator for faster moving prey.

Divers found 68% of *Phacellophora camtschatica* fishing with short vertical excursions of between 1 and 12 m, at rates of 0–2 m/s (Strand and Hamner, 1988). They may remain motionless at the top and bottom of each excursion, becoming ambush predators. However, while swimming vertically they behave like *Pelagia noctiluca*, being cruising predators for their slow-moving prey and ambush predators for faster species. The largest prey, *Aurelia aurita*, moves primarily horizontally outside aggregations, so vertical movement by *Phacellophora camtschatica* would maximize contact.

Bailey and Batty (1983) examined the predation of *Aurelia aurita* on herring larvae in 5-litre jars. They found the swimming speeds of 5–25 mm *A. aurita* to be approximately 6–16 mm/s, whereas the average speed for first stage herring larvae was 3.7 mm/s. There was also adaptation to the prey following contact. The swimming pattern of *A. aurita* changed markedly from horizontal to more vertical, with a greater number of turns, and the encounter rates increased (Figure 3.5). Although Bailey and Batty also stated that swimming speed increased following first prey capture, their tabulated data does not support that statement.

The model of Madin (1988) examines more closely the encounter probabilities of a medusa, based on the dimensions and arrangement of its tentacles and its swimming behaviour. The model assumes that medusae do not use sensory means to orient toward individual prey prior to contact. It distinguishes between the 'encounter zone' around the medusa in which tentacles can be found, and the 'tentacle density' or fraction of that space actually filled with tentacles. For a medusa such as *Pelagia noctiluca*, which swims trailing the tentacles behind, the encounter zone is a cone, up to 30 times bell diameter in length, in which the average spacing between tentacles ranges from centimetres at the bell to tens of centimetres at the tentacle tips. With only eight tentacles, tentacle density within that cone is low. (The latter fact would intuitively indicate larger prey, rather than the range actually seen in Table 3.2.)

When swimming rapidly the encounter zone of *Phacellophora camtschatica* is also a cone (Strand and Hamner, 1988). However when swimming vertically the medusa may ascend in a slow spiral approximately twice the bell diameter, which causes the numerous tentacles to

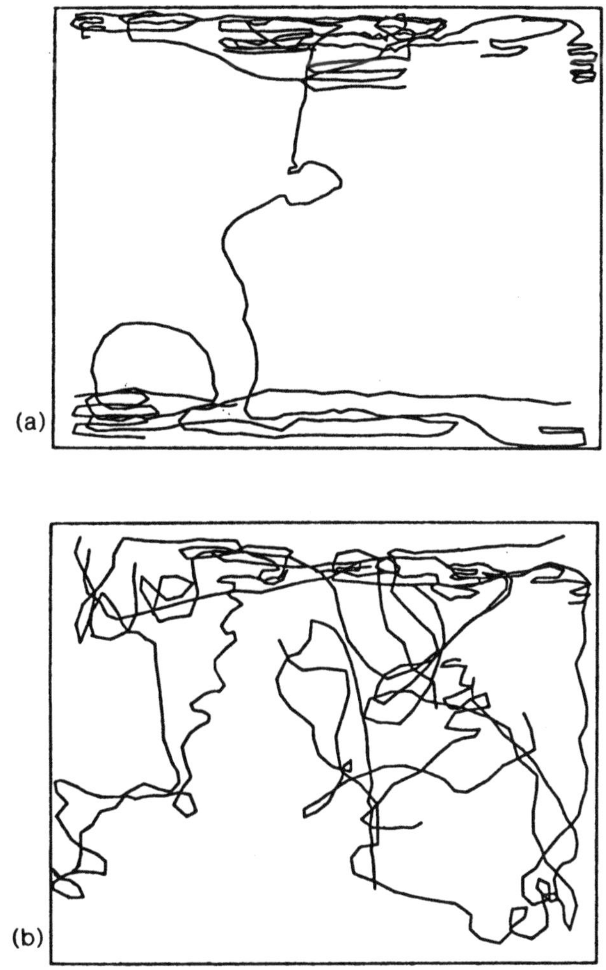

Figure 3.5 Feeding behavior of 12–14 mm *Aurelia aurita* medusae in a 6.6 litre glass tank of standing sea water. (a) Swimming tracks of five medusae without attached herring larvae; (b) swimming tracks of five medusae with one herring larva attached to the oral arm of each medusa. (Source: Bailey and Batty, 1983, with permission of Springer-Verlag.)

swirl out. When sinking with the exumbrella upward, the medusa may drop through the tentacles spreading them outward. When the medusae (more rarely) swim horizontally, they may reverse direction and swim back through an area of already deployed tentacles. All these manoeuvres have the effect of increasing the size of the encounter zone.

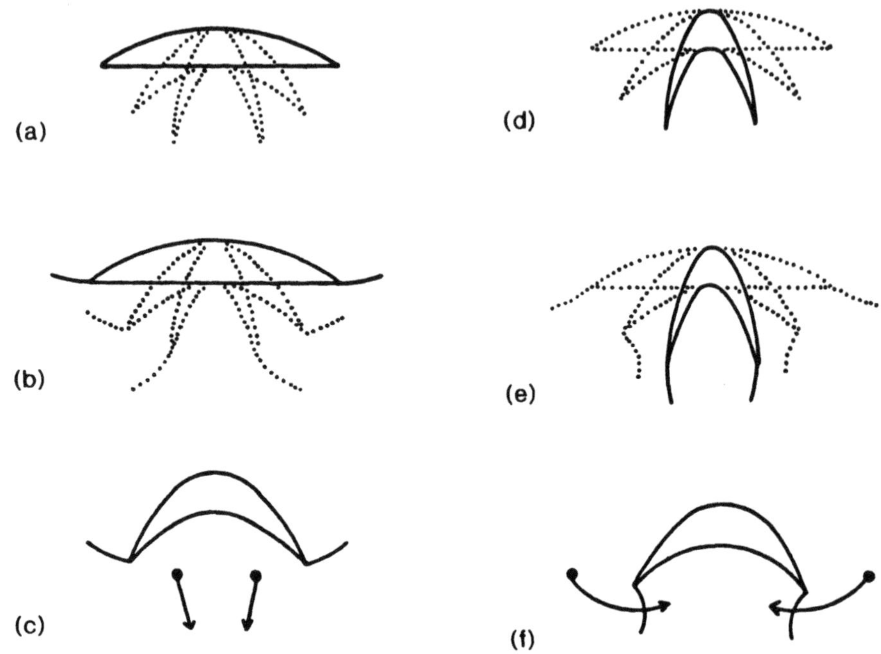

Figure 3.6 Relationship between bell pulsation, fluid motion and prey capture in *Aurelia aurita*. All drawings represent cross-sections. (a) Change in bell form during power stroke; solid form represents initial position while stippled forms represent successive bell positions. (b) Identical to (a) except for the addition of tentacle position during the power stroke. (c) Bell and tentacle position in mid-power stroke; arrows represent motion of fluid and entrained particles. (d) Change in bell form during recovery stroke; solid form represents initial position while stippled forms represent successive bell positions. (e) Identical to (d) except for addition of tentacle position during the recovery stroke. (f) Bell and tentacle position in mid-recovery stroke; arrows represent motion of fluid and entrained particles (including prey). (Source: Costello, 1992, with permission of J. Costello and Scientia Marina.)

Contact with the prey also depends on the fluid motion immediately around the medusa as it swims. The pulsating forward movement of *Aurelia aurita* is characterized by a rhythmic fore-and-aft waving of the short fringing tentacles (Fraser, 1969; Gamble and Hay, 1989). Nematocysts are present on the exumbrellar surface as well as the tentacles (Heeger and Möller, 1987). Contact with small or weak prey will depend on the pattern of water displacement and turbulence created during movement. Eddies circulate over the bell margin, through the tentacles, and into the subumbrellar cavity (Costello, 1992; Costello and Colin, 1994) (Figure 3.6). Prey encounter with the capture surfaces will be a function of the marginal flow velocity compared with the prey escape velocity. Slow prey such as hydromedusae can be captured by small medusae, whereas copepods with fast escape responses are captured primarily by larger medusae with higher marginal flow velocities (Sullivan, Garcia and Klein-MacPhee, 1994).

Rhizostomes lack marginal tentacles so contact is primarily with the manubrium and oral arms. The massive and complicated manubrium has numerous mouth openings on scapulets (leaf-like structures) and four pairs of oral arms. Medusae such as *Stomolophus meleagris* are active swimmers. Contact with the short manubrium is probably dependent on water forced around it by the contracting umbrella during swimming (Larson, 1991). In other species with oral arms extending farther beyond the bell, such as *Pseudorhiza haeckeli*, water may be pumped downwards through the arms (Fancett and Jenkins, 1988). In the sessile *Cassiopea xamachana* the peripheral coronal muscles contract first in the beating sequence. This raises the outer edge of the umbrella so that the subsequent main jet of water is directed through the oral arms (Passano, L.M., 1973).

3.4.2 Medusae attraction to prey

Scyphomedusae, being nonvisual predators, have often been assumed to make random contact with their prey according only to the principles discussed in section 3.4.1. However, attraction to prey does occur. *Aurelia aurita* were tested in a flow-through aquarium with inflow at each end and a central outflow (Arai, 1991). They were attracted to either end of the chamber if *Artemia* prey were present in a screened compartment (Figure 3.7). This was a response to one or more chemicals since they were also attracted to water conditioned by *Artemia* and to ammonium chloride added to one or other of the inflow currents. Medusae have not been observed to make directed movements toward their prey (Oiestad, 1985; Strand and Hamner, 1988) so they may simply move less or turn more after encountering prey or water containing chemicals.

It is not known to what extent such attraction is present in the field. It would obviously be advantageous for the medusae to be able to feed on aggregations of their prey. *Phacellophora camtschatica* is found most often in or close to locations with large numbers of *Aurelia* (see Strand and Hamner, 1988), but this is not necessarily a direct response to the *Aurelia*.

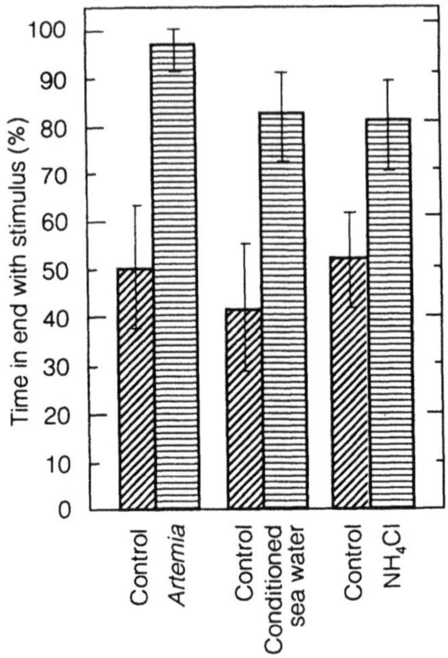

Figure 3.7 Position of *Aurelia aurita* medusae in a flow-through aquarium with inflow at each end and a central outflow. Attraction to the end of the aquarium with *Artemia* in a screened compartment, or with water conditioned by previous presence of *Artemia* or ammonium chloride. (Source: Arai, 1991, with permission of Kluwer Academic Publishers.)

3.5 FEEDING BEHAVIOUR

3.5.1 Medusae prey capture

When contact has been made with the prey, it becomes attached to the scyphomedusa due to either mucus or cnidae action. Prey are then moved towards the mouth by ciliary tracts or by contraction of the tentacles, lappets or oral arms. More rarely a medusa may simply contract around a prey animal and engulf it.

Small particles including microzooplankton may be trapped in mucus and transported by ciliary tracts to the stomach. Adult *Aurelia aurita* capture particles on their exumbrella or subumbrella and transport them by ciliary currents and boundary layer flow to a marginal groove between the row of tentacles and an inner fold of ectoderm (the velarium). Mucus and particles accumulate in adradial widenings of the groove (food pouches) (Orton, 1922; Southward, 1955) (Figure

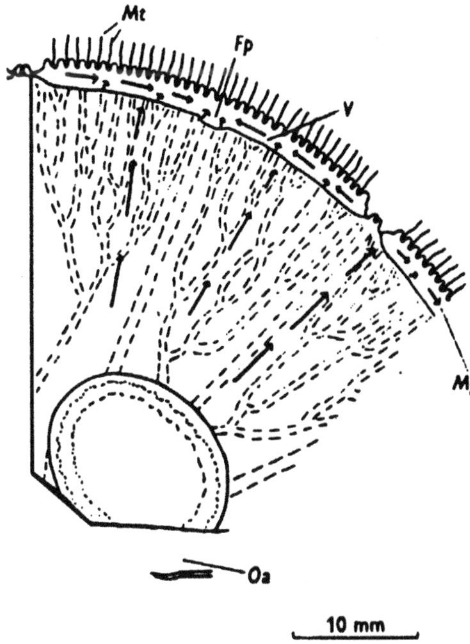

Figure 3.8 Currents on the subumbrellar surface and at the margin of an *Aurelia aurita* medusa. Currents indicated by arrows; underlying canals by dotted lines. Fp = food pouch; Mg = marginal groove; Mt = marginal tentacle; Oa = oral arm; V= velarium. (Redrawn after Southward, 1955.)

3.8). The masses of food and mucus are then transferred by the ciliary tracts of the oral arms to the stomach. During this transfer, inert particles and phytoplankton such as dinoflagellates are rejected (Stoecker, Michaels and Davis, 1987).

Macrozooplankton trigger the action of the nematocysts (section 3.2). When small animals contact medusan epithelia with nematocysts, they rarely escape, but larger animals may break free. *Phacellophora camtschatica* captures all the small hydromedusae and ctenophores that contact single tentacles, but has to entangle *Aurelia aurita* in a number of tentacles to capture it. Most *A. aurita* escape. Larger *P. camtschatica* capture more and larger *A. aurita* but even a 40 cm *P. camtschatica* rarely captures *A. aurita* over 18 cm in diameter (Strand and Hamner, 1988).

Tentacles of many semaeostome medusae are highly contractile, bringing prey quickly into the vicinity of the oral arms and bell. In *Pelagia noctiluca* an individual tentacle of up to 60 cm can contract within 3 seconds to less than 4 cm (Bozler, 1926b; Rottini-Sandrini

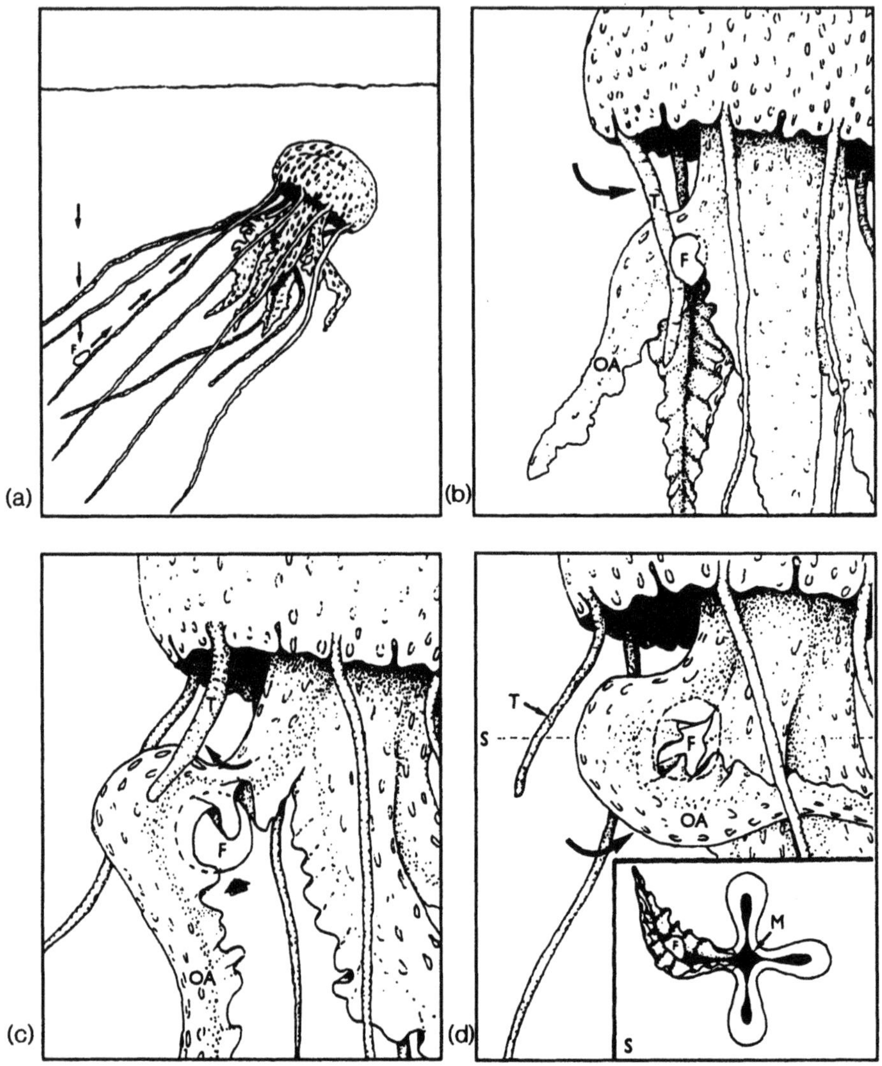

Figure 3.9 *Pelagia noctiluca* feeding behaviour on motile prey (from laboratory and open sea video-recordings). (a) When the prey touches a marginal tentacle, there is an immediate nematocyst discharge, followed by a tentacle contraction after 2–3 s. (b) The stiff tentacle bends towards the nearest oral arm; at the same time the oral arm moves upwards, turns slightly and draws its inner layer near the food. (c) The stiff tentacle releases the prey and moves upwards away from the oral arm. (d) The oral arm grasps the prey completely and starts the peristaltic and mucous movements which drive the food to the oral arm groove, then to the manubrium, and finally to the gastric cavity. (s) Inset: transverse section of (d). F = food; M = mouth opening; OA = oral arm; T = tentacle. (Source: Rottini-Sandrini and Avian, 1989, with permission of L. Rottini-Sandrini and Springer-Verlag.)

and Avian, 1989) (Figure 3.9). Other tentacles remain extended and fishing (Malej, 1989a). The fishing tentacles of *Chrysaora quinquecirrha* can contract to a thirtieth of the resting length (section 2.3.2) (Perkins, Ramsey and Street, 1971). A tentacle of *Phacellophora camtschatica* pulled a 1.5 cm long ctenophore 2 m to the bell margin in 70 seconds (Strand and Hamner, 1988).

Transfer of prey from the contracted tentacles to the oral arms requires coordination of the movement of the muscles of that quadrant of the umbrella, and of the oral arms which bend toward the tentacle or bent umbrella margin. The details of the transfer vary with the extent of the tentacles and oral arms. This feeding behaviour is illustrated for *Pelagia noctiluca* in Figure 3.9. The oral arms of *P. noctiluca* may also collect non-motile prey directly, without the intervention of the tentacles (Rottini-Sandrini and Avian, 1989).

When prey reaches the oral arms it is enveloped by ciliary creeping aided by muscular contraction. The oral arms of *Cyanea capillata* can spread over the surface of the prey to form a thin, closely adhering film (Plotnikova, 1961). Isolated oral arms will also spread over a petri dish or a surface film (Seravin, 1991). A 1 g piece of oral arm of *Drymonema dalmatinum* can spread over an area of 25cm^2 in the bottom of a dish, the total potential surface area of the oral arms being several square metres (Larson, 1987c). This movement may be rapid. *Chrysaora quinquecirrha* can envelop a 10 cm ctenophore in about 5 minutes and move it toward the stomach at 3–7 mm/minute (Larson, 1986a). When feeding rapidly, prey may collect in a temporary 'bag' of oral arm tissue (Lebour, 1923). In semaeostome medusae, prey may be either passed into the stomach or digested within the oral arms (section 4.2.1).

In most coronate medusae, such as *Nausithoe punctata* and *Periphylla periphylla*, the tentacles are rigid, bending without contracting (Larson, 1979). They may be held in front of the umbrella as the medusa swims (Child and Harbison, 1986). The central disc of the umbrella is surrounded by a coronal groove and a peripheral zone with radial thickenings of mesoglea (pedalia) and peripheral marginal lappets. Both tentacles and lappets with prey bend inwards to close off a subumbrellar cavity and transfer the prey to the simple lips of the mouth.

Swimming (section 2.6) may be reduced or stopped during prey transfer. In *Pelagia noctiluca*, bell pulsations are inhibited on the side of the umbrella nearest the food and weak but rapid on the opposite side (Larson, 1987d). As a result, while pulsation rates may increase, motility decreases (Rottini-Sandrini and Avian, 1989). In *Aurelia aurita* there is a reduced frequency of bell contractions after uptake of 8–15 herring larvae by medusae 20–21 mm in diameter (Heeger and Möller, 1987). Similarly in *Nausithoe punctata* the rhythmical discharge of the marginal ganglia that control swimming is inhibited during feeding (Horridge, G.A., 1956a).

In Rhizostomeae there are numerous digitata (small finger-like structures with nematocyst concentrations at their tips) along the edges of the oral arms. The digitata are involved with prey capture, and bend to pass prey inward into ciliated grooves leading to a canal system extending to the stomach (Smith, H.G., 1936; Thiel, M.E., 1964; Larson, 1991). In the rhizostome species which have been examined so far, digestion has not been observed in the oral arms.

Prior to the development of the oral arms and tentacles, the feeding of the ephyrae of *Aurelia aurita* differs from the adult. Ephyrae can catch prey on the lappets at the tips of the ephyra arm. They then use the mobile manubrium to pick small arthropods from the bent ephyra arm (Gemmill, 1921; Southward, 1955; Horridge, G.A., 1956b; Sveshnikov, 1963).

Deepstaria enigmatica and *D. reticulum* are meso-bathypelagic semaeostome medusae with large, thin umbrellas and no tentacles. The method of feeding may be very unusual. The coronal muscle can rapidly contract the margin of the umbrella to purse it shut (Figure 3.10). Larson *et al.* (1988) suggest that upwardly swimming prey may enter the subumbrellar chamber, and that contact will stimulate contraction of the muscle to trap the prey. The prey could then be stung by subumbrellar nematocysts and eventually grasped by the short oral arms.

3.5.2 Polyp prey capture

Prey of stauromedusae are captured and transported to the mouth by the arms and short tentacles (Hyman, 1940; Berrill, M., 1962) (Figure 3.11). Some species, such as *Kishinouyea corbini*, can also perform a somersaulting manoeuvre, trapping prey between the oral surface and the substrate (Larson, 1980).

Semaeostome and rhizostome polyps capture planktonic food on their tentacles. Individual tentacles then shorten and bend, bringing the prey to the mouth (Chapman, D.M., 1965; Cargo, 1971). The tentacle may enter the mouth so that the prey is wiped off as it withdraws from the tightened lips (Loeb and Blanquet, 1973) (Figure 3.12). On *Aurelia aurita* and *Chrysaora quinquecirrha* polyps, currents pass up the column carrying mucus and particles to the tips of the tentacles (Percival, 1923; Southward, 1955; Blanquet and Wetzel, 1975). It is not known whether this also results in feeding if the tentacles bend to the mouth.

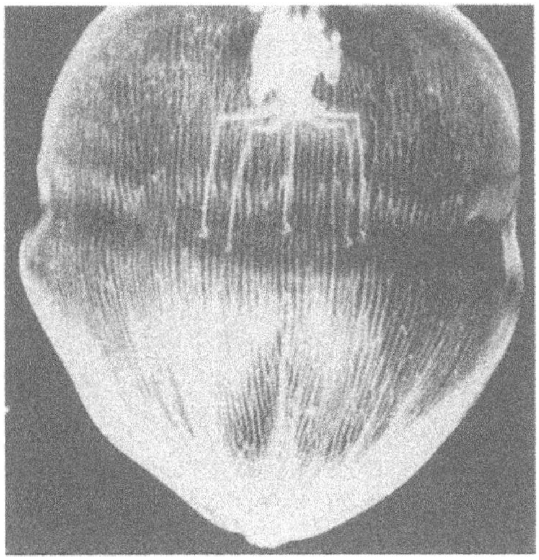

Figure 3.10 *Deepstaria enigmata* with umbrella margin pursed shut. Note absence of tentacles and the presence of a peristaltic wave moving upwards on the umbrella. (Source: Larson *et al.*, 1988, with permission of R.J. Larson and Cambridge University Press.)

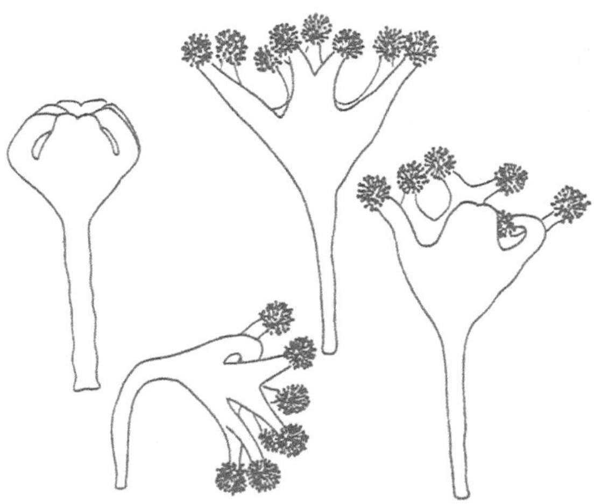

Figure 3.11 *Lucernaria quadricornis* showing fully expanded and fully contracted states, together with browsing and general feeding postures. (Source: Berrill, 1962, with permission of National Research Council of Canada.)

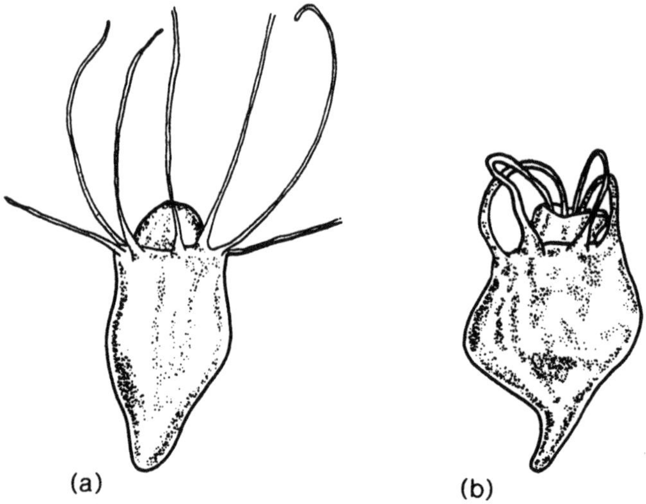

Figure 3.12 Feeding behavior of *Chrysaora quinquecirrha* polyps. (a) Polyp with closed mouth and outstretched tentacles prior to introduction of feeding stimulant; (b) polyp after exposure to 10^{-5} Molar reduced glutathione. Most of the tentacles have been omitted from the drawings for clarity. (Source: Loeb and Blanquet, 1973, with permission of *Biological Bulletin*.)

3.5.3 Chemical induction of feeding

Chemical stimuli are important in controlling feeding behaviour in coelenterates. Following puncture by nematocysts, prey release chemicals that trigger the sequence of events leading to ingestion. The naturally occurring chemicals known to have effects on scyphozoan behavior are summarized in Table 3.3. They include organic acids such as pyruvate and lactate, urea, some fatty acids and lipids, a number of amino acids and some peptides, including the tripeptide reduced glutathione (GSH). No carbohydrates tested to date have been effective for initiation of the feeding response.

Chemicals differ in the minimum effective concentration to elicit a response. Also different chemicals may stimulate one or more of the responses in the sequence of events leading to actual ingestion of the prey. These effects have been most extensively investigated for *Chrysaora quinquecirrha* polyps where the responses include tentacle writhing, gaping of the mouth, and stuffing of the tentacles into the mouth (Loeb and Blanquet, 1973) (Figure 3.12). The most effective chemical tested was reduced glutathione, which elicits all three responses at concentrations down to 10^{-12} Molar.

Table 3.3 Feeding activators

Species	*Feeding activators*	*Source*
Aurelia aurita medusae	asparagine glycine leucine tyrosine oleic acid palmitic acid triolein	Henschel, 1935
Aurelia species	alanine	Kauffman and Muscatine in Lenhoff, 1971
Aurelia species polyps	reduced glutathione	Muscatine in Loeb and Blanquet, 1973
Aurelia aurita medusae	cephalin	Seravin *et al.*, 1979; Seravin, 1995
Chrysaora quinquecirrha polyps	19 amino acids (not lysine) reduced glutathione glycylglycine urea α-ketoglutarate pyruvate lactate	Loeb and Blanquet, 1973
Cyanea capillata medusae	asparagine glycine leucine tyrosine	Henschel, 1935
Cyanea capillata medusae	proline	Seravin *et al.*, 1979 Seravin, 1995
Linuche unguiculata medusae	proline reduced glutathione	Larson, 1979
Lucernaria quadricornis medusae	cephalin	Seravin *et al.*, 1979
Nausithoe punctata medusae	proline reduced glutathione	Larson, 1979

The receptors for the chemicals have not been identified. In *Chrysaora quinquecirrha* at least some of the receptors are present on the tentacles, as isolated tentacles can be induced to respond.

3.6 FEEDING RATES

Feeding rates of field-caught medusae may be determined by examining the gut contents (summarized in section 3.3.1) and measuring the digestion rates of the same size and type of prey at the same temperature (section 4.2.3). This approach requires minimum maintenance of these delicate organisms in the laboratory. Medusae can be dipped from surface layers with minimum loss of gut contents, although those subjected to net capture may lose gut contents or feed in the nets. Although there may be errors if prey do not leave identifiable remains, the method emphasizes natural diets. Feeding rates measured in this way are usually expressed as predation rates, i.e. the number of food animals eaten per predator per day as shown in Table 3.4.

Alternatively feeding rates may be measured in containers in the laboratory or field enclosures. These experiments do not usually present the broad range of alternative prey present in the field, and they require longer maintenance of the medusae in good condition. They are also subject to the effects of reduced turbulence, and of size and shape of the enclosure on both predator and prey (de Lafontaine and Leggett, 1987; Toonen and Chia, 1993). They therefore only produce approximations of the feeding rates in nature, but they are useful in allowing experimental manipulation of factors affecting

Table 3.4 Field predation rates of scyphomedusae based on stomach contents and digestion rates; number of prey per medusa per day

Species	*Prey*	*Rate*	*Source*
Aurelia aurita			Möller, 1980b
(6–25 mm)	fish larvae	1.6	
(16–40 mm)	fish larvae	0.6	
(36–50 mm)	fish larvae	15.9	
Aurelia aurita			Matsakis and Conover, 1991
(10–150 mmm)	mixed prey	3.8	
Chrysaora quinquecirrha			Purcell, 1992
(18–120 mm)	copepods	10–18682	

Table 3.5 Daily clearance rates of scyphomedusae based on laboratory or enclosure experiments; litres cleared per medusa per day

Species	*Volume (litres)*	*Prey*	*Rate*	*Source*
Aurelia aurita	0.5	ciliates	1–13	Stoecker, Michaels and Davis, 1987
(40–56 mm)		rotifers	1	
(66–118 mm)		ciliates	1–134	
		copepod nauplii	2–67	
		dinoflagellates	39	
Aurelia aurita (35–88 mm)	260–6350	fish larvae	204–548	de Lafontaine and Leggett, 1987
Aurelia aurita (12–85 mm)	5000	fish larvae	0–1347	Gamble and Hay, 1989
Aurelia aurita (4–9 mm)	4.6	rotifers	0.4	Olesen, Frandsen and Riisgard, 1994
Aurelia aurita (60 mm)	90	various prey	0–600	Båmstedt, Martinussen and Matsakis, 1994
Chrysaora quinquecirrha (40 mm)		*Artemia*	240	Feigenbaum and Kelly, 1984
Chrysaora quinquecirrha (64–130 mm)	3000	fish eggs fish larvae	32–45 12–23	Cowan and House, 1993
Chrysaora quinquecirrha (40–109 mm)	3200	ctenophore	< 200 –6180	Purcell and Cowan, 1995
Cyanea capillata (40 mm)	25	various prey	< 140	Fancett and Jenkins, 1988
Pelagia noctiluca (14 mm)	5	*Artemia*	6–11	Morand, Carré and Biggs, 1987
Phyllorhiza punctata		copepod		Garcia and Durbin, 1993
(30–40 mm)	14	nauplii	< 200	
(200–240 mm)	350		<9600	
Pseudorhiza haeckeli (40 mm)	25	various prey	< 400	Fancett and Jenkins, 1988

feeding. Feeding rates measured in this way are often expressed as clearance rates, i.e. as the volume of water swept clear of prey by the predator per day as presented in Table 3.5.

Absolute feeding rates vary widely depending on a variety of factors. Some of these factors such as selection of prey types and medusan size will be discussed in the following two sections.

3.6.1 Selection of prey types

As noted in section 3.3 scyphozoa are primarily carnivorous animals. Presumably coelenterates are unable to exploit macrophytes as food sources because they have no mechanisms for mechanically disrupting cell walls. It is less clear why phytoplankton are selected against. *Aurelia aurita* does not utilize flagellates or diatoms when microzooplankton such as ciliates are present (Stoecker, Michaels and Davis, 1987; Båmstedt, 1990). Post-capture selection of particles occurs during this microphagous feeding. For example, the dinoflagellate *Heterocapsa* is expelled by the oral arms during transport, remaining in good enough condition that it may be able to swim after release.

Selection is also shown for or against particular animal prey. In some cases this is a lack of predation on particular animal groups. Giorgi *et al.* (1991) found that *Pelagia noctiluca* is mainly a non-selective predator of meso- and macrozooplankton, but that they do not prey on adults and ephyrae of scyphomedusae. However, Larson (1987d) found that *P. noctiluca* would eat pieces of *Cassiopea* in the laboratory.

Within the diet prey may be proportionately more or less abundant than in the plankton. The most often used measure of prey selection is the index C which ranges from −1 to +1, zero valued for no selection, and can be derived from the chi-square formulation (Pearre, 1982). Calculations indicate for example that for adult *Chrysaora quinquecirrha* there is a positive selection for anchovy eggs and larvae, for copepods and for cladocera but negative selection for copepod naupli (Purcell, 1992; Purcell *et al.*, 1994). Similarly *Stomolophus meleagris* is a largely non-selective predator but selects against calanoid and cyclopoid copepods and their nauplii (Larson, 1991). *Cyanea capillata* and *Pseudorhiza haeckeli* show positive selection for fish eggs and yolk-sac larvae, and varied responses to particular arthropods (Fancett, 1988). *Aurelia aurita* show positive selection for hydromedusae and barnacle nauplii (Sullivan, Garcia and Klein-MacPhee, 1994). Other investigators have not used the Pearre index. *A. aurita* medusae show selection for large, non-loricate ciliates over most metazoan microzooplankton (Stoecker, Michaels and Davis, 1987). Among the metazoans copepod nauplii are selected over rotifers and polychaete larvae (Stoecker, Michaels and Davis, 1987), and mollusc larvae over copepods (Hamner, Gilmer and Hamner, 1982).

Selection of animal prey may be due to differential contact rates (section 3.4), or to differential capture rates following contact (section 3.5). Post-capture sorting may also release some potential prey together with inanimate particles. Tentacles of *Pelagia noctiluca* release ephyrae of the scyphozoan *Cotylorhiza tuberculata* after killing them, whereas most prey cause a fast contraction of the tentacle (Rottini-Sandrini and Avian, 1989). Some animals such as closed bivalve larvae may even be released unharmed from the stomach of *Chrysaora quinquecirrha* (see Purcell *et al.*, 1991).

3.6.2 Factors affecting feeding rates

Feeding rates are affected by contact rates (section 3.4) and by selectivity (section 3.6.1). Feeding rates are also strongly affected by the sizes of both the medusan predator and the prey.

The effect of medusan size has been demonstrated for *Aurelia aurita* by Möller (1980b), Bailey and Batty (1983; 1984), Stoecker (1987), de Lafontaine and Leggett (1987; 1988) and Gamble and Hay (1989). When tanks or enclosures were stocked with a fixed prey concentration, predation rates of *A. aurita* increased with increased diameter of the medusae (Figure 3.13). Similar increases in feeding rates with increasing medusan size have been found for *Pelagia noctiluca* (see Morand, Carré and Biggs, 1987), *Cyanea capillata, Pseudorhiza haeckeli* (see Fancett and Jenkins, 1988), *Chrysaora quinquecirrha* (see Purcell, 1992; Purcell and Cowan, 1995) and *Phyllorhiza punctata* (see García and Durbin, 1993). At the largest predator sizes, feeding rates may again decrease (de Lafontaine and Leggett, 1988). This is partly due to deterioration of the largest animals. In the later part of the annual cycle, most of the *Cyanea* sp. in the Niantic River estuary have empty gastrovascular cavities, although prey are still available (Brewer, 1989) (Figure 3.4).

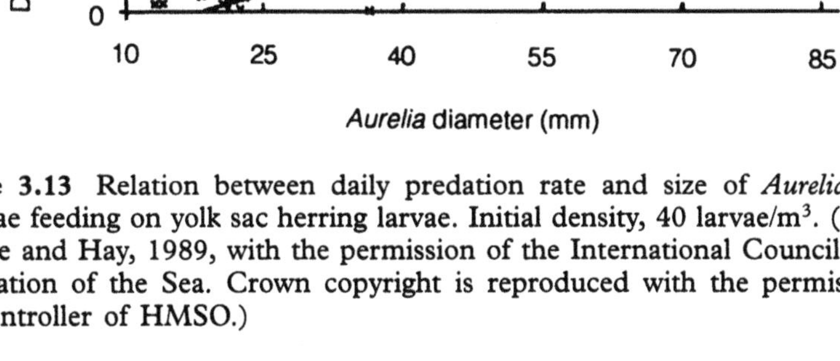

Figure 3.13 Relation between daily predation rate and size of *Aurelia aurita* medusae feeding on yolk sac herring larvae. Initial density, 40 larvae/m^3. (Source: Gamble and Hay, 1989, with the permission of the International Council for the Exploration of the Sea. Crown copyright is reproduced with the permission of the Controller of HMSO.)

As fish larvae increase in size they usually become less vulnerable to predation. As goby larvae grow from 3 to 10 mm standard length, predation by *Chrysaora quinquecirrha* decreases at a rate that, when extrapolated, would predict no mortality at over 11.4 mm length (Cowan and Houde, 1992, 1993). Predation by *Aurelia aurita* on the smaller and weaker yolk-sac fish larvae is greater than that on the larger feeding larvae (Möller, 1984b; Bailey, K.M., 1984; Bailey, K.M. and Batty, 1984; Gamble and Hay, 1989). Decreased predation is again correlated with increased larval length. The actual cause of the decreased predation is unknown. It may be due to decreased contact rates, increased larval escape speed, increased response to mechanical touch or water movement, or to pain, or decreased susceptibility to nematocyst stings. If feeding-stage fish larvae are starved, they become more vulnerable to predation by *A. aurita*, but only when the larvae have almost reached the point of irreversible starvation (Bailey, K.M., 1984; Gamble and Hay, 1989).

If feeding rates depend only on contact rates, for any one predator–prey combination clearance rates should remain constant as prey density changes, and predation rates should increase linearly with prey density. Predation rates do increase with prey density for *Aurelia aurita* feeding on copepods (Anninsky, 1988a), on capelin larvae (de Lafontaine and Leggett, 1988), on herring larvae (Gamble and Hay, 1989), on the ciliate *Strombidium sulcatum*, on the rotifers *Synchaeta* sp. and *Brachionus* sp., on mixed zooplankton (Båmstedt, 1990; Olesen, Frandsen and Riisgard, 1994) and on *Artemia* nauplii (Båmstedt, Martinussen and Matsakis, 1994). Predation rates also increase with prey density for *Pseudorhiza haeckeli* and *Cyanea capillata* vs copepods (Fancett and Jenkins, 1988) and for *Chrysaora quinquecirrha* vs *Artemia* (see Clifford and Cargo, 1978), vs copepods (Purcell, 1992), and vs anchovy eggs (Cowan and Houde, 1993).

When satiation (i.e. reduced clearance rates at high prey concentration) occurs it is probably only at prey densities above those occurring in nature. Bailey and Batty (Bailey, K.M. and Batty, 1983) found satiation of *Aurelia aurita* feeding on herring larvae only at larval concentrations of over 7000 larvae/m^3. Clearance of rotifers by *A. aurita* ephyrae decreased only 30–50% with increasing prey concentration from 7 to 13 000 individuals per litre (Olesen, Frandsen and Riisgard, 1994). Clearance rates of *Phyllorhiza punctata* feeding on copepods remain constant up to prey densities above 200 prey per litre, i.e. to densities higher than patch densities in the field (Garcia and Durbin, 1993).

It is an advantage to a predator to be able to feed at rates greater than the average prey density, in order to be able to fully utilize small-scale aggregations of prey. It is unclear to what extent scyphomedusae can utilize such aggregations. Caution must be used in extrapolating from the above experiments, with high prey density in tanks or enclosures, to turbulent field conditions.

Predation on one species of prey is in some cases reduced by the presence of another prey species. *Aurelia aurita* feeding rates on yolk-sac capelin larvae were not affected by the presence of wild zooplankton <1 mm in length, even at five times natural densities (de Lafontaine and Leggett, 1988). Nor was the predation of *Chrysaora quinquecirrha* on goby larvae affected by the presence of zooplankton < 1 mm (Cowan and Houde, 1992). However, the predation was reduced 20–25% when the ctenophore *Mnemiopsis leidyi* was present as an alternative prey.

Little is known about the effects of the physical environment on feeding rates. The predation rates of *Aurelia aurita* and *Chrysaora quinquecirrha* increase with increased temperature (Anninsky, 1988a; Purcell, 1992). This may be due to increased swimming rate as discussed in section 2.6.2.

Surprisingly predation rates have not been shown to be significantly affected by light or dark conditions. This point has been examined for *Aurelia aurita* (see Möller, 1980b; Bailey, K.M. and Batty, 1983), *Cyanea capillata* and *Pseudorhiza haeckeli* (see Fancett and Jenkins, 1988). It is reasonable that since these are non-visual predators, the medusae should feed similarly in light or dark. However, light may influence avoidance reactions, or vertical migration, by prey with eyes, and hence would be expected to influence predation rates.

4 Nutrition

4.1 INTRODUCTION

Chapter 3 considered the acquisition of prey by scyphozoa. The present chapter will discuss processing of that material; intake, digestion, and distribution. It will also consider the acquisition of nutrients from alternative sources. Organic material is obtained by some scyphozoa through uptake of dissolved organic material (DOM) from sea water or by translocation from endosymbiotic algae. Little is known about the relative importance of these alternative sources. The presence of symbionts has a restricted distribution among scyphozoan species, but uptake of DOM may prove to be more widespread.

4.1.1 Units of intake

The ingestion rate (or daily ration) is the amount of food ingested per animal per day. This may be expressed in the same units as predation rates, i.e. prey per medusa per day. Daily ration may also be expressed as ash-free dry weight (AFDW), carbon or nitrogen (also intake per animal per day). Daily ration is affected by a number of factors already discussed in section 3.6.2 on feeding rates.

The daily ration may be converted to a specific daily ration for comparison between different predator–prey situations or experimental treatments. The specific daily ration expresses the daily intake as a percentage of the predator content using the same units of measurement (AFDW, C, or N). Table 4.1 summarizes the highly variable

Table 4.1 Specific daily rations of scyphomedusae

Genus	*Prey*	*Experimental conditions*	*Specific daily ration*	*Source*
Aurelia		L		Båmstedt, 1990;
3–4 mm	ciliate < 50/ml	AFDW	0.42	Båmstedt, Martinussen and Matsakis, 1994
2–5 mm	rotifer < 600/1		1.5–13	
4–18 mm	mixed 28/1		< 10	
6–7 mm	mixed 323/1		82	
> 45mm	mixed > 100/1		5–40	
55–85 mm	mixed (300–1000 μm) 25/1		2.7–66	
55–85 mm	mixed (>1000 μm) 25/1		262	
Aurelia		F		Matsakis and
1–8 cm	mixed	C	400–750	Conover, 1991 (see text)
Chrysaora		F		Purcell, 1992
< 45 mm	mixed	N	8.6	
> 45 mm	(hard body only)		4.0	
Cyanea		L		Båmstedt, Martinussen
150 mm	mixed (300–1000 μm) 25/1	AFDW	< 0.5	and Matsakis, 1994
65–220 mm	*Aurelia* 50–260 mm		650–700	
Pelagia		L		Morand, Carré and
14 mm	*Artemia* 5/1	N	13	Biggs, 1987
14 mm	*Artemia* 20/1		35	
Phyllorhiza		L		Garcia and Durbin,
5–27 cm	Mixed 24–188/1	AFDW	1–16	1993

C, calculated on the basis of carbon contents; F, field; L, laboratory; N, calculated on the basis of nitrogen contents; AFDW, calculated on the basis of ash-free dry weight.

data available to date. Matsakis and Conover (1991) were unable to explain why their values were much higher than those obtained by other investigators. Values were also higher for hydrozoa they examined than for hydrozoa examined by Larson (1987b) under very similar field conditions. It is very unlikely that these values are correct.

The specific daily ration is potentially useful in considering whether the intake is proportionately greater for smaller medusae. Little data is available. Using mixed plankton in laboratory experiments on *Aurelia aurita*, Båmstedt (1990) found that, over a broad size spectrum, the specific daily ration decreased with increased medusa size. Matsakis and Conover (1991) also found this to be true.

4.1.2 Dietary requirements

The daily ration as defined in section 4.1.1 refers to the total intake of food, not what is actually assimilated or what would be optimal for body processes. Assimilation, the physiologically useful ration, is difficult to measure in scyphozoa due to difficulty in retrieval of eliminated waste products. Assimilation efficiencies, i.e. [(ingestion − egestion)/ingestion] × 100%, measured by Anninsky (1988c) for *Aurelia aurita* feeding on mixed prey or copepods varied from 36 to 86%. These are the only data for scyphozoa.

Lack of accurate information on assimilation rates, of acquisition of nutrients from alternative sources, and of many other terms in carbon and energy budgets, makes calculation of daily ration from organic requirements for metabolism, growth and reproduction highly speculative (Mironov, 1967; Shushkina and Musayeva, 1983; Larson, 1987e; Malej, 1989a; Båmstedt, 1990; Arai, in press). However, it is possible to observe experimentally whether a given daily ration is adequate for growth (Chapter 7).

Little is known about the ability of scyphozoa to utilize particular organic compounds. The presence of appropriate digestive enzymes indicates the ability to digest proteins, carbohydrates and lipids (section 4.2.2), but not the extent to which they contribute to metabolism. Also little is known about scyphozoan requirements for particular vitamins, amino acids, fatty acids or minerals. The requirement of iodine-containing compounds for strobilation of *Aurelia* will be discussed in section 6.4.3.

4.2 DIGESTION

4.2.1 Extracellular and intracellular digestion

Digestion in scyphozoa involves an extracellular phase as well as a following intracellular phase (Chuin, 1929a,b, 1930). The gastrovascular system of scyphomedusae consists of a central stomach surrounded by stomach pouches and/or a canal system (Figure 4.2). In most species the stomach contains gastric cirri (gastric filaments), tapering structures each with a mesogleal core covered with gastrodermal cells. These structures are concerned with extracellular digestion within the stomach. Some species can also digest prey before it reaches the stomach, using the oral arms. The polyp lacks gastric cirri but cells releasing enzymes may be concentrated on the longitudinal septa.

The gastrodermis of the gastric cirri contains ciliated mucous and serous secretory cells (Figure 4.1). In medusae of *Aurelia aurita* mucous cells are concentrated in the apical region of the cirri (Heeger and Möller, 1987). Serous cells, presumed to produce the digestive enzymes, are concentrated in the basal region of the cirri where their discharge can be directly applied to entangled prey. Similar cells are present on the septa of the scyphistoma (Hentschel and Hündgen, 1980). As noted in section 3.2.5, microbasic eurytele nematocysts have been found in the gastric cirri of medusae such as *Rhopilema verrilli* and *Deepstaria reticulum* (see Calder, 1972; Larson, Madin and Harbison, 1988). In these species nematocysts may be involved in the delivery of digestive enzymes as well as attachment to the food.

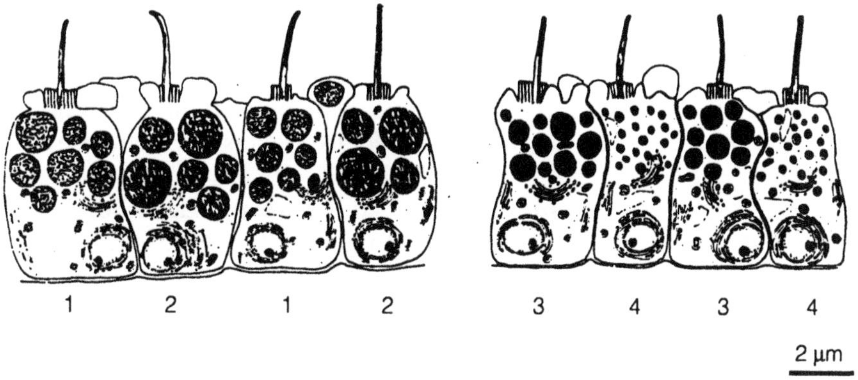

Figure 4.1 Mucous (1,2) and serous (3,4) cells of *Aurelia aurita*, indicating the four types of glandular cells in the gastric cirri. (Source: Heeger and Möller, 1987, with permission of T. Heeger, H. Möller and Springer-Verlag.)

Digestion by the oral arms allows utilization of large prey. Although *Chrysaora quinquecirrha* medusae possess gastric cirri, prey which are too large to enter the stomach are also digested in the oral arms. Pieces of excised oral arm tissue are capable of protein digestion and must therefore be capable of releasing enzymes (Larson, 1986a). Similarly a *Cyanea capillata* can digest an *Aurelia aurita* of equal size using the oral arms (Plotnikova, 1961; Seravin, 1991). In a few species of medusae, such as *Drymonema dalmatinum* over 10 cm, gastric cirri are absent and the oral arms completely envelop the prey during digestion (Larson, 1987c).

Intracellular digestion occurs in the gastrodermal cells following endocytosis of particles derived from the food. Particles of *Artemia* nauplii homogenate injected into the stomachs of scyphistomae of *Cassiopeia xamachana* are engulfed phagocytically by the gastrodermal cells (Fitt and Trench, 1983). Lysosomes fuse with the resulting phagosomes and empty their contents into the phagosomes. Similar digestion within food vacuoles occurs in *Aurelia aurita* medusae (Heeger and Möller, 1987).

4.2.2 Enzymes

Two early papers on the extracelluar digestion of scyphozoa examined symbiotic species. Ohtsuki (1930) obtained fluids from the stomach and body cavity of *Mastigias papua*. He found digestion of starch, glycogen and olive oil, as well as a substance tentatively translated as rennin by the present author, i.e. probably digestion of protein. Extracts of the gastric cirri and stomach wall contained other enzymes including a cellulase. Smith (1936) fed *Cassiopea frondosa* with mollusc meat. The pH of the coelenteric fluid dropped from 7.8 to 7.2, and a proteolytic enzyme with an optimum pH of 7.0 was present. A similar protease was present in tissue extracts of the gastric cirri. A glycogenase was also present in the fluid of the stomach, but no lipase, amylase, sucrase, lactase or cellulase was detected.

Other workers have not distinguished between extracellular and intracellular digestion. Bodansky and Rose (1922) used tissue suspensions of the gastric cirri of *Stomolophus meleagris* to demonstrate the presence of proteolytic enzymes that could digest gelatin in both acid and alkaline media, a lipolytic enzyme that could digest ethyl butrate, an amylase that could digest starch, and a maltase. Similarly Båmstedt (1988) demonstrated the presence of a proteolytic enzyme and an amylase in a whole animal homogenate of *Aurelia aurita*. Manchenko and Zaslavskaya (1980) found a leucine aminopeptidase in *Cyanea capillata*. Stewart and Lakshmanan (1975) mention the presence of an alkaline phosphatase in the gastric fluid of *Chrysaora quinquecirrha*, but used tissue homogenates to characterize alkaline and acid phosphatases capable of hydrolysing a wide variety of substrates.

There is a need for further characterization of known enzymes and for an extensive search for others. For example, there has been no search for possible scyphozoan chitinases. It is likely that they would be present since many scyphozoa utilize arthropod prey and sea anemones do possess such enzymes (Shick, 1991). It would also be interesting to investigate whether cellulase and other enzymes digesting plant polysaccharides are present in non-symbiotic species. It has often been assumed that scyphomedusae are restricted to a primarily carnivorous diet by inability to produce appropriate enzymes. However, anthozoa are known to produce ß-glucuronidase, amylase and laminarinases potentially capable of digesting cell walls of marine algae (Shick, 1991).

4.2.3 Digestion rates

There has been a good deal of interest in the rates of extracellular digestion by scyphomedusae in connection with derivation of feeding rates from stomach contents (section 3.6). Table 4.2 summarizes the data available.

A number of factors affect these rates including temperature, comparative sizes of prey and predator, and type of prey. Digestion of *Calanus heligolandicus* by *Aurelia aurita* is faster at 23.6°C than at 6.7°C (Q_{10} = 1.2–1.4) and the rate is also somewhat increased with increased size of the medusa (Anninsky, 1988c). A greater effect of temperature was seen in digestion of herring larvae by *A. aurita* where 19 hours passed from the capture of 10 herring larvae to the shedding of faeces at 5°C, but less than 4 hours at 22°C (Heeger and Möller, 1987). Rates of copepod digestion by *Chrysaora quinquecirrha* were strongly related to temperature over a range of 20–27°C, less strongly related to the number of prey in each medusa, and not shown to be related to medusa size (Purcell, 1992).

With reference to type of prey, the rate of digestion by *Cyanea capillata* and *Pseudorhiza haeckeli* is significantly slower for fish eggs than copepods (Fancett, 1988). Some potential prey may not be digested at all. *Aurelia aurita* does not digest amphipods (Fraser, 1969). *Chrysaora quinquecirrha* medusae and ephyrae capture but rarely digest live oyster veliger larvae; 98% of the larvae survived for 24 hours after egestion (Purcell *et al.*, 1991). However, the scyphistoma stage of *C. quinquecirrha* digested 48% of the veligers ingested.

Table 4.2 Digestion times of scyphomedusae

Species	*Prey*	*Time (h)*	*Temp. (°C)*	*Source*
Aurelia aurita	FL	2	12	Fraser, 1969
Aurelia aurita	FL	4–10	10–12	Möller, 1980b
Aurelia aurita	FL	4–6	10–13	Bailey, K.M. and Batty, 1983
Aurelia aurita	FL	19	5	Heeger and Möller, 1987
	FL	12	10	
	FL	8	15	
	FL	4	22	
Aurelia aurita	CO	7–24	7–24	Anninsky, 1988c
Aurelia aurita	MP	3–4	4	Matsakis and Conover, 1991
Chrysaora quinquecirrha	CT	< 10	26–29	Larson, 1986a
Chrysaora quinquecirrha	CO	2–6	20–27	Purcell, 1992
Chrysaora quinquecirrha	FE	4	26	Purcell *et al.*, 1994
	FL	1		
Cyanea capillata	CT	8	7–12	Plotnikova, 1961
	ME	10		
	SI	26		
Cyanea capillata	ME	6–25		Loginova and Perzova, 1967
Cyanea capillata	CO	2		Fancett, 1988
Cyanea capillata	CT	3–6		Seravin, 1991
	ME	24–32		
Drymonema dalmatinum	ME	6–8		Larson, 1987c
Pelagia noctiluca	ME	8–10	27–29	Larson, 1987d
Pelagia noctiluca	MP	42–54	15	Rottini-Sandrini and
		20–28	19	Avian, 1989
Pseudorhiza haeckeli	CO	2		Fancett, 1988
	FE	3		
Stomolophus meleagris	BA	2	28–30	Larson, 1991
	CO	1.5–2		
	FE	3		
	LA	1.5		
	TI	1.5		
	VE	2–4		

BA, barnacle larvae; CO, copepods; CT, ctenophora; FE, fish eggs; FL, fish larvae; LA, larvacea; ME, medusae; MP, mixed prey; SI, siphonophores; TI, tintinnids; VE, veligers.

4.3 CIRCULATION AND TRANSLOCATION

4.3.1 Circulatory canals and ciliary currents

As extracellular digestion proceeds, the products are distributed through the gastrovascular system for uptake by gastrodermal cells. The ciliary tracts leading to the mouth were described in section 3.5.1. Further ciliary tracts move fluid, mucus and food to the stomach, and then beyond to the rest of the gastrovascular system. Spatially or temporarily separated centripetal currents move waste products back through the stomach to the mouth for discharge.

In scyphomedusae the central stomach is usually four-sided or extended into four pouches containing the gastric cirri. The stomach is surrounded by a peripheral marginal zone extending to the umbrella margin. In coronate and some semaeostome medusae the cavity of this zone is divided by radial septa into a series of pouches. In the remaining semaeostome medusae (including *Aurelia*), and all rhizostome medusae, flow is restricted to a system of peripheral canals.

Circulation has been most extensively examined in *Aurelia aurita* medusae (Widmark, 1911, 1913; Gemmill, 1921; Wetochin, 1930; Southward, 1955). The manubrium, leading from the mouth at the base of the oral arms to the stomach, is cruciform with four V-shaped grooves. Currents move outward at the peripheral base of the groove and inward along the proximal walls of the groove. The inward currents then pass along the roofs of the stomach pouches, over the gastric cirri, the gonads and the surrounding gastrocircular groove and along the roofs of the eight adradial canals to the periphery. Outward currents move back beneath the inward currents in the adradial canals, as well as through a network of other canals, from the periphery to the floor of the gastrocircular grooves, along the floor of the gastrocircular grooves and up the manubrium (Figure 4.2).

Although the configurations of the stomach and peripheral gastrovascular system vary widely in other scyphomedusae, the main principles of circulation seen in *Aurelia* apply. In the manubrium the outward currents are peripheral to the inward currents (Smith, H.G., 1936; Larson, 1976, 1979). Outward currents are also separated from the inward currents in the remainder of the gastrovascular system, either in separate canals or flowing beneath the inward currents in the same structures (Thiel, M.E., 1964; Larson, 1987d).

The gastrovascular system of stauromedusae and scyphistomae of other orders consists of a tubular cavity divided in its upper portions by four radial septa. The open nature of this system makes maintenance

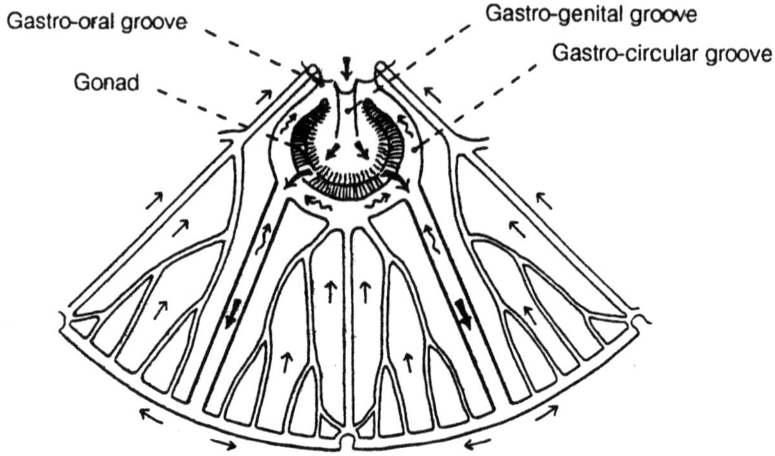

Figure 4.2 Circulation in the gastrovascular system of *Aurelia aurita*. Dark arrows indicate movement toward the periphery along the roofs of chambers and large adradial canals; wavy lines indicate passage back along the floor of these same canals. Lighter straight arrows indicate return toward the mouth along smaller perradial and interradial canals. Simplified by omission of many branches of the perradial and interradial canals. (Source: Russell, 1970, with permission of Cambridge University Press.)

of simultaneous inward and outward currents more difficult to maintain. In polyps of five coronate species, ciliary currents carry food down along the edges of the septa. These currents can reverse, allowing defecation (Chapman, D.M., 1973). The reactions in semaeostome polyps are variable. Blanquet and Wetzel (1975) found only outward currents in polyps of *Chrysaora quinquecirrha*, even when *Artemia* extract or solutions of reduced glutathione were applied (Figure 4.3). These solutions did cause tentacles to bend into the mouth. The relatively large prey may only be delivered by the tentacles, and the currents may be needed only for removal of waste products and silt. Chapman (1973) observed reversal of ciliary currents in *Aurelia aurita* scyphistomae he had turned inside-out with forceps and a blunt glass rod, i.e. some semaeostome polyps may resemble the coronate polyps.

4.3.2 Endocytosis

Internalization of extracellular particles by endocytosis is carried out by gastrodermal cells lining much of the gastrovascular cavity. Coated vesicles are often associated with endocytosis. Vesicles of this type are present at the gastrovascular cavity end of nonsecretory gastrodermal

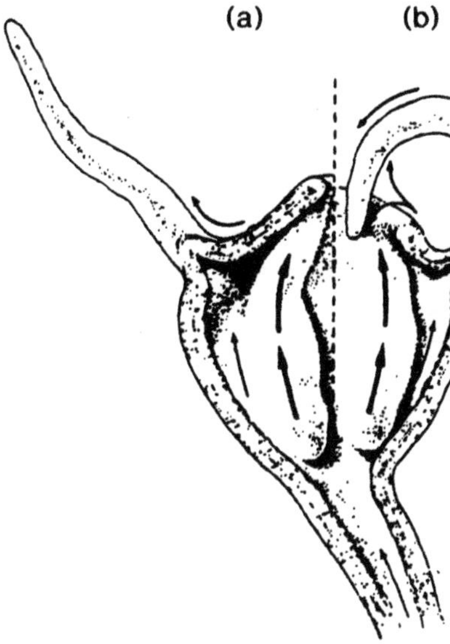

Figure 4.3 Water currents generated by the gastrodermal cilia in (a) non-feeding and (b) feeding polyps of *Chrysaora quinquecirrha*. Currents are particularly rapid along the septal edge. (Source: Blanquet and Wetzel, 1975, with permission of *Biological Bulletin*.)

cells in *Cyanea capillata* (see Anderson and Schwab, 1981). They have been observed in peri-rhopalial tissue at the periphery of the gastrovascular system.

The sequence following presentation of digested food has been examined in *Aurelia aurita* polyps and medusae (Hentschel and Hündgen, 1980; Heeger and Möller, 1987) and, in greater detail, in *Cassiopea xamachana* scyphistomae (Fitt and Trench, 1983). Particles of *Artemia* were engulfed phagocytically by cells of *C. xamachana*. Phagosomes containing groups of *Artemia* particles were formed by overlapping pseudopods. Lysosomes (which had been previously labelled with ferritin) then fused with the phagosomes to deliver digestive enzymes.

4.3.3 Translocation

It is not known how organic compounds are moved from the gastrodermal cells lining the gastrovascular cavity to other cells, such as the muscle layers, or to the mesoglea. Muscle cells are always in contact with mesoglea but may be some distance from the lining of the gastrovascular cavity. Also tentacles may have a solid core of gastroderm so that the tentacle tips are a great distance from the circulating fluid of the cavity. It is possible that sufficient free water is present in the mesoglea to allow adequate diffusion of organic compounds.

Cells are present in the mesoglea of some species (section 2.2). Amoebocytes of *Aurelia* contain many vesicles and granules which could contain storage products (Chapman, D.M., 1974). However, since cells are not present in all species, they cannot be the only agent for distribution of organic compounds.

4.4 UPTAKE OF DISSOLVED ORGANIC MATERIAL

Sea water contains dissolved organic material (DOM) which may reach concentrations of 3 mg carbon/litre in the open ocean, and much higher values in areas near organic detritus. Many anthozoa are able to absorb these compounds, especially glucose and amino acids (Schlichter, 1980; Sebens, 1987). It is not known how widespread this ability may be in the Scyphozoa. In particular information is lacking about the importance of DOM to larval stages such as the planula.

Uptake of DOM has been demonstrated in four species. Autoradiographs of *Pelagia noctiluca* medusae which have been incubated in a ^{14}C amino acid mixture show label incorporation into the epidermis of the subumbrellar surface, oral arms, outer portions of the tentacles and tentacular bulbs (Ferguson, 1988). However, analysis of the stable isotope composition of the medusae indicates that DOM is probably an insignificant food source (Malej, Faganelli and Pezdic, 1993). *Lucernaria quadricornis* can concentrate ^{14}C-labelled glucose (Erokhin, 1979). *Linuche unguiculata* with symbiotic zooxanthellae can take up ^{15}N-labelled glycine, ^{15}N-labelled leucine and ^{14}C-labelled alanine from seawater (Wilkerson and Kremer, 1992). The only quantitative observations on azooxanthellate forms were made by Shick (1973, 1975) on ^{14}C-glycine uptake by the polyps and ephyrae of *Aurelia aurita*.

Since *Aurelia aurita* polyps have high intracellular concentrations of free amino acids, especially glycine (Webb, Schimpf and Olmon, 1972; Shick, 1976), the uptake of glycine from sea water must occur against a strong concentration gradient. The variation in rate with substrate concentration for a large number of enzyme-catalysed reactions is described by the Michaelis–Menton equation. The uptake of radiolabelled glycine does follow Michaelis–Menton kinetics (Shick, 1975), although this does not necessarily prove that the transport process is enzyme mediated. The values of the two kinetic parameters K_T and V_{MAX} are directly related to acclimation temperature between 12°C and 32°C. (K_T is the substrate concentration at which uptake is experimentally determined to be half the maximal uptake rate V_{MAX}.) The uptake decreases with decreasing salinity, but is unaffected by the presence or absence of bacteria (a potential complication in such measurements), or by starvation for up to 12 days (Shick, 1973).

It should be noted that the previous paragraph refers to the uptake of radiolabelled glycine. It is not known to what extent the intracellular free amino acids may show efflux to the environment by diffusion or excretion. If the efflux of glycine is significant then the net influx of glycine will be less than that indicated by the uptake of the radiolabelled glycine.

Whatever the net rate of uptake, the actual utilization of the radiolabelled glycine is demonstrated by the production of $^{14}CO_2$ (Shick, 1973). Also Shick (1975) demonstrated that uptake could benefit strobilizing *Aurelia aurita*. Eight weeks of starvation in Millipore-filtered artificial sea water at 20°C result in a 78% reduction in the number of polyps strobilating in response to temperature increase and exposure to iodide. This effect of starvation can be prevented by exposing the starved polyps to environmental concentrations of glycine or alanine during the starvation period.

4.5 SYMBIOSIS

Endosymbiotic algae carry out photosynthesis and are sources of organic compounds for some scyphozoa. These symbionts, commonly referred to as zooxanthellae, are present in scyphistomae and medusae such as the coronate medusa *Linuche unguiculata* and several rhizostomeae including species of *Cassiopea* and *Mastigias*. Most research has concentrated on species of *Cassiopea* and their symbionts.

4.5.1 Identity and location of algal symbionts

Zooxanthellae are dinoflagellates. Within the host they are coccoid (spherical) in form and lack flagella and surface grooves (Figure 4.4), but if cultured they can regain the two flagella, grooves and motility characteristic of dinoflagellates (Figure 4.5) (Freudenthal, 1962; Loeblich and Sherley, 1979). Cultures include a smooth-walled, coccoid and non-motile vegetative phase, which divides to liberate the motile gymnodinioid phase. The coccoid phase in culture is similar to that in the host although the complex cell covering (periplast),

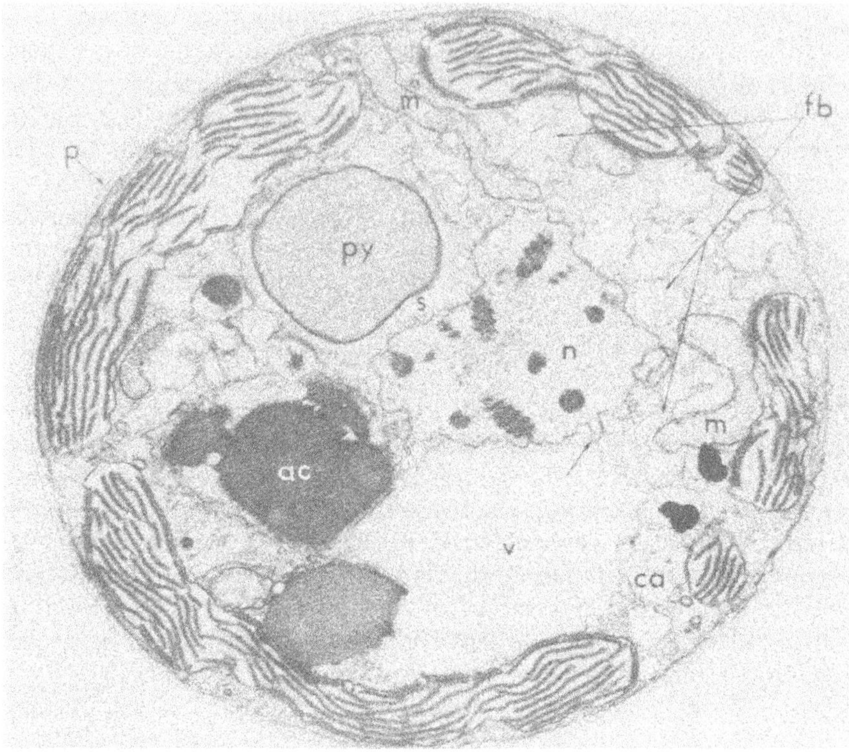

Figure 4.4 Electron micrograph of a zooxanthella, *Symbiodinium microadriaticum*, isolated from a *Cassiopea*. The single but multilobed chloroplast is seen as peripheral sections enclosing lamellae. ac = accumulation body; ca = calcium oxalate crystals; fb = fibrous bodies; m = mitochondria; n = nucleus; p = periplast; py = pyrenoid; s = starch; v = vacuole. ×14 600. (Source: Kevin *et al.*., 1969, with permission of *Journal of Phycology*.)

consisting of several apposed membranes, is thinner in the host than in culture. The cell organelles present include the nucleus, one or more chloroplasts with attached pyrenoid, mitochondria, vacuoles, granules, and an accumulation body (Figure 4.4).

The systematics of the zooxanthellae present in scyphozoa is not clear. Nevertheless, the dinoflagellate *Symbiodinium microadriaticum* was first described as a new species using cultures of zooxanthellae derived from *Cassiopea* sp., presumably *C. xamachana* (see Freudenthal, 1962). The binomium *Symbiodinium microadriaticum* therefore refers to the symbionts of *Cassiopea* (see Kevin *et al.*, 1969; Trench and Blank, 1987; Blank and Huss, 1989). However, morphologically

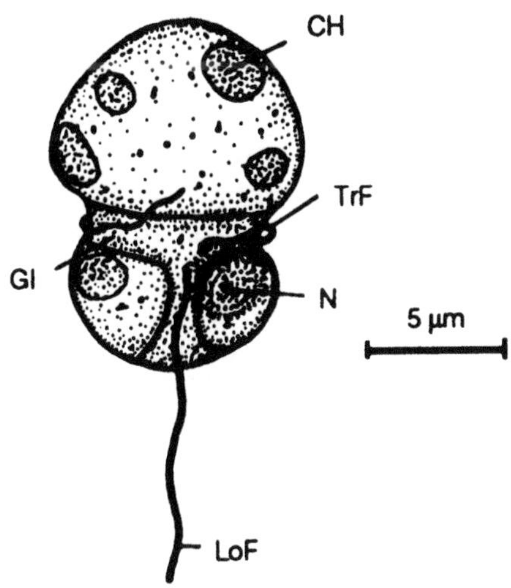

Figure 4.5 Gymnodinioid zoospore of *Symbiodinium microadriaticum*. CH = chloroplast; GI = girdle; N = nucleus; LoF = longitudinal flagellum; TrF = transverse flagellum. (Source: Freudenthal, 1962, with permission of *Journal of Protozoology* and the Society of Protozoologists.)

similar algae are symbiotic with a phyletically broad range of hosts (Taylor, 1974).

There has been extensive discussion using a wide variety of criteria as to whether strains of *Symbiodinium* isolated from different hosts represent different species or not. In order to prevent differences due to direct interaction with host tissue, strains have been compared in culture. Differences between strains cultured from different hosts have been found in morphological (Schoenberg and Trench, 1980b; Blank, 1986), biochemical (Schoenberg and Trench, 1976, 1980c; Chang and Trench, 1982, 1984) and physiological and behavioural studies (Schoenberg and Trench, 1980a; Fitt, Chang and Trench, 1981; Iglesias-Prieto and Trench, 1994). Although data on sexual recombination are lacking, differences between strains in chromosome number and volume and DNA composition and sequences indicate that multiple species are probably present (Blank and Trench, 1985; Blank, Huss and Kersten, 1988; Blank and Huss, 1989; Rowan and Powers, 1991a,b). Zooxanthellae obtained from different individuals of the same host species belong to the same strain according to a variety of criteria, but depending on the criteria used (even DNA sequences) indistinguishable algae may be isolated from taxonomically very divergent hosts. By many criteria *Cassiopea xamachana* and *C. frondosa* appear to contain identical algae (Trench and Blank, 1987), but the zooxanthellae differ in DNA sequences (Rowan and Powers, 1991b).

Recently the symbiotic algae from *Linuche unguiculata* have been examined in the medusa and in culture (Trench and Thinh, 1995). A new species, *Gymnodinium linuchae*, has been described based on morphology. This dinoflagellate genus already contained both symbiotic and freeliving forms, but this is the first species to be described from a coelenterate.

In *Cassiopea* and *Mastigias* medusae most zooxanthellae are found in cells of the mesoglea. In *C. xamachana* medusae they are present in narrow bands beneath the exumbrellar and subumbrellar epithelia of the bell, especially near the muscle bands, and more broadly in the mesoglea of the oral appendages (Blanquet and Riordan, 1981; Blanquet and Phelan, 1987). Similarly in *Mastigias* sp. they are clustered in the mesoglea immediately beneath the epidermis of the bell, especially near the coronal muscle, and of the oral lobes and arms (Muscatine and Marian, 1982).

In *Linuche unguiculata* medusae the algae are present in endodermal cells (Trench and Thinh, 1995; Montgomery and Kremer, 1995). Patches of zooxanthellae in the subumbrella (Figure 4.6) expand and become thinner during the day and contract at night (Costello and Kremer, 1989). The mechanism of this complex movement of zooxanthellae and endodermal host cells is not known. Unlike similar rhythms in anthozoa it is partly endogenous rather than being directly cued by ambient light intensity. Motility of *Gymnodinium linucheae* in culture also shows a 24-hour rhythmicity, but this may be coincidental and does not provide direct proof that the algae control the rhythm *in situ* (Crafts and Tuliszewski, 1995).

4.5.2 Metabolic exchange between symbiont and host

Zooxanthellae, like free-living algae, photosynthesize and fix carbon dioxide into organic compounds. A small *Cassiopea* sp. medusa with its algal symbionts can photosynthesize at a rate of at least 75 μg C/cm^2 per hour in the light (Drew, 1972). On a dry-weight basis, rates of photosynthesis of *Cassiopea andromeda* are approximately 80 μmol C/g per hour in the medusae and 15 μmol C/g per hour in the polyps (Hofmann and Kremer, 1981). The difference in rates is largely due to differential density of algae in the two stages of the life cycle, and average rates of photosynthesis based on content of chlorophyll *a* are very similar (5 μmol C/mg Chl *a* per hour). Rates of ^{14}C-fixation in the dark do not exceed 5% of the photosynthetic rates at the same temperature.

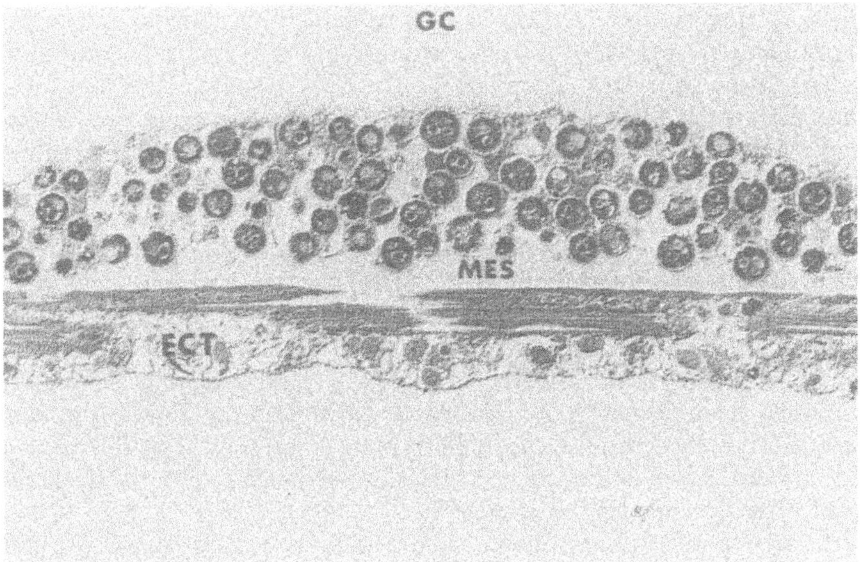

Figure 4.6 Light micrograph of a cross-section through the lower bell of *Linuche unguiculata* showing a contracted patch of zooxanthellae. Note the position of the zooxanthellae close to the striated circular myofibrils of the subumbrellar ectoderm. ECT = ectoderm; GC = gastrovascular cavity; MES - mesoglea. ×450. (Source: Costello and Kremer, 1989, with permission of Inter-Research.)

Carbon dioxide is the primary substrate for photosynthetic carbon assimilation in zooxanthellae, which can fix metabolic CO_2 produced by the host scyphozoan. This is facilitated by the frequent location of the zooxanthellae in the vicinity of actively metabolizing host tissues such as muscle. However, metabolic CO_2 may be insufficient to maintain the high photosynthetic rates, and it may be necessary to utilize carbon from the sea-water bicarbonate pool. At pH 8.2 to 8.3, most of the inorganic carbon in sea water is in the form of HCO_3^-, and the uncatalysed conversion to CO_2 is a relatively slow process. The enzyme carbonic anhydrase, which catalyses this conversion, is present in animal tissue from the zooxanthellate cnidarian *Cassiopea xamachana* but absent in the azooxanthellate form *Aurelia aurita* (see Weis, Smith and Muscatine, 1989). It is not known whether this carbonic anhydrase has been transported out of the algae or has been synthesized by the host, nor how movement of carbon dioxide and bicarbonate is influenced by the membranes and possible pH differences of host and algal cells.

Algae also require sources of nitrogen, phosphorus and other elements. One source of nitrogen is ammonium produced by the host metabolism. Symbiotic *Cassiopea* sp. medusae excrete less ammonium into the surrounding sea water than do aposymbiotic animals (Cates and McLaughlin, 1976). There may also be uptake of dissolved nutrients from sea water. *Mastigias* sp. medusae show a net uptake of ammonium from sea water both day and night (Muscatine and Marian, 1982). Comparison with uptake rates of isolated zooxanthellae suggests that most ammonium taken up by the medusae is used by the zooxanthellae. *Linuche unguiculata* can take up phosphate, ammonium, nitrate and at least three amino acids from sea water (Wilkerson and Kremer, 1992). However, the rate of nitrate uptake is low and nitrogen from amino acids is primarily incorporated into the host tissue. Therefore it is likely that the main source of algal nitrogen in this system is also ammonium. No ammonium excretion was measured even after several days in the dark.

A portion of photosynthate produced by the algae is translocated to the host. If the association of the *Cassiopea* sp. medusa and its algae is exposed to an aqueous solution containing ^{14}C-labelled carbon dioxide in the light, labelled products of photosynthesis can be detected autoradiographically in both the zooxanthellae and the host tissue (Balderston and Claus, 1969). Freshly isolated zooxanthellae from *C. frondosa* and *Rhizostoma* sp. medusae, when incubated in the light in a solution containing $NaH^{14}CO_3$ and fresh host tissue homogenate, liberate 20–23% of the photosynthates into the medium as water-soluble organic compounds (Trench, 1971a). For *C. frondosa* zooxanthellae the principal exudate is glycerol, whereas for *Rhizostoma* zooxanthellae it is fumarate/succinate. Host homogenate was included in the incubation medium because in anthozoa it can greatly enhance the excretion of the photosynthetic products (Trench, 1971b). *Symbiodinium microadriaticum* cultured from *C. xamachana* also release large molecular weight glycoproteins (Markell, Trench and Iglesias-Prieto, 1992; Markell and Trench, 1993). It is not known to what extent the metabolic responses of scyphozoan zooxanthellae isolated from the host may differ from those *in hospite*.

In addition to transfer of organic compounds from algae to host, there may also be uptake of host organic compounds by the algae. Zooxanthellae isolated from *Cassiopea xamachana* can take up ^{14}C-alanine, ^{14}C-glucose and ^{14}C-glycerol and incorporate them into a variety of organic compounds (Carroll and Blanquet, 1984b; Macon McDermott and Blanquet, 1991). The rate of uptake of alanine is not affected by the rate of algal photosynthesis, but is strongly inhibited by a low molecular weight (2–10 kilodaltons) fraction of the host tissue (Carroll and Blanquet, 1984a) or of sea anemone tissue (Blanquet, Emanuel and Murphy, 1988). The uptake of glucose is also inhibited by a different low molecular weight (< 2 kD) fraction of host tissue but the uptake of glycerol is unaffected (Macon McDermott and Blanquet, 1991). There is a Na^+-dependent, active transport system for glucose but simple or facilitated diffusion of glycerol. It is not known what uptake rates may occur when the zooxanthellae are enclosed in vacuoles within host cells.

4.5.3 Establishment and control of algal numbers

In sexual reproduction of most of the scyphozoa so far examined, algae must be acquired from the environment during the scyphistoma stage. Thus algae are absent in the eggs and planulae of the otherwise symbiotic species *Cassiopea andromeda* (see Gohar and Eisawy, 1961b), *C. frondosa* (see Smith, H.G., 1936), *C. xamachana* (see Trench, Colley and Fitt, 1981), *Cotylorhiza tuberculata* (see Kikinger, 1986), and *Mastigias papua* (see Sugiura, 1963, 1964). The zooxanthellae are then transmitted from the scyphistoma to buds and further polyps, or to ephyrae and hence to the medusae. However, in *Linuche unguiculata* eggs are released in mucus strands that also contain dinoflagellates. The developing embryos or planulae are infected in the 24 hours post-fertilization (Montgomery and Kremer, 1995). Algae are at first incorporated into the ectoderm, but are found increasingly in the endoderm at the planulae age.

In the laboratory infection with algae may occur by direct interaction or by uptake of prey containing algae. *Mastigias papua* scyphistomae may be infected by feeding pieces of a medusa containing zooxanthellae or by adding motile algae to the culture dish (Sugiura, 1964). Brine shrimp nauplii ingest zooxanthellae. Capture and infection of these brine shrimp by aposymbiotic scyphistomae of *Cassiopea xamachana* leads to infection of the scyphistomae (Fitt, 1984). However, motile algae also enter the coelenteric cavity of *C. xamachana* directly and establish a symbiosis. This is aided by responses of both the algae and the scyphistoma. The algae are attracted to aposymbiotic scyphistomae and to fed symbiotic individuals, but not to starved symbiotic scyphistomae. The water surrounding the attracting hosts contains high levels of ammonia which may be the attractant compound. Scyphistomae also ingest algae using responses similar to those in feeding. The presence of algae, particularly motile forms, increases the frequency with which tentacles are moved into the mouth.

The sources of the algae infecting scyphistomae in the wild are unknown. Free-living *Symbiodinium microadriaticum* have rarely been found. This may be because, as noted in culture, motility is limited to short light periods and motile algae remain close to the bottom (Fitt, Chang and Trench, 1981). Zooxanthellae released from other hosts or by predators of those hosts may also be sources (Fitt, 1984; Trench, 1987).

Once in the coelenteric cavity the algae are endocytosed by the gastrodermal cells lining the cavity. The algae must trigger the necessary reactions for algal sequestration and persistence in appropriate positions in the host, and must avoid exocytosis, or extracellular or intracellular digestion, by the host. Different strains of algae differ in their ability to infect hosts successfully (Trench, Colley and Fitt, 1981; Colley and Trench, 1983; Fitt, 1984, 1985; Trench, 1987), although it is difficult to evaluate the degree of specificity due to the taxonomic problems discussed in section 4.5.1. For example, Fitt (1984) fed cultures of algae isolated from 17 hosts to brine shrimp. When the *Artemia* sp. were fed to scyphistomae of *Cassiopea xamachana* all algal strains were taken up by the digestive cells after 4 hours. Twelve of the isolates of zooxanthellae remained in the host to establish a permanent symbiosis and others disappeared within 24 hours.

The rate of endocytosis by *Cassiopea xamachana* of algae isolated from the same host species is influenced by algal history. Freshly isolated algae are endocytosed at higher rates than those which have been cultured (Trench, Colley and Fitt, 1981; Colley and Trench, 1983). This may be because animal membranes are associated with the freshly isolated algae even after repeated washing, i.e. there may be recognition by host cells. Treatment with Triton-X-100 to remove membranes does cause lower rates of phagocytosis, but addition of animal homogenates to cultured algae does not enhance phagocytosis. Also the non-symbiotic scyphystomae of *Aurelia aurita* did not phagocytize cultured algae although they phagocytized freshly isolated algae from *C. xamachana* (which persisted only four days). In addition to lack of animal membranes, cultured algae also differ from the recent isolates in other possibly pertinent respects such as increased thickness of the periplast and in decreased release of photosynthate.

During endocytosis of live *Symbiodinium microadriaticum* by an gastrodermal cell of *Cassiopea xamachana*, pseudopods surround the algae and each is sequestered in an individual tight-fitting vacuole (Fitt and Trench, 1983). Phagosomes containing food particles fuse with lysosomes containing the intracellular digestive enzyme acid phosphatase near the apex of the cell. A portion of the vacuoles containing live or experimentally heat killed *S. microadriaticum* also fuse with lysosomes. However, within 8 hours after phagocytosis about 25% of the live algae are transported basally away from zones of high lysosome

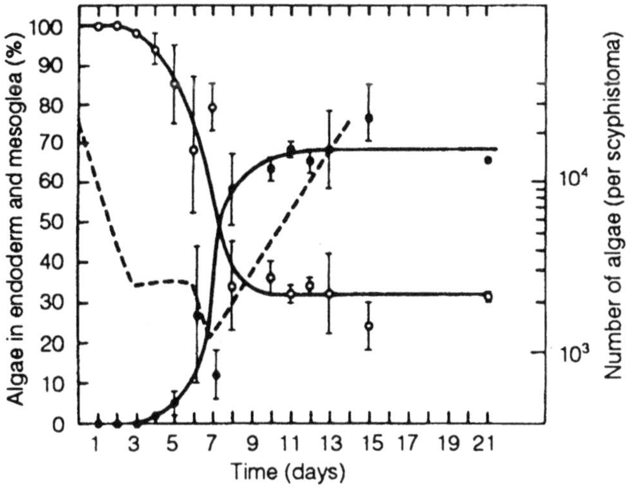

Figure 4.7 Relative distribution of algae in gastrodermal cells (open circles) and in mesogleal amoebocytes (closed circles) of *Cassiopea xamachana* at different times after infection with *Symbiodinium microadriaticum*. The broken line represents numbers of algae per scyphistoma. (Source: Colley and Trench, 1985, with permission of Springer-Verlag.)

density (Colley and Trench, 1985). Most gastrodermal cells with algae cease to be phagocytically active within three days. They migrate into the mesoglea to form 'amoebocytes', and then the contained algae proliferate (Figure 4.7).

In both scyphistomae and medusae, numbers of zooxanthellae depend partly on the ambient light level. The number of zooxanthellae decrease in *Cassiopea* sp. medusae held in continuous dark for 14 days (Zahl and McLaughlin, 1959). Aposymbiotic scyphistomae can be produced by four weeks of darkness including two weeks of starvation (Ludwig, 1969). However, even if *Cephea cephea* scyphistomae have been maintained in darkness for 16 months until bleached, zooxanthellae reappear when they are illuminated.

There may be stress-related expulsion of zooxanthellae. Expulsion of zooxanthellae has been observed in starved *Cassiopea xamachana* and *C. frondosa* (see Mayer, 1914b; Smith, H.G., 1936) as well as in *Mastigias* sp. (see Muscatine, Wilkerson and McCloskey, 1986). In anthozoa bleaching may also occur due to changes in temperature or salinity (Spencer Davies, 1992) and it is probable that scyphozoa would show similar reactions.

Under favourable conditions, the population densities of zooxanthellae may remain relatively constant or decrease with the growth of their host medusae. For *Cephea cephea* density of algae decreases greatly as the ephyrae develop into young medusae (Sugiura, 1969). However, in a *Mastigias* sp. population in the Western Caroline Islands, weight-specific algal population density is independent of host size from 2 to 16 cm in diameter (Muscatine, Wilkerson and McCloskey, 1986). There are no reported examples of the algae overgrowing their hosts.

It is not known how algal numbers or division rates are controlled. As noted in section 4.5.2, factors produced by the host may decrease the uptake of nutrients by the zooxanthellae. The mitotic index of zooxanthellae in *Mastigias* sp. and *Linuche unguiculata* medusae is lower than that typical of free-living dinoflagellates (Wilkerson, Muller-Parker and Muscatine, 1983; Kremer *et al.*, 1990). Muscatine, Wilkerson and McCloskey (1986) compared the algal growth rates in one population of *Mastigias* with medusan growth rates in another population and tentatively concluded that the small medusae may grow faster than their zooxanthellae whereas larger medusae may grow marginally more slowly. When medusae were held in the laboratory, the algae in small medusae were capable of transient increases of up to seven times in the mitotic index compared with newly captured medusae. They speculated that maintenance of population density depends on facultative increase in algal growth rates in small medusae and on expulsion or digestion in large medusae.

4.5.4 Ecological significance of symbiosis

Algal symbiosis is not widespread in scyphozoa, even in near-surface waters where light is present. It is not clear why some closely related putative host species possess symbionts when others do not. It is also not known whether the complex association, when all effects are taken into consideration, is of benefit to either or both alga and host. As was discussed in section 4.5.2, both alga and host gain some nutrients and supply others. The toxic effects of molecular oxygen and light must be prevented, and the behaviour of the host modified to allow light to reach the algae.

Photosynthetic carbon fixation by the algae can be of great importance to the host scyphozoan. For example, zooxanthellate *Mastigias* sp. medusae in Eil Malk Jellyfish Lake, Western Caroline Islands, have not been observed feeding holozoically (Muscatine and Marian, 1982). The contribution of photosynthesis to the symbiotic association is often assessed by comparing the oxygen production in daylight (P) to the 24-hour oxygen consumption of algae and medusae (R) to give a P/R ratio. The 24-hour oxygen consumption is extrapolated from dark measurements, involving the possibly erroneous assumption that respiration is unaffected in light conditions. If P/R exceeds a value of 1 then excess photosynthate, over the needs of symbiont and host for respiration, is available toward meeting their combined requirements for growth and reproduction. P/R ratios measured to date include: *Cassiopea* sp. 0.95–2.5 (Cates, 1975), *Cassiopea andromeda* 1.3–1.5 (Mergner and Svoboda, 1977; Svoboda, 1978), *Cassiopea xamachana* 1.8–2.0 (Kikinger, 1992), *Cotylorhiza tuberculata* 0.4–1.2 (Kikinger, 1992), *Linuche unguiculata* 1.5–1.8 (Kremer *et al.*, 1990) and *Mastigias* sp. 1.1–1.8 (McCloskey, Muscatine and Wilkerson, 1994).

As described in section 4.5.2 it is not known what proportion of the photosynthate is actually translocated to the host, nor the rate of uptake of organic compounds by the algae *in hospice*. Nevertheless Kremer *et al.* (1990) calculated that if the photosynthate not used for algal growth or respiration were all translocated to host *Linuche unguiculata*, it could provide all the carbon required for medusan respiration and somatic growth, although not for female egg production. For further discussion see section 5.2.6.

It has been stated in reviews of this field that zooxanthellae are necessary for strobilation of some symbiotic species. Symbiosis enhances the rate of strobilation, but it is not clear that there is an obligate dependence of the host on the alga. Strobilation may occur also in a temperature dependent manner and the effects of temperature have not been fully explored for most species. Sugiura (1964) reared aposymbiotic polyps of *Mastigias papua* at 23°C and was unable to induce strobilation by temperature increase to 30°C or abundant feeding of *Artemia*. If polyps reared at 15°C were infected by zooxanthellae, strobilation occurred after a latent period at 24°C. Similarly Kikinger (1992) caused strobilation of *Cotylorhiza tuberculata* by reinfection of aposymbiotic scyphistomae, but not by a temperature increase from 19°C to 25°C. On the other hand, aposymbiotic scyphistomae of *Cephea cephea* were induced to strobilate by raising the temperature from 20°C to 29°C.

The above differences could be attributed to generic differences in adaptation to symbiosis, but conflicting results have been obtained within *Cassiopea* sp. Trench, Colley and Fitt (1981) and Fitt (1984) stated that they had not observed strobilation in many cultures of aposymbiotic scyphistomae of *Cassiopea xamachana* maintained at constant temperature. On the other hand Rahat and Adar (1980) induced strobilation of aposymbiotic scyphistomae of *C. andromeda* at temperatures below 25°C, albeit at a lower rate than found using symbiotic scyphistomae. On the same species Ludwig (1969) and Hofmann and Kremer (1981) stated that the presence of at least a small population of zooxanthellae is indispensable for strobilation of *C. andromeda* even at 24°C. Strobilation was possible at a lower rate even after prolonged cultivation in darkness or following inhibition of photosynthesis with DCMU (3-(3,4-dichlorophenyl)-1,1-dimethylurea). As stated by Hofmann and Kremer (1981): 'Strobilation in the polyps thus seems to be significantly supported, but not definitely triggered by algal photosynthetic activity.'

It is also often claimed that algal symbiosis is highly advantageous in oligotrophic environments where radiant energy is abundant but organic input is scarce (see for example the model of symbiosis by Hallock, 1981). Efficient recycling of nutrients between the host and symbionts would increase production of organic matter by the association. This may be true in some nutrient-poor environments (Wilkerson and Kremer, 1992), but in stress, such as starvation, medusae may blanch, expelling their symbionts (Mayer, 1914b; Smith, H.G., 1936). It is possible that with complete lack of organic input from food there is nutrient limitation of the zooxanthellae, and they become a net drain on the resources of the host. The cycling of organic material between host and symbiont is extremely complex. Whereas non-symbiotic species of scyphomedusae can survive long periods of starvation, simply growing smaller (section 7.4.1), such responses to unfavourable food supplies may be hindered with zooxanthellae present.

Molecular oxygen, produced as an obligatory by-product of photosynthesis, is potentially damaging to the host when in excess of what is required for host consumption. Symbiotic scyphozoa must develop protective enzymes. Molecular oxygen generates free radicals and hydrogen peroxide (H_2O_2). The enzyme superoxide dismutase (SOD) removes superoxide radicals but generates hydrogen peroxide. Catalase in turn removes hydrogen peroxide, minimizing the toxic effect of these products. In *Cassiopea xamachana* the activity of these two enzymes is in direct proportion to the chlorophyll content of the tissues (Dykens, 1984).

In order to carry out photosynthesis, algae must receive light which is potentially damaging to both the host tissue and the algae. Ultraviolet wavelengths of light (below 400 nm) are damaging to biological materials. Wavelengths of visible light may also be biologically disruptive, particularly if synergistically mediated by the presence of molecular oxygen. In culture, the growth of zooxanthellae from *Cassiopea* medusae is severely impaired in ultraviolet light, although not by visible light (Jokiel and York, 1982; Read, 1986). *Cassiopea xamachana* contains a blue pigment, Cassio Blue, diffused within the acellular portion of the mesoglea in the same areas of the bell where

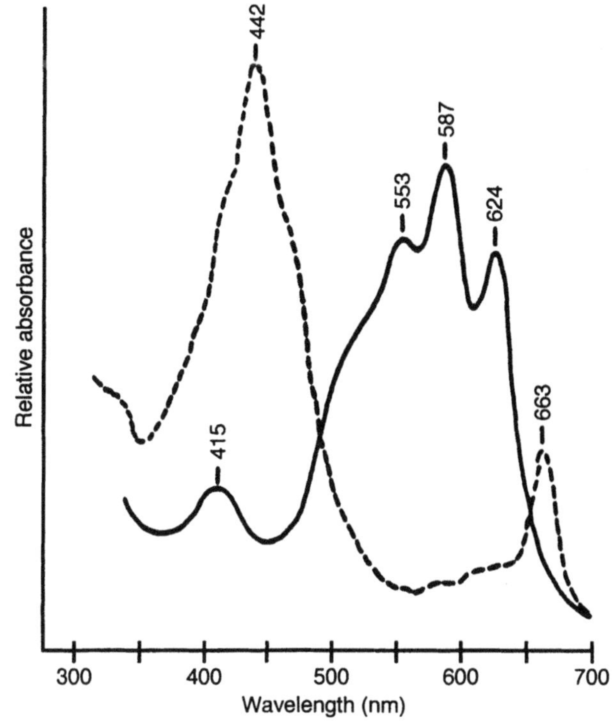

Figure 4.8 Absorption spectra of purified Cassio Blue pigment (solid line) and methanol-extracted photosynthetic pigments of zooxanthellae (broken line) from *Cassiopea xamachana*. (Source: Blanquet and Phelan, 1987, with permission of R.S. Blanquet and Springer-Verlag.)

cells containing zooxanthellae are concentrated (Blanquet and Phelan, 1987). The pigment is a polymeric glycoprotein with light absorption maxima at 624, 587, and 553 nm, whereas the photosynthetic pigments of the zooxanthellae have an absorption maximum of 442 nm (Figure 4.8). Cassio Blue probably acts as a visible light attenuator for injurious solar radiation other than the photosynthetically active wavelengths. *Rhizostoma pulmo* contains a similar pigment, but the absorption spectrum of its zooxanthellae has not yet been examined (Christomanos, 1954). Compounds protecting against UV have not been identified in scyphozoa.

Finally, there has been no evaluation of the costs of behavioural modifications of the host necessary to maintain a light-dependent population of symbionts. *Cassiopea* medusae spend much time pulsing in an inverted position on the bottom of shallow pools (Bigelow, 1900;

Mayer, 1906). This allows exposure to light of the symbionts in the oral arms, but modifies the currents bringing food particles. *Mastigias* sp. migrate to follow incident light (Hamner and Hauri, 1981). *Linuche unguiculata* possesses patches of zooxanthellae which expand and contract with circadian regularity (Costello and Kremer, 1989). Modifications such as these may not only represent energetic costs for maintenance, but may also decrease efficiency of obtaining particulate food.

5 *Metabolism*

Once organic compounds have been distributed to the tissues, they are excreted as waste, or utilized for reproduction, somatic growth, or as sources of energy. Production of adenosine triphosphate (ATP), the cell's energy currency, is coupled with processes utilizing it. Waste products, particularly nitrogenous compounds derived from protein metabolism, must be excreted. In coelenterates the net uptake of oxygen and excretion of ammonium has been measured, but less is known of the cellular metabolism. In addition to carbon containing organic compounds, the concentrations of water and inorganic ions must also be controlled in each cell. Organic compounds and ions in turn affect the buoyancy of the animals. Reproduction and growth will be considered in Chapters 6 and 7. The remaining associated topics will be discussed in this chapter.

5.1 INTRODUCTION

5.1.1 Definitions

The terms 'respiration' and 'metabolism' are defined several ways. For the purposes of this chapter respiration will be defined as the sum of the processes by which the respiratory gases, oxygen and carbon dioxide, are transferred between environment and tissues (Burggren and Roberts, 1991). Metabolism will be defined as the intracellular process that consumes substrates and produces by-products in the course of generating chemically stored energy as ATP. The metabolic rate is the rate at which that chemical energy is consumed by an animal in growth and maintenance, i.e. the amount of energy consumed per unit time. The production of ATP may be anaerobic (i.e. oxygen independent) or aerobic (i.e. oxygen dependent).

It should be noted that the rate of oxygen consumption is a measure of only aerobic metabolism and should not be equated with total metabolic rate. The extent of anaerobic metabolism is unknown in scyphomedusae. The reason why oxygen consumption is often used as an index of the metabolic rate is its relative ease of measurement. Total metabolic rate could, in theory, be obtained by measuring the total production of heat, the form of energy to which the chemical energy consumed is converted. Unfortunately this method is so far impractical for scyphomedusae.

5.1.2 Aerobic and anaerobic metabolism

The metabolic pathways producing ATP in coelenterates are similar to those of higher animals. The main pathways for vertebrates are described in great detail in many textbooks: phosphagen mobilization, glycolysis, the pentose shunt, the Krebs cycle, the electron-transport system and ß-oxidation. Invertebrates follow the same general pathways although they may differ in such ways as alternative anaerobic pathways, i.e. fermentations (Hochachka, 1991). For coelenterates most is known about anemones (Shick, 1991), but the very scattered data on scyphozoa also fit this general picture (Figure 5.1).

A number of key enzymes for aerobic carbohydrate metabolism via the glycolytic pathway and the Krebs cycle have been identified in *Aurelia aurita*, *Chrysaora quinquecirrha*, and *Cyanea capillata* (see Raymont, Krishnaswamy and Tundisi, 1967; Lin, A.L. and Zubkoff, 1973, 1976a, 1977; Zubkoff and Linn, 1975; Manchenko and Zaslavskaya, 1980; Thuesen and Childress, 1994). These include hexokinase, pyruvate kinase, citrate synthase, isocitrate dehydrogenase, succinic dehydrogenase and malate dehydrogenase.

The requirement of some enzymes for nucleotides as cosubstrates is indicative of the presence of an electron-transport system. Some enzymes, such as isocitrate dehydrogenase, differ from higher animals in their specificity for nicotinamide adenine dinucleotide phosphate (NADP+) rather than nicotinamide adenine dinucleotide (NAD+) as their requisite nucleotide cosubstrate (Lin, A.L. and Zubkoff, 1977; Hoffmann, Bishop and Sassaman, 1978).

Workers found glucose-6-phosphate dehydrogenase (G6PDH), the first enzyme of the pentose shunt, but little or no 6-phosphogluconate dehydrogenase (6PGDH), another enzyme of the shunt, in *Cassiopea*

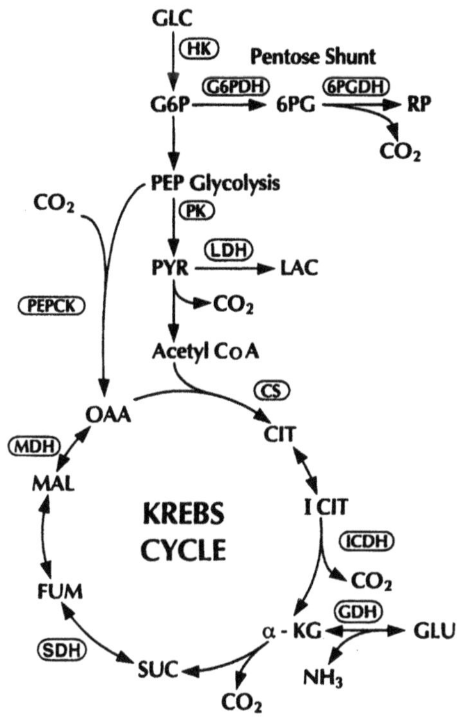

Figure 5.1 Pathways of metabolism for which there is fragmentary evidence among several species of scyphozoa. Identified enzymes catalysing key reactions are shown in ovals. See text for discussion. Sites of NAD or NADP oxireduction are not shown, nor are possible sites and yields of ATP production. α-KG, α-ketoglutarate; AcetylCoA = acetylcoenzyme A; CIT = citrate; CS = citrate synthase; FUM = fumarate; GDH = glutamate dehydrogenase; GLC = glucose; GLU = glutamate; G6P = glucose-6-phosphate; G6PDH = glucose-6-phosphate dehydrogenase; HK = hexokinase; ICDH = isocitrate dehydrogenase; ICIT = isocitrate; LAC = lactate; LDH = lactate dehydrogenase; MAL = malate; MDH = malate dehydrogenase; OAA = oxaloacetate; PEP = phosphoenolpyruvate; PEPCK = phosphoenolpyruvate carboxykinase; 6PG = 6phospho-D-gluconate; 6PGDH = 6-phosphogluconate dehydrogenase; PK = pyruvate kinase; PYR = pyruvate; RP = D-ribulose-5-phosphate; SDH = succinic dehydrogenase; SUC = succinate.

sp., *Mastigias* sp., *Chrysaora quinquecirrha*, and *Cyanea capillata* (see Powers, Lenhoff and Leone, 1968; Blanquet, 1972b; Zubkoff and Linn, 1975; Lin, A.L. and Zubkoff, 1976b; Manchenko and Zaslavskaya, 1980). In some other invertebrates, more molecules of CO_2 produced in the shunt are derived from the C-1 of glucose rather than the other structural carbon. For *C. quinquecirrha* the amount of

$^{14}CO_2$ produced by the *in vivo* oxidation of [1-^{14}C]-glucose is greater than for [6-^{14}C]-glucose, supporting the presence of a functional pentose shunt (Lin, A.L. and Zubkoff, 1976b).

The ß-oxidation of fatty acids has not been examined. The possibility of transamination of amino acids forming glutamate, followed by deamination of glutamate and oxidation in the Krebs cycle, is indicated by the presence of the enzyme glutamate dehydrogenase (GDH) (Hoffmann, Bishop and Sassaman, 1978).

One common difference in anaerobic metabolism between invertebrates and vertebrates is in the terminal dehydrogenases of glycolysis. As the Krebs cycle does not operate, in anaerobic conditions complete oxidation is suppressed and intermediate end-products may accumulate. In vertebrates, lactate is the anaerobic glycolytic end-product after reduction of pyruvate. In invertebrates, lactate dehydrogenase may be replaced by functionally analogous imino acid dehydrogenases so that an imino acid (octopine, alanopine, strombine or tauropine) replaces lactate. Although imino acid dehydrogenases are present in anthozoa they were not found in *Aurelia aurita* (see Sato *et al.*, 1993). Lactate dehydrogenase was also not detected in *A. aurita* or *Chrysaora quinquecirrha* (see Lin, A.L. and Zubkoff, 1977; Sato *et al.*, 1993). However, it has been recently been found in *Atolla vanhoeffeni*, *Atolla wyvillei*, *Nausithoe rubra*, *Paraphyllina ransoni*, *Periphylla periphylla* and *Pelagia colorata*, especially in the coronal muscles (Thuesen and Childress, 1994).

An alternative pathway for anaerobiosis of some invertebrates is production of succinate (via oxaloacetate) from phosphoenol pyruvate (PEP) rather than production of pyruvate. The presence of pyruvate kinase in *Aurelia aurita*, *Chrysaora quinquecirrha*, *Atolla wyvillei* and *Periphylla periphylla* indicates the production of pyruvate (Lin, A.L. and Zubkoff, 1977; Thuesen and Childress, 1994). However, the presence of the enzyme phosphoenolpyruvate carboxykinase (PEPCK) in *A. aurita* and *C. quinquecirrha* indicates that the alternative pathway may also be active (Lin, A.L. and Zubkoff, 1977).

Other possible fermentation pathways have not yet been identified. There has also not been any examination of quantities of anaerobic end-products.

Whatever anaerobic pathways may be present in scyphozoa, it is unlikely that they are quantitatively as important as aerobic pathways for the production of ATP. The biochemical efficiency (i.e. ATP yield per mol glucose or other energy source) is low for all known anaerobic pathways compared with aerobic pathways. Although the total rate of ATP production by fermentation may be increased by increasing the flux in the pathways, they are employed by most animals primarily in situations of environmental or physiological hypoxia. For example, anaerobic pathways may be important in actively contracting muscle of scyphozoa, as they often are in active muscle of higher animals. High levels of lactate dehydrogenase are present in the coronal muscle of *Periphylla periphylla* and *Pelagia colorata* (see Thuesen and Childress, 1994).

It is not known what proportions of fat, carbohydrate or protein are oxidized in aerobic metabolism, or whether carbohydrate or protein is utilized in anaerobic metabolism. Lipid cannot be utilized in known anaerobic metabolic pathways. Since scyphozoa are carnivorous, and also themselves contain a higher percentage of protein (Table 7.2), it is probable that the main substrate is protein. Unfortunately this uncertainty about substrate limits the ability of researchers to convert respiration rates to carbon turnover rates useful in carbon budgets (see also section 5.2).

In some other animals the substrate for aerobic metabolism can be determined from molar ratios of excreted carbon dioxide and nitrogen to the oxygen consumed, i.e. the respiratory quotient (RQ) and nitrogen quotient (NQ) (Gnaiger, 1983a). In scyphozoa these data are not yet available. The chemistry of carbon dioxide in sea water is complex due to the high solubility, and the large amounts of carbon dioxide, carbonate and bicarbonate normally found there. Carbon dioxide as a measure of metabolic rate must be measured under steady state conditions (Burggren and Roberts, 1991).

5.2 FACTORS AFFECTING OXYGEN CONSUMPTION

The rates of aerobic metabolism are affected by a variety of factors discussed below. Comparative respiration rates *per se* may be used as measures of relative changes in aerobic metabolism due to various activities or environmental changes.

To incorporate respiration into energy or carbon budgets in absolute terms it is necessary to determine the respiratory substrates. The substrates are not yet known for scyphozoa (section 5.1.2). For carbon budgets, the respiratory quotient (RQ) (CO_2 produced/O_2 consumed) values for carbohydrate and lipid are 1 and 0.72 respectively. That for protein varies as a function of the excretory product from 0.84 to 0.97 (Gnaiger, 1983a). In practice various authors building carbon budgets have assumed RQ values of 0.8–1.0 (Larson, 1987e; Schneider, 1989b; Kremer *et al.*, 1990).

Rates of oxygen consumption may be reported in volumetric, gravimetric or molar terms. Theoretically molar terms are preferred for calculations of molecular relationships (Gnaiger, 1983a,b), but the majority of data on scyphozoa has been expressed in volumetric terms. Also the oxygen consumption rates measured have been expressed with reference to wet weight (WW), dry weight (DW) and weight of protein of the respiring animal. There are difficulties in measuring dry weight of gelatinous animals, as will be discussed in section 5.4.1, but most earlier authors expressed their measurements with reference to dry weight. Oxygen consumption rates of scyphozoa in Tables 5.1 and 5.2 are presented in the units used by the original investigators except for conversion of gravimetric to molar terms (1 mg O_2/h = 31.251 μmol O_2/h). Not included in Table 5.1 are the first measurements on scyphomedusae, those of Vernon (1895) on *Rhizostoma pulmo*, because his units differ from all subsequent authors. Other excluded data based on fewer than four specimens may be found in Biggs (1977), Kuzmicheva (1980) and Smith (1982). Thuesen and Childress (1994) measured respiration rates of 1–8 specimens of five species of coronate medusae.

Respiration measurements are also subject to various experimental effects. Activity of animals can be affected by handling, agitation, or flow of the water, and by being confined in small chambers (Yakovleva, 1964; Larson, 1987e). This has been minimized by recent workers who have monitored rates of pulsation in the chambers versus that outside, and used larger chambers as needed (e.g. Larson, 1987a).

In spite of the above deficiencies and variations, there is now a large enough body of data on scyphozoan respiration to make generalized comparisons with other animals possible. When oxygen consumption is expressed with reference to wet or dry weight, it is much lower in gelatinous coelenterates than in most other animals. However, when consumption is compared on the basis of organic material present the rates are comparable (Larson, 1987e; Schneider, 1992).

5.2.1 Body size

The respiration rate of medusae increases with increased size of the animal (Figures 5.2 and 5.4).

The relationship between respiration and size may be expressed by the allometric equation $R = \mathrm{a}W^{\mathrm{b}}$, where R = O_2 consumption, W = weight of the individual, a = constant for the species and temperature, and b = exponent for size; ‘b’ is the slope of the regression line in a log–log plot of respiration against weight as in Figures 5.2 and 5.4. If weight-specific respiration remains constant as size increases, the value of b is 1.0. The ratio between surface and volume changes as size increases is such that the value of b approximates 0.667 for

Table 5.1 Oxygen consumption rates of nonsymbiotic semaeostome and rhizostome scyphomedusae

Species (specimens)	*Temp. (°C)*	*A*	*B*	*C*	*D*	*Source*
Aurelia aurita (86)	16	.003–.005				Thill, 1937
Aurelia aurita (15)	10–21	.002–.007† .02–.14*	.09–.36†			Yakovleva, 1964
Aurelia aurita (7)	20	.014				Pavlova, 1968
Aurelia aurita (142)	12–14	.002–.004				Kerstan, 1977
Aurelia aurita					.1	Svoboda, 1978
Aurelia aurita (18)	20	.003–.011				Kuzmicheva, 1980
Aurelia aurita (46)	10–15		.14–.24			Larson, 1987e
Aurelia aurita ephyrae (50)	22	.11				Mangum, Oakes and Shick, 1972
Aurelia aurita ephyrae (25+)	15				.10–.19	Olesen, Frandsen and Riisgard, 1994
Chrysaora hysoscella (12)	15	.0066				Krüger, 1968
Cyanea capillata (6+)	10		.15			Mangum, Oakes and Shick, 1972
Cyanea capillata (24)	10–15		.48–.88			Larson, 1987e
Cyanea capillata (4)	6	.003	.07			Bailey, T.G., Youngbluth and Owen, 1995
Cyanea sp. (23)	15	.016				Krüger, 1968
Pelagia noctiluca (4)	23–29			16		Biggs, 1977
Pelagia noctiluca (4)	18		1.02			Davenport and Trueman, 1985
Pelagia noctiluca						Malej and Vukovic, 1986
(12)	16			0.1–1.5		
(4)	25			4.0–8.6		
Pelagia noctiluca (42)	12–25		.05–.68		.003–.07	Malej, 1989b; Malej, 1991; Malej, Faganelli and Pezdic, 1993
Poralia rufescens (6)	4–7	.0004–.012	.01–.29			Bailey, T.G., Youngbluth and Owen, 1995
Rhizostoma pulmo (25)	19–25	.004–.013† .004–.009*	.27–.62†			Yakovleva, 1964
Rhizostoma pulmo (37)	15	.014				Krüger, 1968
Stomolophus meleagris (68)	30	.05				Larson, 1987a

Expression of oxygen consumption rate:
A, as µl 0_2/hr per mg WW
B, as µl 0_2/hr per mg DW
C, as µl 0_2/hr per mg protein
D, as µmol 0_2/hr per mg DW
† = respiration in still water
* = respiration in running water

Table 5.2 Oxygen consumption rates of scyphozoan polyps and planulae

Species (specimens)	*Temp. (°C)*	*A*	*B*	*C*	*Source*
Aurelia aurita strobilae (97)	22	.095			Mangum, Oakes and Shick, 1972
Aurelia aurita polyps	20				Shick, 1975
(200 fed)			.9		
(400+ starved)			.25		
Aurelia aurita planulae	20–22		3.5		Schneider and Weisse, 1985
Chrysaora quinquecirrha polyps	22	.08			Mangum, Oakes and Shick, 1972
Chrysaora quinquecirrha polyps					Black, 1981
starved 7 days				2.1	
starved 70 days				1.8	

Expression of oxygen consumption rate:
A, as µl 0_2/hr per mg WW
B, as µl 0_2/hr per mg DW
C, as µl 0_2/hr per mg protein

Table 5.3 Respiration rate weight exponent values for scyphomedusae (see text for explanation)

Species	*Mass exponent*	*Source*
Aurelia aurita	0.82–0.83	Yakovleva, 1964
Aurelia aurita	0.86	Kuzmicheva, 1980
Aurelia aurita	0.91–0.92	Larson, 1987e
Aurelia aurita	0.94*	Schneider, 1989b
Chrysaora hysocella	0.98	Krüger, 1968
Cyanea capillata	1.00–1.04	Larson, 1987e
Cyanea sp.	0.91	Krüger, 1968
Rhizostoma pulmo	0.91–0.97	Yakovleva, 1964
Rhizostoma pulmo	0.99	Krüger, 1968
Stomolophus meleagris	0.99	Larson, 1987a

* Calculated from data in Thill (1937) and Kerstan (1977)

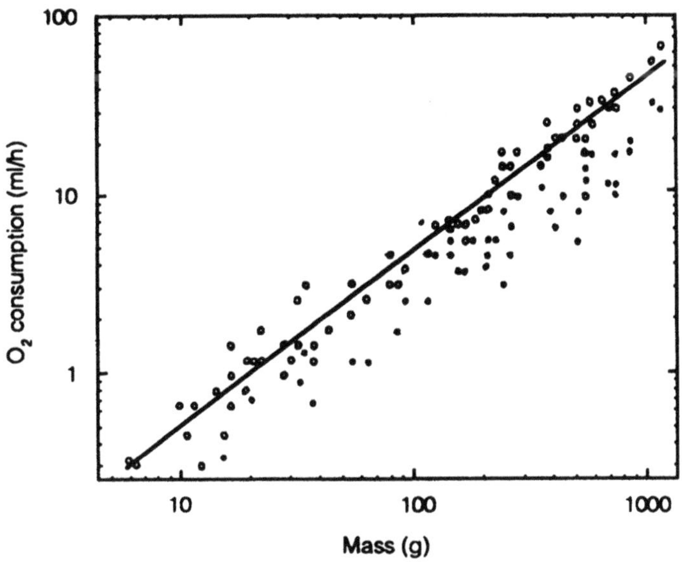

Figure 5.2 Plot and regression line of oxygen consumption of individual medusae of *Stomolophus meleagris* vs their mass (wet weight). Open circles and regression line = routine (active) rate; solid circles = lower standard (inactive) rate following crushing of marginal ganglia. (Source: Larson, 1987a, with permission of R.J. Larson and National Research Council of Canada.)

surface-dependent respiration. Values of b for most plankton lie between 0.67 and 1.0. Values for scyphozoa, summarized in Table 5.3, show that there is a small decrease in weight-specific respiration with increased size of medusae. Data on size effects for one developmental stage should not be extrapolated to other stages of the life cycle with different shapes, types of activity and composition. No study has been reported on the polyps.

For medusae there is some doubt that this relationship is a size effect *per se*. Within the medusa stage it is not possible to distinguish between the effects of size and age. Most studies have been done on moderately small to medium sized, actively growing individuals. Change in measured respiration rate with size may merely reflect change in somatic growth rate with size. Pulsation rates may also be higher in smaller individuals (section 2.6.1). Nevertheless the size effect is possibly real in that metabolism is limited in larger animals of most phyla that have been examined (Burggren and Roberts, 1991).

It is not known why this decrease in weight-specific respiration of larger animals occurs, although there is a large body of work on the phenomenon in other animals. For some species it may be due to reduced delivery of oxygen or substrate to the tissues of larger animals. However, *in vitro* studies show that the intrinsic metabolic rate of isolated tissues of larger vertebrates may be lower than that of smaller animals, i.e. that changes in cellular metabolism are involved.

5.2.2 Muscular activity

When measuring the respiration rate of scyphomedusae it is possible to control the swimming level only by abnormal treatment. It is therefore not possible to distinguish between 'standard' respiration (the respiration rate when an animal is inactive) and the component due to activity. The rate ideally measured is of 'routine' respiration, when the animal is spontaneously active in the absence (as much as possible) of external stimuli. Unfortunately, as noted above, respiration is often measured in conditions where handling, small chambers or agitation of the medium have increased metabolic rate toward the 'active' rate at maximum sustainable swimming speed.

Studies have attempted to measure the component due to swimming. Davenport and Trueman (1985) measured the respiration rate of four *Pelagia noctiluca* medusae before and after they were anaesthetized with methanol. The respiration rate of swimming medusae was 2.17 times that of the anaesthetized animals. It is uncertain what side effects methanol might have on the animals, but they were able to resume normal swimming when returned to clean sea water. It is possible that anaesthetized rates would be lower than standard rate due to decreased rates of activities other than swimming.

Larson (1987a) measured respiration of 42 *Stomolophus meleagris* before and after immobilization by crushing the marginal ganglia. With this experiment also it is unclear whether there may be other side effects of the treatment, but rates for inactive medusae were again approximately 50% less (Figure 5.2). Assuming that the above respiration rates of immobilized medusae represent standard rates, the ratio of active to standard respiration is lower in medusae than in higher animals such as insects, fish and mammals (Burggren and Roberts, 1991).

5.2.3 Food

Food has multiple effects on respiratory rates of animals. The mechanical activity of feeding increases metabolic rate. The increase in oxygen consumption shortly after feeding, the 'specific dynamic effect' (SDE), is imperfectly understood but probably relates in part to deamination of amino acids. Finally, starvation often depresses metabolic rate. It is this latter point that has been most investigated in scyphozoa.

Vernon (1895) kept *Rhizostoma pulmo* medusae in captivity for up to 5 weeks without feeding. The weight loss (both organic and inorganic constituents) was about 8% per day at temperatures of 11–13°C. Subsequent authors have starved scyphomedusae for periods ranging from the actual period needed for respiration measurements to many previous days. Thill (1937) found that decrease in oxygen uptake of *Aurelia aurita* medusae paralleled decrease in body volume over 10 days of starvation. His first measurements were made after one day without food. Other experiments over shorter time spans, (e.g. Malej, 1991), have not distinguished between the effects of short-term starvation and of decreasing oxygen availability during longer residence in the experimental apparatus (compare section 5.2.5).

The polyps of scyphozoa are also able to survive long periods without food. Shick (1975) starved *Aurelia aurita* scyphistomae for 56 days at 20°C, greatly decreasing the rate of strobilation and budding. Weight-specific oxygen consumption declined to 26% of the value in fed polyps, over the first two weeks of food deprivation (Figure 5.3). Exposure of the starving polyps to environmental levels of dissolved glycine restored strobilation to normal, but respiration was still less than 30% of normal.

5.2.4 Temperature

The respiration of most marine invertebrates rises about 2.5 times per 10°C increase in their temperature within their thermal lethal limits. The standard (inactive) metabolism increases continuously with temperature up to lethal levels. The active metabolism may either increase to a plateau or pass through a peak above which the animals are incapable of sustaining higher levels of metabolism (Burggren and Roberts, 1991). Routine measurements of respiration will lie between standard and active.

The Q_{10} term is the factor by which respiration is increased or decreased by a rise of 10°C. Q_{10} values of 2–3 indicate thermal effects on biochemical reactions, whereas higher or lower values indicate other processes such as permeability. A Q_{10} value of 1 indicates temperature insensitivity. Q_{10} values are not constant over the normal temperature range so the range over which they are calculated must be stated.

It is pertinent at this point to state that reactions of animals to a particular temperature will vary with their thermal history. Within their lethal range, animals acclimate to temperatures of their environment

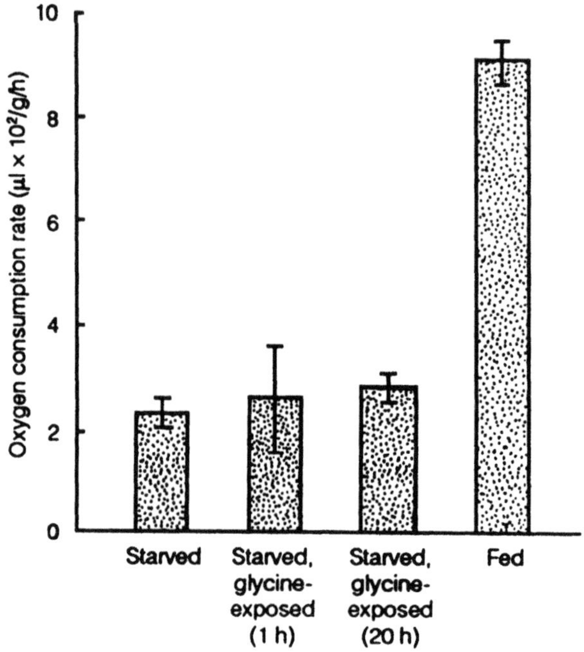

Figure 5.3 Oxygen consumption rates in four groups of 100–150 starved, starved/glycine-exposed (1 hour) and starved/glycine-exposed (20 hours) *Aurelia aurita* scyphistomae, and in four groups of 50 fed scyphistomae. Values are means ± SD. The respiration of starved or glycine-exposed polyps is greatly decreased compared with fed animals. (Source: Shick, 1975, with permission of *Biological Bulletin.*)

by metabolic and behavioural adjustments. They then respond to acute changes from the acclimation temperature. More details of the effects of temperature will be discussed in section 8.2.1.

As expected, the respiration rates of scyphozoa increase with increased temperature. Figure 5.4 shows the increase in respiration of *Aurelia aurita* and *Cyanea capillata* medusae held for 24 hours and tested at 15°C, compared with those held and tested at 10°C. The Q_{10} values were 2.9 and 3.4 respectively. Similar Q_{10} values for other species and measurements by other workers are summarized in Table 5.4. Although variable, they lie predominately in the expected range. Acclimation also occurs. *A. aurita* and *Chrysaora quinquecirrha* polyps that have been acclimated to 22°C show less increase in respiration when tested at higher temperatures than did those acclimated to 12°C (Mangum, Oakes and Shick, 1972).

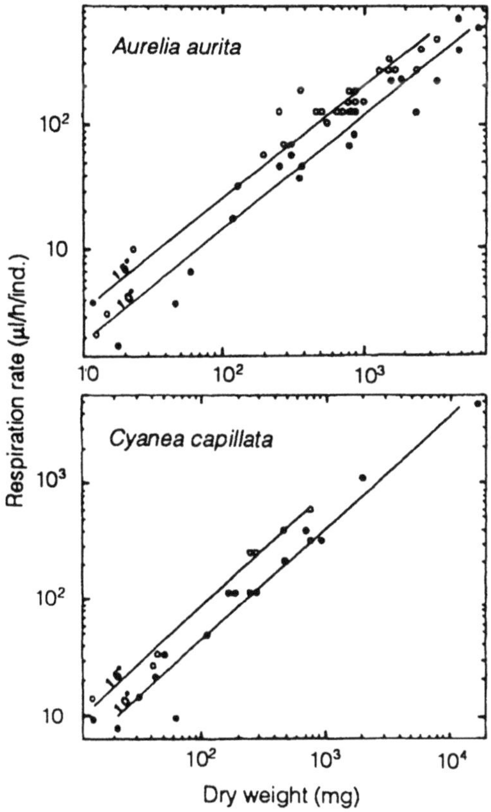

Figure 5.4 Plot of respiration vs dry weight for *Aurelia aurita* and *Cyanea capillata*. Closed circles and lower regression lines = measurements at 10°C; open circles and upper regression lines = measurements at 15°C. Note increased respiration at the higher temperature. (Source: Larson, 1987e, with permission of R.J. Larson and Elsevier Science Ltd, The Boulevard, Langford Lane, Kidlington, OX5 1GB, UK.)

5.2.5 Oxygen availability

The relatively small decrease in weight-specific metabolic rate with increased size of medusae (section 5.2.1) indicates that there can be little restriction on supply of oxygen to the tissues of larger animals.

There have been no studies on movement of oxygen in scyphozoa. Oxygen enters all animal cells by diffusion. Epidermal cells, directly in contact with sea water, could receive oxygen directly. Gastrodermal cells, in contact with the gastrovascular system, would also be supplied directly, provided the fluid in the system is circulated actively enough to maintain adequate oxygen levels as it is used by respiring cells.

Table 5.4 Q_{10} values of scyphozoan respiration

Species	Q_{10}	*Temp. (°C)*	*Source*
Aurelia aurita medusae	2.9	10–15	Larson, 1987e
Aurelia aurita polyps	1.7–6.4	12–32	Mangum, Oakes and Shick, 1972
Chrysaora quinquecirrha polyps	1.1–18.9	12–32	Mangum, Oakes and Shick, 1972
Cyanea capillata medusae	1.2–2.6	5–15	Mangum, Oakes and Shick, 1972
Cyanea capillata medusae	3.4	10–15	Larson, 1987e
Pelagia noctiluca medusae	2.6–3.4	16–23	Malej, 1989b; Malej, 1991
Periphylla periphylla medusae	2.6	5–10	Thuesen and Childress, 1994
Rhizostoma pulmo medusae	2.6	10–24	Vernon, 1895 (calculated in Larson, 1987)

Circulation in the gastrovascular cavity was described in section 4.3.1 but there have been no measurements of oxygen in these fluids.

Problems may arise in supply of oxygen to cells away from the surfaces. Diffusion time increases with the square of the diffusion path length. At the metabolic rates of most tissues in other animals and at usual partial pressures of oxygen a diffusion distance of 0.5–1 mm is considered maximal (Burggren and Roberts, 1991). Without a cardiovascular system, increase in size of coelenterates is therefore restricted to increased area of the epidermal and gastrodermal tissue layers, or to increase in amount of extracellular products.

Respiration rates are affected by the partial pressure of oxygen (P_{O_2}) in the surrounding sea water. If a specimen of *Aurelia aurita* is allowed to respire in a chamber so that it depletes the oxygen, respiration decreases after the oxygen levels have dropped by approximately 20% (Thill, 1937). Animals which are able to maintain their respiration rate at ambient P_{O_2} levels down to a critical lower level before the uptake begins to fall are termed oxygen regulators. For *A. aurita* contraction continues at normal rates until oxygen is much lower than the critical level. Greater decrease in medusa volume in oxygen-depleted water than in aerated water and increased oxygen uptake after return to oxygenated water indicate the possibilty of anaerobic metabolism and an oxygen debt, a term that refers to the aerobic metabolism of glycolytic end-products produced during a previous anaerobic period.

Scyphozoa may also encounter decreased oxygen in the sea (section 8.2.4). Thus, anaerobic metabolic pathways may become important to these animals (Thuesen and Childress, 1994).

5.2.6 Effects of symbionts

In scyphozoa with symbionts the rate of respiration depends on the zooxanthellae, as well as the medusan tissue. It is usually assumed that the respiration in the light is the same as that which can be measured in the dark. The rate of production of oxygen by the zooxanthellae in the light is then calculated from the measured oxygen production of the intact association in the light plus the dark respiration rate of the intact association. Oxygen production usually exceeds diel respiration (section 4.5.4) indicating that excess photosynthate is available for growth and reproduction of host and/or algae.

Carbon budgets have been used to calculate CZAR – the percentage contribution of carbon translocated from the zooxanthellae to the daily respiratory carbon requirements of the host animal. In order to calculate the oxygen consumption of host tissue only, it is assumed that respiration measured can be partitioned between host and zooxanthellae in the same ratio as protein or carbon is partitioned. Calculation of carbon budgets also involves assumptions of substrates used in respiration of host and alga (as discussed in the introduction to section 5.2) and of percentage of photosynthate translocated. In spite of all the assumptions involved, CZAR is commonly calculated for anthozoa with symbionts (Sebens, 1987; Shick, 1991). For scyphozoa, CZAR is 160% for *Linuche unguiculata* (see Kremer *et al.*, 1990), and 97% and 176% for *Mastigias* sp. lake and lagoon forms respectively (McCloskey, Muscatine and Wilkerson, 1994). (Only three specimens of the lagoon form were examined.) The low CZAR value for *Mastigias* sp. lake form is due to the high respiration rates of this actively migrating medusa, rather than to low production by the algae.

5.3 NITROGEN EXCRETION

Most marine invertebrates excrete ammonium as the main nitrogenous compound resulting from protein and nucleic acid catabolism. Other possible compounds include urea, uric acid, amino acids and purines. Excretion measurements on scyphozoa have concentrated on measurements of ammonium (Biggs, 1977; Smith, K.L., 1982; Muscatine and Marian, 1982; Schneider and Weisse, 1985; Malej and Vukovic, 1986; Morand, Carré and Biggs, 1987; Schneider, 1989a;

Malej, 1989b, 1991; Nemazie, Purcell and Glibert, 1993). With reference to other possible compounds, amino acids are excreted by 'jellyfish', possibly scyphozoa, of the Sargasso Sea (Webb and Johannes, 1967). There are to date no measurements of total nitrogen excretion, and therefore no indication of what percentage of excretory products ammonium represents. This is a major lack in knowledge of the metabolism of these animals.

Like oxygen consumption (section 5.2), rates of ammonium excretion are expressed in various terms and are subject to experimental effects. When expressed with reference to dry weight ammonium excretion by scyphomedusae is low, but when expressed with reference to carbon content of the medusa it is similar to that of other marine taxa (Schneider, 1990).

Excretion of ammonium by zooplankton contributes to nitrogen requirements of phytoplankton. Ammonium from blooms of scyphomedusae has been found to contribute up to 14% of the nitrogen required by the phytoplankton (Schneider, 1989a; Nemazie, Purcell and Glibert, 1993).

5.3.1 Factors affecting rates of excretion

Most data on metabolic pathways in scyphozoa refer to aerobic metabolism of carbohydrate (section 5.1.2). It is nevertheless reasonable to assume that the rate of excretion of nitrogen depends primarily on the amount of protein that is metabolized.

The ratio of oxygen atoms consumed to nitrogen atoms excreted is indicative of the proportion of protein in organic compounds used in aerobic metabolism. If ammonium is the only nitrogenous end-product of purely protein catabolism, the ratio is approximately 8 (Ikeda, 1974, 1977). Figures for the ratio of oxygen to ammonium nitrogen of scyphozoa range from 5.9 to 25.2 (Biggs, 1977; Smith, K.L., 1982; Schneider and Weisse, 1985; Malej and Vukovic, 1986; Morand, Carré and Biggs, 1987; Schneider, 1989a; Malej, 1989b, 1991). The higher figures indicate either that other substrates, as well as protein, are oxidized, or that other nitrogenous products are also formed in protein catabolism. The O : N ratio decreases in starving *Pelagia noctiluca* (see Malej, 1989b). The proportion of protein metabolized will probably increase in unfed animals, since there is less storage of lipid or carbohydrate in scyphozoa than in most prey (section 7.2).

The relation of excretion to weight may be described by an allometric equation similar to that for respiration (section 5.2.1). For *Chrysaora quinquecirrha* the weight-specific ammonium excretion rate was almost the same for all sizes of medusae tested ($b = 1.0$) (Nemazie,

Purcell and Glibert, 1993), whereas for *Aurelia aurita* it decreased slightly with increased size (b = 0.93) (Schneider, 1989a).

Excretion is also affected by temperature. As expected, ammonium excretion increases with increased temperature (Malej and Vukovic, 1986; Morand, Carré and Biggs, 1987; Malej, 1989b, 1991; Nemazie, Purcell and Glibert, 1993).

As discussed in section 4.5.2 algae require sources of nitrogen, and symbiotic algae may utilize ammonium produced by the host metabolism. Symbiotic medusae therefore excrete less ammonium, or may show a net uptake from surrounding sea water.

5.4 OSMOTIC AND IONIC REGULATION

Scyphozoa are believed to be osmoconformers, i.e. the proportion of water in intracellular and extracellular fluids of the tissues varies with the salinity of the surrounding sea water according to osmotic principles. Water moves freely from the side of the cell membrane with proportionately fewer dissolved solute molecules or ions to the other side. As salinity of the sea water decreases the percentage of water in the animal and in each cell increases until a lethal level is reached. Table 5.5 lists measurements of water content of medusae with reference to salinity of the surrounding sea water. The percentage of water varies with salinity, but there has been no rigorous test of conformity to osmotic principles.

Within its normal range of salinities, each animal must control the concentrations of particular ions. Most animal cells have higher intracellular concentrations of potassium and organic anions than the surrounding medium, whether that medium is extracellular animal fluids or external bathing media such as sea water. These concentrations are necessary for establishing membrane potentials and appropriate conditions for enzyme function. If the membrane is permeable to the ions, maintenance of concentration gradients requires metabolic energy (Kirschner, 1991).

The cells must also regulate cell volume. Since water is incompressible, net water loss causes cells to shrink while water entry will cause them to swell. Since entry of water cannot be prevented if cells are placed in a hypoosmotic (relatively more dilute) medium, cells must decrease the internal solute to balance the reduced external solute and prevent swelling. This can be done by decreasing inorganic ions like potassium, or by decreasing the concentration of free amino acids (FAA).

Scyphomedusae have been shown to change volume with changes in the salinity of the sea water surrounding them. *Aurelia aurita*

Table 5.5 Water content (as percentage of wet weight) of nonsymbiotic scyphomedusae

Species	*Portion*	*Method*	*Salinity (‰)*	*Water content*	*Source*
Atolla wyvillei	W	O 60°	33.5	95.1	Clarke, A., Holmes and Gore, 1992
Aurelia aurita	W	O 80°	7.3	98	Thill, 1937
Aurelia aurita	W	O 60°–110°	31.5–32.6	95.9–96.6	Hyman, 1938
Aurelia aurita	W	FD	28–30	96.2	Larson, 1986d
Chrysaora fuscescens	W	O 75°	31–33	96.4–96.7	Shenker, 1985
	U			96.5	
	OA			95.9	
	G			98.1	
Chrysaora melanaster	U	O 110°	31.5	96.4	Koizumi and Hosoi, 1936
Cyanea capillata	U	O 110°	31.5	96.5	Koizumi and Hosoi, 1936
Cyanea capillata	W	FD	28–30	95.8	Larson, 1986d
	U			96.3	
	OA			96.2	
	T			94.8	
Manania atlantica	W	FD	28–30	91.1	Larson, 1986d
Periphylla periphylla	W	O 50°	33.1–33.3	95.7–96.6	Fosså, 1992
Phacellophora camtschatica	W	FD	28–30	96.3	Larson, 1986d
	U			96.7	
	T			95.4	
Rhizostoma pulmo	W	VD 60°	13.5	98.2	Gubareva *et al.*, 1983

G, gonad; OA, oral arm; T, tentacle; U, umbrella; W, whole specimen; FD, freeze drying; O, overdrying at indicated temperature; VD, vacuum drying at indicated temperature.

medusae collected from salinity of 7.3% in the Baltic Sea can regulate their volume down to salinities of approximately 6.4%, but below that salinity the volume increases (Thill, 1937). Scyphistomae and planulae appeared normal down to approximately 5%, but planulae swelled at lower salinities.

That volume regulation is related to free amino acids (FAA) is shown by linear relations between the FAA concentrations and salinity in polyps of *Aurelia aurita*, *Chrysaora quinquecirrha* and *Cyanea capillata* (see Webb, Schimpf and Olmon, 1972). Glycine is the most concentrated free amino acid in *A. aurita*, *C. quinquecirrha* and *Cyanea* sp.

although ornithine and taurine are more concentrated in *Rhizostoma pulmo* (see Daumas and Ceccaldi, 1965; Severin, Boldyrev and Lebedev, 1972). Recently fed scyphistomae of *A. aurita* show reduced uptake of glycine from sea water as salinity is decreased (Shick, 1973). Complex changes in the rate of RNA and protein synthesis occur in the course of salinity acclimation (Lukanin, 1976; Berger, 1977; Khlebovich and Lukanin, 1980; Weiler and Black, 1991). The concentrations of FAA will also be affected by the rates of protein and amino acid catabolism.

5.4.1 Water content

Many workers have been interested in the water content of gelatinous animals such as scyphomedusae. Some early workers claimed concentrations of water of over 99.8% of wet weight (Bateman, 1932). However, measurements made using modern methods, tabulated in Table 5.5, have not found more than 98% in whole animals. The exact value varies with the external salinity, as described above. Other than in very low salinity situations, such as the Baltic, workers have rarely found more than 97%.

Water content of other stages in the life cycle may be less than that in the medusae. For example, planulae of *Chrysaora hysocella* from the English Channel at Roscoff contain only 65–71% water (Teissier, 1926).

In order to determine water content, dry weight is measured and the result subtracted from the wet weight. The dry weight varies with the method used, particularly with the drying temperature. Bound water of hydration remains after drying, unless temperatures used are high enough also to cause oxidation of organic compounds (section 7.1.2). In practice most recent measurements are made by freeze drying, or by oven drying at 60°C, which retains some bound water. Table 5.5 includes only measurements for which the salinity and drying method are stated in the relevant publication. Further measurements are included in tables in Vinogradov (1953) and Larson (1986d).

5.4.2 Buoyancy

The density of known scyphomedusae is identical or slightly heavier than surrounding sea water. The density of *Aurelia aurita* is identical to sea water in which it has been caught (Lowndes, 1942, 1943; Aleyev and Khvorov, 1980). Anaesthetized *Pelagia noctiluca* sink only 0.3 cm per second (Davenport and Trueman, 1985). Thus these animals use little energy in swimming in order to maintain a given depth in the water column.

Table 5.6 Sulphate and chloride composition of nonsymbiotic scyphomedusae, presented as percentage of concentration in sea water

Species	*Position*	*Cl^-*	*SO_4^{2-}*	*Source*
Aurelia aurita	MF	104	47	Robertson, 1949
Aurelia aurita	W		56	Bidigare and Biggs, 1980
Chrysaora melanaster	M	108	51	Koizumi and Hosoi, 1936
Cyanea capillata	M	109	42	Koizumi and Hosoi, 1936
Cyanea capillata	MF		40	Newton and Potts, 1993
	G		135	
Pelagia noctiluca	W		62	Bidigare and Biggs, 1980
Rhizostoma pulmo	MF		52	Newton and Potts, 1993
	G		129	Newton and Potts, 1993

G, gastrovascular fluid
M, mesoglea
MF, mesogleal fluid
W, whole specimen

Buoyancy requires lift to balance the heavy proteinaceous tissue. Some lift may be provided by the presence of lipid, but concentrations of lipid are low in scyphomedusae (see Table 7.2). Lift may also be provided by the partial exclusion of heavier ions and their isosmotic replacement with lighter ions. Changes in ion concentrations also affect the density of the water due to changes in its structure. In many gelatinous plankton, sulphate exclusion is coupled with an isosmotic replacement with lighter chloride ions. The replacement of all the sulphate by chloride in the sea water would produce a lift of 2.1 mg/ml (Newton and Potts, 1993).

Concentrations of sulphate ions in scyphomedusae relative to sea water are presented in Table 5.6. As expected from its greater buoyancy, *A. aurita* is able to exclude a greater proportion of its sulphate than *P. noctiluca*. Comparison with the protein concentration shows that *P. noctiluca* is only capable of offsetting 66% of its protein mass by exclusion of 38% of its sulphate content (Bidigare and Biggs, 1980).

The metabolic basis for this ion transport and the energy expenditure involved are not known. Sulphate is excluded from the mesogleal fluid (Table 5.6). It is moved into the gastrovascular fluid, as shown by increased concentrations in that fluid (Newton and Potts, 1993).

6 *Reproduction*

6.1 SYNOPSIS

As noted in Chapter 1, in the typical scyphozoan life cycle the fertilized egg develops into a planula and thence into a polyp. The scyphopolyp produces one or more medusae asexually, which then reproduce sexually (Figures 1.1, 6.1 and 6.6). Either planula or scyphopolyp may also reproduce asexually by budding, or may form cysts. Polyp or medusa may be reduced or absent. In the latter case the polyp becomes the stage that reproduces sexually. The life cycles appearing in each order were briefly outlined in section 1.2. The development of each stage is described in the present chapter. More detailed descriptions of the microanatomy of each stage may be found in Lesh-Laurie and Suchy (1991).

6.1.1 Types of reproduction and trade-offs

The many possible asexual modes of reproduction, even within a single species, enable the potentially extremely complex life cycles of scyphozoa. For example, *Aurelia aurita* may reproduce with a life cycle which sequentially includes a fertilized egg, planula, scyphopolyp (scyphistoma), strobila, ephyra and adult as shown in Figure 6.1 and many textbooks. However, the planula may also develop directly into an ephyra shortly after settling, without formation of a scyphistoma (Kakinuma, 1975; Yasuda, 1975b). In addition the scyphistoma may produce further polyps by longitudinal fission, by direct budding, or via formation of stolons, cysts or planuloid buds (Thiel, H., 1963b; Chapman,

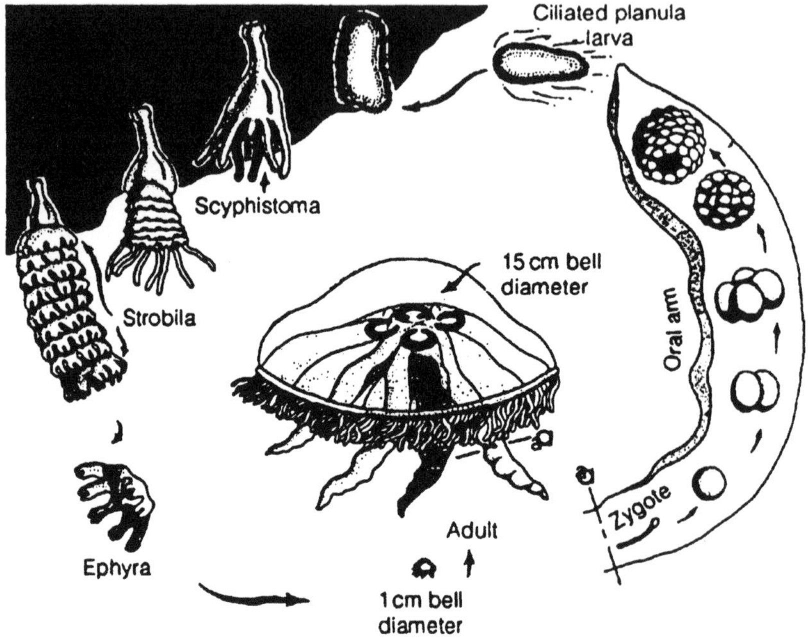

Figure 6.1 Life-cycle of *Aurelia aurita* emphasizing sexual reproduction. See text for variations in asexual reproduction. (Source: Hamner and Jensen, 1974, with permission of W.M. Hamner and American Society of Zoologists.)

D.M., 1968; Kakinuma, 1975). To date there is little data on the relative significance of these various possibilities in field situations.

Similar morphological stages, such as types of cysts or types of polyps, may differ physiologically or ecologically when produced differently. For example, *Cyanea* sp. commonly produces two kinds of cysts: podocysts produced by the polyps and planulocysts produced by the planulae. In the Niantic River, Connecticut, planulocysts all excyst within a given season, whereas a high proportion of the podocysts remain encysted until the following season (Brewer and Feingold, 1991) (Figure 6.13). Planuloid buds of *Cotylorhiza tuberculata* differ from planulae in selection of position for settling and in time required for development into scyphistoma (Kikinger, 1992). Much of the earlier field data must be critically re-evaluated, as more data on the factors which trigger production of particular stages, and their physiological properties, are provided by laboratory work.

Given the wide plasticity in possible scyphozoan life cycles, there has been much speculation, but little data, on why particular sequences of stages have been maintained. It is assumed that sexual reproduction is necessary for genetic variability (Tardent, 1984). However, sexual reproduction could be carried out by polyps, as is done by the stauromedusae, without the complex morphogenetic processes involved in medusa production.

Medusae may be of importance for greater dispersal of the species than is possible by the short-lived planula. The ciliated swimming planula, when present, may be more important in selection of favourable environment for the benthic stages than in dispersal. Medusae may also exploit food sources unavailable to the benthic polyps, allowing the high energy expenditure involved in sexual reproduction (Mackie, 1974; Cornelius, 1992).

Virtually nothing is known either about why benthic stages are necessary, or about the trade-offs between particular types of asexual reproduction. Benthic stages may aid survival of temperate and arctic species during periods when there is little planktonic food available. However, medusae may also survive starvation, particularly at low temperatures, simply by absorbing their own substance and growing smaller (section 7.4).

Cysts may be formed as a protection against temperature change or other physical factors, but they are often present during seemingly favourable summer months. Brewer and Feingold (1991) have suggested that they may also serve as protection against predation, or maintain the animal against seasonal encroachment on space. As suggested above, planulocysts and podocysts may not serve the same functions, podocysts being more important in long-term resistance to unfavourable physical conditions.

6.1.2 Genetics

Nothing is known of inheritance in scyphozoa, although the comparative structure of RNA and DNA has been examined in some species to determine phylogenetic relationships (section 1.3). In order to study inheritance, it will be necessary to find intraspecific characters that can be used to identify populations or individual medusae or polyps, and to distinguish between genetic and environmental effects. Isozyme patterns of the malate dehydrogenase enzyme of *Aurelia aurita*, as well as morphological characters such as types of cnidae present, differ between cultured polyps from different geographical locations (Zubkoff and Linn, 1975). These differences are probably genetic.

Fautin and Lowenstein (1992) suggested that proteins characterized by radioimmunological methods may be of use in distinguishing species. They were able to identify relatively stable proteins distinguishing

Aurelia aurita from *Pelagia colorata* through several developmental stages. However, they were investigating species from two different families and the method may not be applicable to more closely related forms.

6.2 GAMETOGENESIS

Most scyphozoa are gonochoristic (or dioecious), i.e. they have separate sexes. However, medusae of *Chrysaora hysocella* are protandrous hermaphrodites, i.e. they produce first sperm and then ova (Claus, 1877; Widersten, 1965). Other than in the gonads, sexual dimorphism is rare. One example is *Cotylorhiza tuberculata*, where brood-carrying filaments near the mouth are developed only by the females (Kikinger, 1986, 1992).

Sex ratios vary from approximately equal numbers to 1.7 (females : males) in pelagic populations of coronates. This has been observed in samples of more than 100 individuals each in the coronate species *Atolla parva*, *A. vanhoeffeni* and *Paraphyllina ransoni* (see Repelin, 1965, 1966). Similar sex ratios have been observed in some inshore species (e.g. *Cassiopea andromeda*) (Gohar and Eisawy, 19601b). However, the sex ratio in populations with a marked annual cycle may vary during the season. For example, in *Cyanea capillata* the sex ratio is 1 at maturity, but female medusae with planulae outlive the males (Brewer, 1989). As mentioned above, in the hermaphroditic *Chrysaora hysocella* sperm production precedes that of the ova.

6.2.1 Gonad formation

Gonads of the scyphozoa arise from the gastrodermis. In medusae the gonads are usually situated on the floor of the gastrovascular cavity peripheral to the gastric cirri (Figures 1.5, 4.2, and 6.2). In the coronate medusae eight gonads are usually present, in the semaeostome and rhizostome medusae usually four. In the medusae *Cyanea capillata*, *Aurelia aurita* and *Rhizostoma pulmo* the gastrodermal cells form evaginations containing mesoglea that become the gonads (Widersten, 1965; Kon and Honma, 1972). Subgenital sinuses appear formed by overgrowth of the gonad (Figures 6.2 and 6.3). In the hermaphroditic *Chrysaora hysocella* the female gonads are similar to those in other species, but the sperm develop in small vesicles scattered over the gastrodermal endothelium, or on genital filaments near the ovary (Widersten, 1965).

In the stauromedusae there are eight gonads, each extending from near the mouth into an arm (Figure 1.4). The coronate polyps may

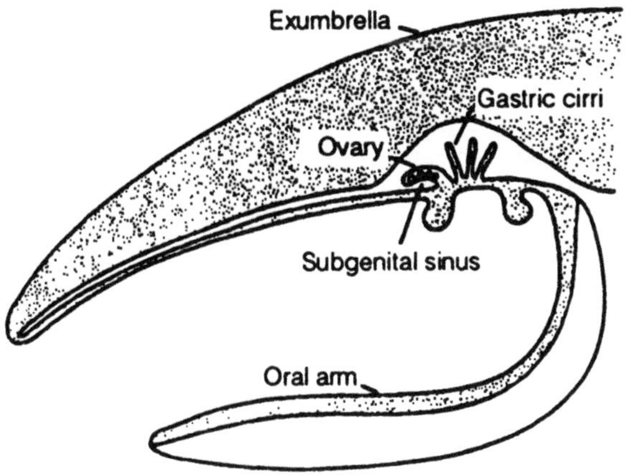

Figure 6.2 Location of ovary on floor of gastrovascular cavity peripheral to gastric cirri in *Aurelia aurita*. (See also Figure 4.2.) (Source: Eckelbarger and Larson, 1988, with permission of K.J. Eckelbarger, R.J. Larson and Springer-Verlag.)

form gametes in the mesenteries prior to strobilation, form hermaphroditic eumedusoids, or lack gametes entirely (section 6.4.3) (Komai and Tokuoka, 1939; Werner, 1971b, 1974; Werner and Hentschel, 1983).

6.2.2 Gamete production

The cells that will develop into ova arise from the gastrodermal epithelium of the ovary. In coronate medusae, such as *Linuche unguiculata*, the oocytes migrate into the mesoglea and remain solitary as they grow (Eckelbarger and Larson, 1992). They must, therefore, receive nutrients through the mesoglea as they synthesize yolk. Yolk formation is associated with elaboration of Golgi complexes and a rough endoplasmic reticulum as well as with invagination of the oolemma to form intraooplasmic channels.

As the oocytes of semaeostome and rhizostome medusae grow they gradually bulge into the mesoglea of the gonad but maintain contact with specialized cells in the epithelium. Examples include *Aurelia aurita*, *Chrysaora hysocella*, *Cotylorhiza tuberculata*, *Cyanea capillata*, *Discomedusa lobata*, *Pelagia noctiluca*, *Rhizostoma pulmo* and *Stomolophus meleagris* (see Von Lendenfeld, 1882; Tsukaguchi, 1914; Widersten, 1965; Rottini-Sandrini, Bratina and Avian, 1986; Eckelbarger and Larson, 1988; Avian and Rottini-Sandrini, 1991; Kikinger, 1992;

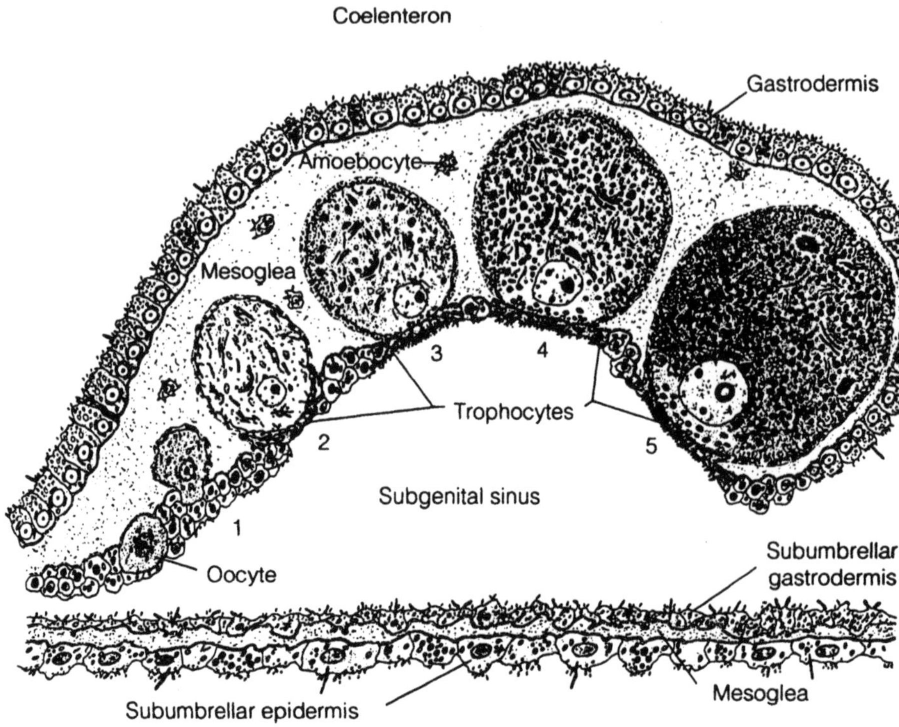

Figure 6.3 Transverse section through *Aurelia aurita* ovary, illustrating various stages of oogenesis as indicated by numbers: 1. Oocyte begins moving into mesoglea from the germinal epithelium. 2. Gastrodermal cells in the germinal epithelium begin differentiating into trophocytes. 3. Oocyte completes movement into mesoglea, retaining association with trophocytes, and yolk synthesis begins. 4. Synthesis of yolk (round dark bodies in cytoplasm) through activity of Golgi complex and rough endoplasmic reticulum, with uptake of precursors through endocytosis. 5. Late-stage oocyte (diameter 175 mm). (Source: Eckelbarger and Larson, 1988, with permission of K.J. Eckelbarger, R.J. Larson and Springer-Verlag.)

Eckelbarger and Larson, 1992) (Figure 6.3). The specialized cells contain membrane-bound inclusions. Some workers have called them 'nurse cells', but they do not maintain cytoplasmic continuity with the oocyte by intracellular bridges as the classical nurse cells do. Eckelbarger and Larson (1988) have applied the term 'trophocyte' to these cells. Trophocytes may be involved in the transfer of yolk precursors to the oocytes, but experiments with labelled precursors have not demonstrated such transfer (Avian *et al.*, 1987). In *D. lobata* and *P. noctiluca* some of the epithelial cells produce mucus which is released

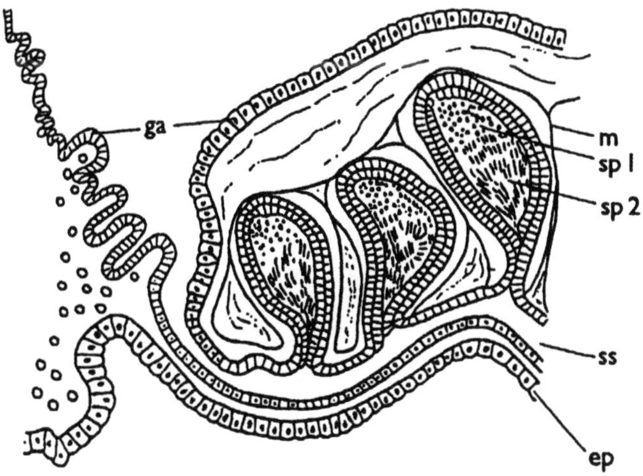

Figure 6.4 Distal part of testes of *Aurelia aurita*. Spermatocytes are produced from the upper walls of the follicles, and mature as they move down the follicles toward release into the subgenital sinus. ep = epidermis; ga = gastrodermis; m = mesogleal fibril; sp1 = spermatocyte; sp2 = spermatid; ss = subgenital sinus. (Source: Widersten, 1965, with permission of Royal Swedish Academy of Sciences.)

during spawning to cover the oocyte (Avian and Rottini-Sandrini, 1991).

In each ovary of stauromedusae such as *Haliclystus auricula* and *H. octoradiatus* a series of follicles is formed (Uchida, 1929). Oocytes move from the peripheral epithelium toward the lumen of the follicle as they grow and form yolk, i.e. they do not differentiate within the mesoglea as in the other orders (Eckelbarger and Larson, 1993). Each oocyte is surrounded by a loosely organized layer of squamous cells.

Sperm of scyphozoa such as those of *Aurelia aurita*, *Craterolophus convolvulus*, *Cyanea capillata* and *Discomedusa lobata* develop in follicles formed by invagination of the epithelium into the mesoglea of the testis (Widersten, 1965; Kon and Honma, 1972; Bratina, Rottini-Sandrini and Avian, 1981; Rottini-Sandrini, Bratina and Avian, 1986; Hedwig and Schäfer, 1986) (Figure 6.4). The cavity of the follicle is open to the gastrovascular cavity. Spermatocytes are liberated from the wall of the follicle and mature in the cavity of the follicle. Sperm may accumulate in the subgenital sinus or in the oral arms prior to spawning. In the rhizostome medusae *Cassiopea andromeda*, *C. frondosa* and *Cotylorhiza tuberculata* this pattern is modified to form spermatozeugmata, sperm packages with somatic cells in the centre

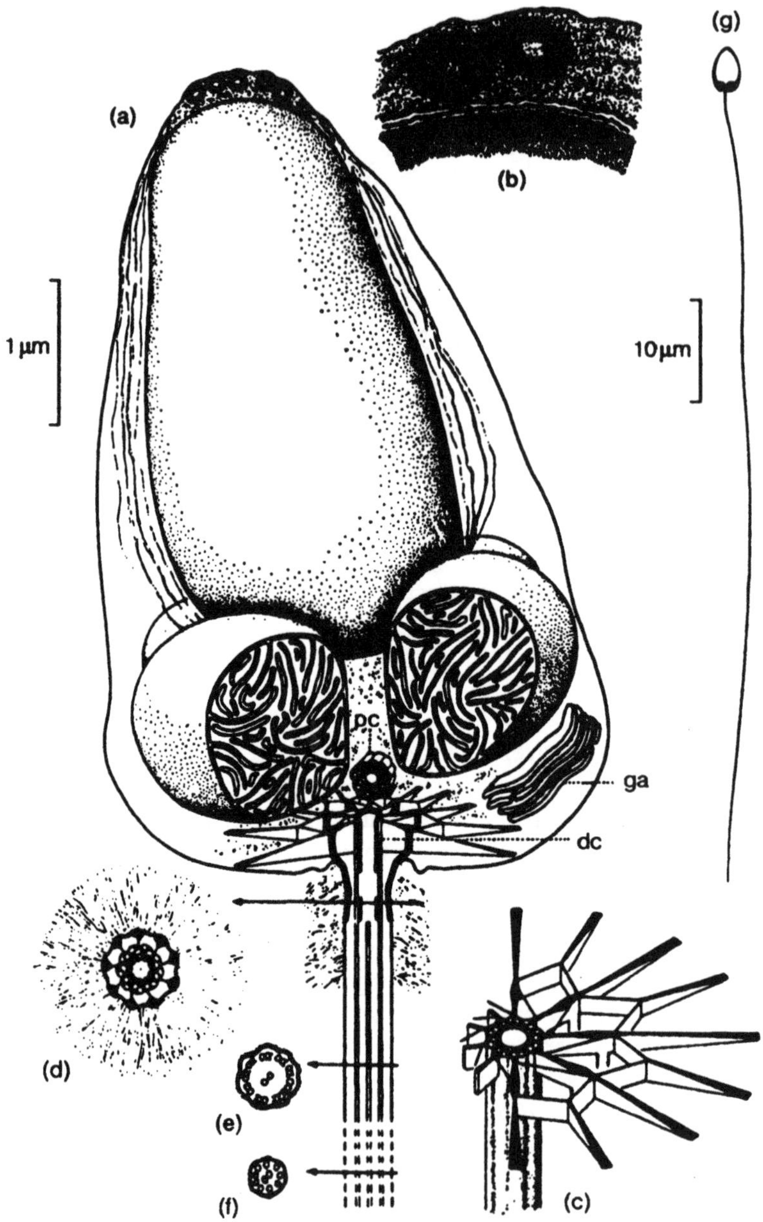

Figure 6.5 Ultrastructure of sperm of *Nausithoe* sp. (a) The short, conical head includes the nucleus and a layer at the tip containing vesicles; the midpiece contains four mitochondria, a proximal centriole and a distal centriole surrounded by pericentriolar processes; the proximal part of the flagellum has a hairy coat of slender fibrils. (b) The anterior end of the sperm head to show the vesicles.

surrounded by sperm with tails oriented outward (Smith, H.G., 1936; Gohar and Eisawy, 1961b; Kikinger, 1992). In these species the spermatozeugmata are each spawned as a unit.

The sperm of the coronate *Nausithoe* sp. has a short conical head, a midpiece containing four large mitochondria and two centrioles, and a long flagellum (Afzelius and Franzén, 1971) (Figure 6.5). The distal centriole is surrounded by pericentriolar processes of striated fibres and is the source of the flagellar microtubules. The longitudinal microtubules are similar to those that power many metazoan flagella. The microtubules show the '9 plus 2' configuration, with nine doublets and a central pair. The sperm are primitive in lacking an acrosomal cap, which in most metazoa contains enzymes for penetration at the anterior tip of the sperm head. However, a layer containing vesicles is present, which may serve the same function. Sperm of semaeostome and rhizostome medusae such as *Aurelia aurita*, *Chrysaora hysocella*, *Pelagia noctiluca* and *Rhizostoma pulmo* are similar to those of the coronates (Hinsch and Clark, 1973; Hinsch, 1974; Rottini-Sandrini, Bratina and Avian, 1983; Hedwig and Schäfer, 1986). However, sperm of the stauromedusa *Craterolophus convolvulus* differ in containing five mitochondria and only one centriole (Hedwig and Schäfer, 1986).

Oceanic scyphomedusae, such as the coronates *Atolla* spp. and the semaeostome *Poralia rufescens*, may produce ova at low rates throughout the year (Russell, 1959a; Mauchline and Harvey, 1983; Larson, 1986b). However, the better known temperate neritic species are strongly seasonal in their production of ephyrae (as will be discussed in section 6.4.3) and of gametes. For example, in *Cotylorhiza tuberculata* of the Mediterranean Sea, gamete production occurs only in late summer and fall (Kikinger, 1986, 1992) (Figure 6.6).

It is difficult to distinguish the factors controlling seasonal gamete production, although it is clearly not due simply to the time elapsed since seasonal ephyrae production. (The factors triggering the spawning behavior will be discussed in the next section.) In *Aurelia aurita*, sexual maturation is correlated with size rather than time since strobilation. Maturation can be started or stopped repetitively by feeding the medusa so that it grows, or starving it so that it 'de-grows' and the gonads regress (Hamner and Jenssen, 1974). (Nevertheless,

(c) The distal centriole with its pericentriolar processes. (d) Cross-section of the flagellum at the level of the hairy coat. (e) Cross-section of the flagellum at the level of the main portion of the flagellum. (f) Cross-section of the flagellum nearer the end. (g) Living spermatozoon drawn from light microscope observations. dc = distal centriole; ga = Golgi apparatus; pc = proximal centriole. (Source: Afzelius and Franzén, 1971, with permission of B.A. Afzelius and Academic Press.)

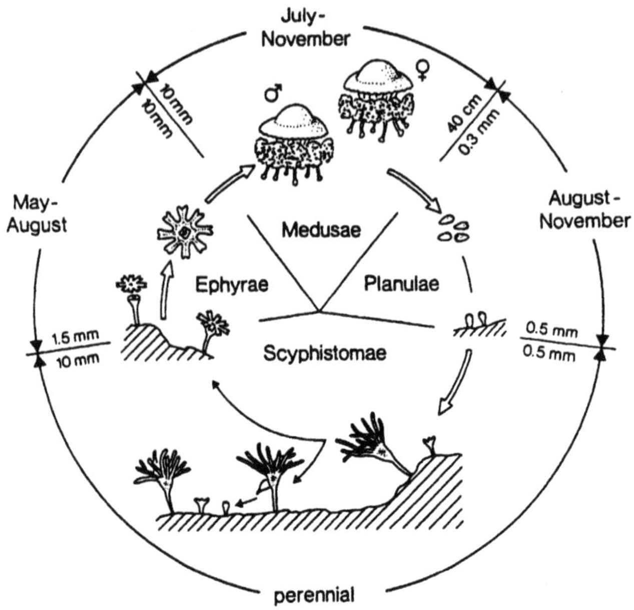

Figure 6.6 Seasonal developmental cycle of *Cotylorhiza tuberculata* in the Bay of Vlyho, Greece. (Source: Kikinger, 1986, with permission of Blackwell Wissenschafts-Verlag.)

spermatogenesis, once initiated, proceeds even if the gonad regresses.) In *Pelagia noctiluca*, both sexes are mature after reaching a diameter of 3.5 cm, although fewer oocytes are produced in summer than spring or autumn (Avian, Giorgi and Rottini-Sandrini, 1991; Rottini-Sandrini and Avian, 1991) and food must be available (Larson, 1987d). In the Niantic River, the largest *Cyanea* sp. reproduce first, although even the smallest medusae eventually reproduce (Brewer, 1989). Unfortunately, these observations do not distinguish clearly between the effects of size *per se* and of nutrition.

Size affects the rates of production, as well as the timing of its commencement. In *Linuche unguiculata* the number of eggs released per day increases as medusa size increases over 9 mm diameter (Kremer *et al.*, 1990). In *Rhopilema esculenta* production increases from 220 × 10^4 to 6700 × 10^4 per day as the umbrella diameter increases from 23 to 53 cm (Huang, M. *et al.*, 1985). This is also probably true of *Aurelia aurita* where the number of brooded eggs and larvae vary linearly with the wet weight of the medusa (Schneider, 1988b).

6.2.3 Fertilization

In *Aurelia aurita* medusae, fertilization occurs in the gastrovascular cavity of the female. The male medusae release sperm from the tips of the oral arms, in mucous strands which are picked up by the female (Southward, 1955; Hamner, Hamner and Strand, 1994). As the strand breaks up the sperm are transported into the gastrovascular cavity, much as food would be (see section 4.3.1 for circulation patterns). Following fertilization, ova move outward much like rejected food material (Widmark, 1913; Southward, 1955; Kon and Honma, 1972). As will be described in section 6.3.2, they become enclosed in pockets in the inner surfaces of the oral arms where development continues.

In other medusae too, fertilization may occur in the gastrovascular cavity. Alternatively, it may occur in the female gonad itself, on the oral arms, or separate from the female in the open sea water. Figure 6.7 shows sperm of *Cyanea capillata* attracted to an ovum still in place in the gonad (Widersten, 1965). Fertilization of *Cotylorhiza*

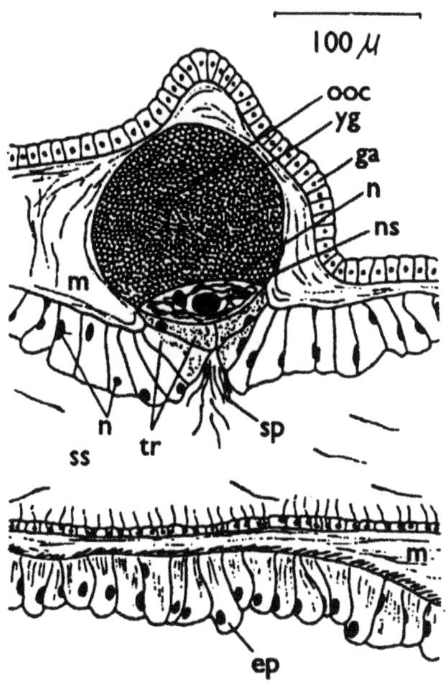

Figure 6.7 Cross-section through the ovary of *Cyanea capillata* showing sperm attracted to a ripe oocyte in the ovarial mesoglea. ep = epidermis; ga = gastrodermis; m = mesoglea; n = nucleus; ns = nucleolus; ooc = oocyte; sp = sperm; ss = subgenital sinus; tr = trophocyte; yg = yolk grains. (Source: Widersten, 1965, with permission of Royal Swedish Academy of Sciences.)

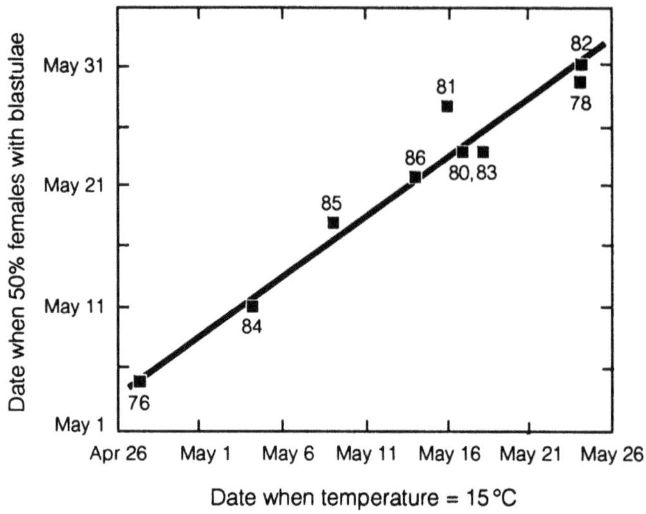

Figure 6.8 Relationship between temperature and the onset of reproduction (appearance of blastulae on the oral folds) in *Cyanea* in the Niantic River estuary, Connecticut. Numbers beside the points indicate the year. (Source: Brewer, 1989, with permission of R.H. Brewer and *Biological Bulletin*.)

tuberculata also occurs in the ovary (Kikinger, 1992). In *Chrysaora hysocella* fertilized ova may even develop to the gastrula stage in the ovarial mesoglea (Widersten, 1965). However, ova of *Chrysaora quinquecirrha* are fertilized in the gastrovascular cavity (Littleford, 1939).

No matter where the ova are fertilized, the sperm is the motile gamete and must detect and reach the ovum. Sperm chemotaxis has been demonstrated in hydrozoa, but attractants have not been investigated in scyphozoa.

The timing of spawning may be correlated with temperature and with light. In *Cyanea* sp. of the Niantic River, the seasonal onset of spawning is not correlated with the first appearance of mature medusae but rather with temperature. The mean time for a female to bear blastulae on its oral folds is 8 days after the surface temperature reaches 15°C (Brewer, 1989) (Figure 6.8).

The diurnal timing of spawning is dependent on light. Scyphozoa, including *Aurelia aurita*, *Chrysaora quinquecirrha* and *Haliclystus octoradiatus*, have been observed to spawn at particular times of day (Wietrzykowski, 1912; Littleford, 1939; Hamner, Hamner and Strand, 1994). *Linuche unguiculata* spawn in the early morning but lose their periodicity if held in complete darkness or in constant light (Figure 6.9) (Conklin, 1908; Kremer *et al.*, 1990). *Haliclystus stejnegeri* spawn

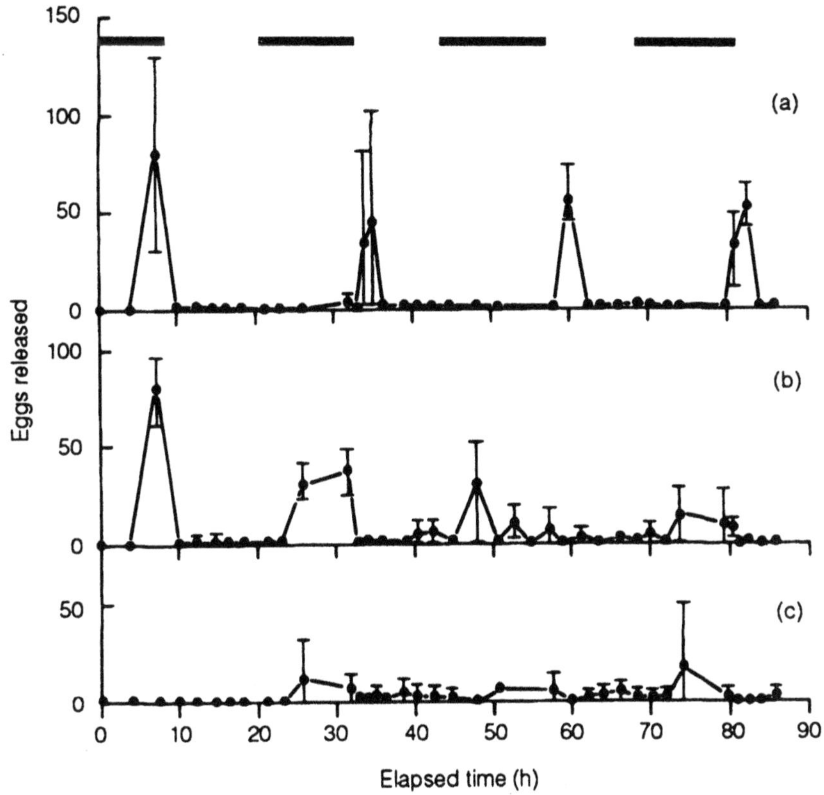

Figure 6.9 Egg production of 18 mm unfed *Linuche unguiculata* held in 200 ml filtered sea water (with transfers every few hours) for 84 hours after collection. (a) Natural light–dark cycle (dark bars denote dark period); (b) constant darkness; (c) constant illumination. Bars show the range of egg release rates for three medusae per treatment. Note the diel periodicity of egg release. (Source: Kremer *et al.*, 1990, with permission of P. Kremer and the American Society of Limnology and Oceanography.)

on exposure to light after being held in the dark for at least 8 hours (Otto, 1976).

It would be advantageous for fertilization if spawning occurred in the presence of conspecifics. Medusae do often form aggregations in which spawning occurs (section 8.4). For example, male *Aurelia aurita* in Saanich Inlet, British Columbia, release sperm when in aggregations but not when isolated (Hamner, Hamner and Strand, 1994). However, evidence that the spawning is directly due to the interactions between individuals is lacking.

6.3 LARVAL DEVELOPMENT

The patterns of development in the Scyphozoa are varied, between species and even within species. The zygote (fertilized ovum) undergoes cleavage. It forms a blastula and then a gastrula, developing into the larval planula. The planulae may then develop into a polyp, form buds, cysts (planulocysts) or scyphorhizae, or develop directly into medusae.

6.3.1 Embryogenesis and planulae

Cleavage of the zygote may be equal or unequal, and of varied patterns which have been interpreted as radial or pseudospiral (Mergner, 1971). In coronate, semaeostome and rhizostome species such as *Aurelia aurita*, *Chrysaora hysocella*, *Cotylorhiza tuberculata*, *Cyanea arctica*, *Linuche unguiculata* and *Mastigias papua*, a hollow blastula (coeloblastula) is formed (Figure 6.10) (Smith, F., 1891; Hein, 1900, 1902; Conklin, 1908; Hargitt and Hargitt, 1910; Uchida, 1926; Teissier, 1929). Some cells may migrate into the blastocoele, but they disappear before gastrulation. In contrast the stauromedusae such as *Haliclystus octoradiatus* form a solid blastula (stereoblastula) (Wietrzykowski, 1912).

Gastrulation (formation of the gastrula from the blastula) may occur by ingression or by invagination. In the stauromedusae, such as *Haliclystus octoradiatus* and *Manania distincta*, the endoderm is formed by 'ingression' of a single column of cells (Wietrzykowski, 1912; Hanaoka, 1934). In modern terminology this expansion of an epithelium along its basal margin is more correctly termed involution (Browder, Erickson and Jeffery, 1991). In *Cyanea capillata*, invagination (inward buckling and folding of the epithelium) occurs (Figure 6.10) (Hyde, 1895; Hargitt and Hargitt, 1910). *Aurelia aurita* may employ multipolar ingression (inward migration of individual cells which lose contact with other epithelial cells) or invagination (Smith, F., 1891; Hyde, 1895; Hein, 1900; Hargitt and Hargitt, 1910; Berrill, N.J., 1949).

Berrill (1949) has correlated the type of gastrulation with the size of the egg. Very small eggs, such as the 0.03 mm egg of *Haliclystus octoradiatus*, lead to gastrulation by involution or ingression, whereas larger eggs such as the 0.3 mm egg of *Pelagia noctiluca* precede gastrulation by invagination. *Aurelia aurita*, which has eggs ranging from 0.15 mm to 0.23 mm in diameter, has varied types of gastrulation.

The planula larva consists of an outer ectoderm separated from an inner endoderm by a thin mesoglea. Semaeostome and rhizostome

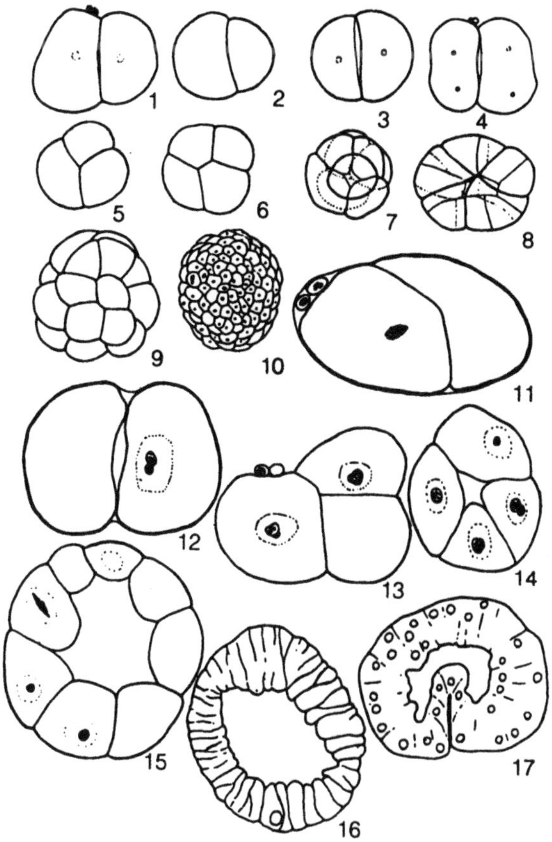

Figure 6.10 Development of *Cyanea capillata*. Drawings 1–10 of entire two-cell stage to blastula, ×360; drawings 11–17 from sections from two-cell stage through blastula to young gastrula, ×715. (Source: Hargitt and Hargitt, 1910.)

planulae are oval or pyrifrom in outline, varying in length from approximately 100 μm to 390 μm (Figure 6.12) (Gohar and Eisawy, 1960; Sugiura, 1963, 1966; Uchida and Nagao, 1963; Korn, 1966; Kakinuma, 1967, 1975; Kühl, 1972; Calder, 1973, 1982; Yasuda, 1979; Neumann, 1979; Ding and Chen, 1981; Martin, V.J. and Chia, 1982; Cargo, 1984; Yasuda and Suzuki, 1992; Kikinger, 1992). These planulae may appear oval or biconcave in cross-section (Brewer, 1976). They may be solid or hollow, but in either case they lack a mouth. Coronate planulae are larger, reaching 300–630 μm, but of similar shape (Werner, 1974; Ortiz-Corp's, Cutress and Cutress, 1987). Planulae of these three orders are ciliated and swim actively. Their locomotion was described in section 2.6.4.

Stauromedusae, such as *Manania distincta*, form nonciliated planulae which creep over the bottom (section 2.6.4; Figure 2.20). In an unusual case of cell constancy in a cnidarian, the planula has a constant number of endodermal cells. For example, *Haliclystus salpinx* and *H. stejnegeri* planulae have 16 cells (Otto, 1976, 1978). These coin-shaped endodermal cells are arranged in a linear stack.

Neurons have not been clearly demonstrated in scyphozoan planulae, and it is not known how sensory and transmittal functions such as those needed for settlement are carried out (Chia and Bickell, 1978). For example, the planulae of *Cassiopea xamachana* contain only four cell types (Martin, V.J. and Chia, 1982). The ectoderm includes supportive cells and nematocytes, whereas the endoderm consists of supportive cells and interstitial cells. Each supportive cell bears microvilli and a single cilium.

6.3.2 Brooding

Brooding is the term used to describe retention of the zygote by the female through at least part of its development. As noted in section 6.2.3, fertilization may occur within the female gonad, or, as the eggs move outward, in the female gastrovascular cavity, or in the open sea water. The zygote may be retained within the gonad, or in specialized brooding chambers in the gastrovascular cavity or on the exterior of the female. It is probable that this provides either protection or nutrition for the developing embryo, although this has not been demonstrated for any scyphozoa.

Retention within the gonad usually extends no further than the planula stage. For example the embryo of *Chrysaora hysocella* develops as far as the gastrula in the ovarial mesoglea (Widersten, 1965). Planulae of *Rhopilema verrilli* are retained within the gonadal tissue until fully developed (Calder, 1973).

Spectacular internal brooding occurs in the semaeostome medusa *Stygiomedusa gigantea.* Only a few specimens of this deep-sea genus have so far been described (Russell, 1959b; Russell and Rees, 1960; Repelin, 1967; Ulrich, 1972; Cornelius, 1972, 1973; Harbison, Smith and Backus, 1973; Larson, 1986b). Much of the bell cavity is occupied by four brood chambers in which development occurs as far as young medusae of over 9 cm in diameter. No male specimens have been described (Larson, 1986b). The poor condition of the known specimens following collection has made the sexual or asexual origin of the embryos a matter of speculation (Hadzi, 1963).

Some common semaeostome and rhizostome medusae retain the embryos to mature planulae in small brood chambers of the oral arms or subumbrella, or associated with brood-carrying filaments near the mouth. Kikinger (1992) summarizes the literature for the rhizostome medusae, as well as describing the brood-carrying filaments of *Cotylorhiza tuberculata*.

Embryos of *Aurelia aurita* develop to planulae in brood pouches formed along the edges of the oral arms (Figures 1.6, 6.1) (Minchin, 1889; Russell, 1970). These pouches may be tightly packed with developing embryos, making the otherwise transparent arms opaque. Embryos or planulae are present for several months but it is not known how long any one embryo takes to develop (Yasuda, 1971). Blastulae develop to planulae on the much-folded margins of the oral folds of *Cyanea* sp. (see Brewer, 1989). Surprisingly, the *Cyanea* oral folds may also retain viable planulae of other species (Lambert, 1935).

6.3.3 Settlement including metamorphosis

Semaeostome and rhizostome planulae larvae such as those of *Aurelia aurita*, *Cassiopea xamachana*, *Chrysaora melanaster*, *Chrysaora quinquecirrha*, *Cyanea* sp., *Rhizostoma pulmo*, *Rhopilema esculenta* and *R. verrilli* remain free-swimming from a few hours up to approximately 10 days (Littleford, 1939; Kakinuma, 1967, 1975; Kühl, 1972; Calder, 1973; Ding and Chen, 1981; Martin, V.J. and Chia, 1982; Cargo, 1984; Brewer, 1984). The larger planulae of the coronate species *Linuche unguiculata* and *Nausithoe eumedusoides* remain planktonic for 3 to 4 weeks (Ortiz-Corp's, Cutress and Cutress, 1987; Werner, 1974). Following the free-swimming period, most scyphozoan planulae must select an appropriate substrate, attach, and undergo metamorphosis into a polyp or cyst. A few develop directly into a medusa as described in section 6.3.4.

It is not known what determines the length of time prior to settlement. That larger planulae tend to remain pelagic longer suggests a role of nutrition. Schneider and Weisse (1985) calculated on the basis of respiration and excretion rates that the maximum survival time of *Aurelia aurita* planulae should be between a few days and a week. However, their calculations assumed no intake during the period. Although planulae do not feed it is possible that they, like the polyps and ephyrae of the same species, are able to absorb dissolved organic matter (compare section 4.5).

The choice of substrate involves active selection by the planula. Prior to settlement the planula may become shorter. Shortened planulae of *Cyanea capillata* approach a substrate with the anterior end, and rotate counter-clockwise (Brewer, 1976, 1984). If the substrate

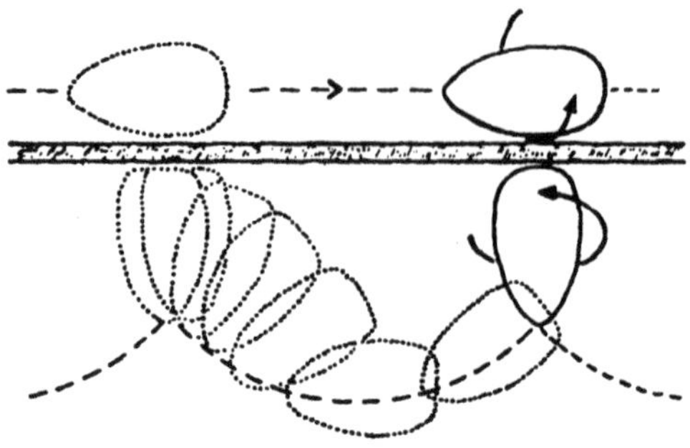

Figure 6.11 Behaviour of planulae of *Cyanea* in contact with upper (top of figure) and under (bottom of figure) surfaces. The blunt end, by which the planula attaches when it settles, is anterior during locomotion. Planulae on the upper surface glide with their long axis parallel to it (dashed arrow), rotating in the direction shown by the solid arrow. The direction of rotation is the same for planulae against the under surface, but they orient perpendicular to it for varied periods of time depending upon the nature of the surface; between these stationary periods of rotation, they move as shown by the dotted lines along a path described by the curved dashed line. (Source: Brewer, 1984, with permission of *Biological Bulletin*.)

is not suitable the planula leaves it, swims briefly and then approaches a surface to 'inspect' another site (Figure 6.11).

Attraction to a substrate depends on several factors, the response presumably enabling selection of an appropriate substrate for the polyp. For example, polyps of *Aurelia aurita* are often found in nature hanging with the oral surface downwards, thus avoiding problems with siltation. If the planulae are offered horizontal objects, such as cover slips, more than 90% fasten to the underside of the objects (Brewer, 1978). Similar selection of the undersurfaces of objects has been demonstrated in *Chrysaora quinquecirrha*, *Cotylorhiza tuberculata* and *Cyanea capillata* (see Cargo and Schultz, 1966, 1967; Brewer, 1976, 1984; Cargo, 1979; Kikinger, 1992). Planulae of *C. capillata* are initially geopositive and remain near the bottom, but after about 50 hours become geonegative and swim upwards in the water column to contact the undersurfaces of objects (Brewer, 1976). This species also settles preferentially in shaded areas (Svane and Dolmer, 1995).

Planulae, such as those of *Aurelia aurita*, *Chrysaora quinquecirrha* and *Cyanea capillata*, settle in greater numbers on rough or grooved surfaces than on smooth ones (Brewer, 1976, 1978, 1984; Cargo, 1979). This may partly be due to the increased wettability of roughened surfaces, since planulae of *Cyanea* sp. attach sooner and in greater numbers to clean plastic (hydrophobic) surfaces than to glass (hydrophilic) surfaces (Brewer, 1984).

It is unclear to what extent aggregation of polyps of *Aurelia aurita* is due to gregarious settlement, i.e. due to attraction to conspecifics. Reciprocal tile transplants by Keen (1987) indicated that concentrations of scyphistomae occur due to response of later settlers to the same physical factors that attracted the initial settlers, rather than to their presence. However, similar transplant experiments by Gröndahl (1988b, 1989) on the same species showed effects of the presence of polyps, which varied with the age of the polyp. Whereas the presence of 10-day-old polyps reduced the number of planulae that settled, in the presence of polyps established for 4 days planula larvae were attracted to the polyps. Ten-day-old polyps of *A. aurita* have well-developed tentacles, whereas 4-day-old polyps do not, and predation by the older polyps on settling planulae may occur. Similarly, planulae of *Cyanea capillata* settle in greater numbers among presettled 4-day-old conspecific scyphistomae than when the substratum is unoccupied or occupied by scyphistomae of *A. aurita* (see Dolmer and Svane, 1993).

Planulae attach by their anterior ends. As shown in Figures 1.1 and 6.12, planulae of *Stomolophus meleagris* develop an elongated stalk from the anterior end (Calder, 1982). The posterior end forms a bulbous calyx with an oral cone and mouth, and the rudiments of the first four tentacles. Similarly, in planulae of *Cassiopea andromeda* attachment and secretion of an adhesive pedal disc occur within 24 hours in culture. This is followed within another 2–3 days by differentiation of a stalk and calyx, the latter with an oral cone and tentacles (Gohar and Eisawy, 1960; Neumann, 1979; Hofmann and Brand, 1987). Similar metamorphic sequences have been described in other semaeostome and rhizostome species, including *Aurelia aurita*, *A. limbata*, *Cephea cephea*, *Chrysaora melanaster*, *Chrysaora quinquecirrha*, *Cotylorhiza tuberculata*, *Mastigias papua*, *Rhizostoma pulmo*, *Rhopilema esculenta*, *Rhopilema nomadica* and *Rhopilema verrilli* (see Hyde, 1895; Hein, 1900, 1902; Uchida, 1926; Littleford, 1939; Uchida and Nagao, 1963; Korn, 1966; Sugiura, 1966; Kakinuma, 1967, 1975; Kühl, 1972; Ding and Chen, 1981; Calder, 1982; Lotan, Ben-Hillel and Loya, 1992).

It is not known how metamorphosis of the planula is controlled, although portions of the planula differ in their ability to form polyps. The anterior portion of the planula is required for development of the normal scyphistoma. If planulae of *Cassiopea andromeda* are cut in half, the anterior portions (corresponding normally to the basal portion of the developing polyp) can reconstitute a smaller-sized

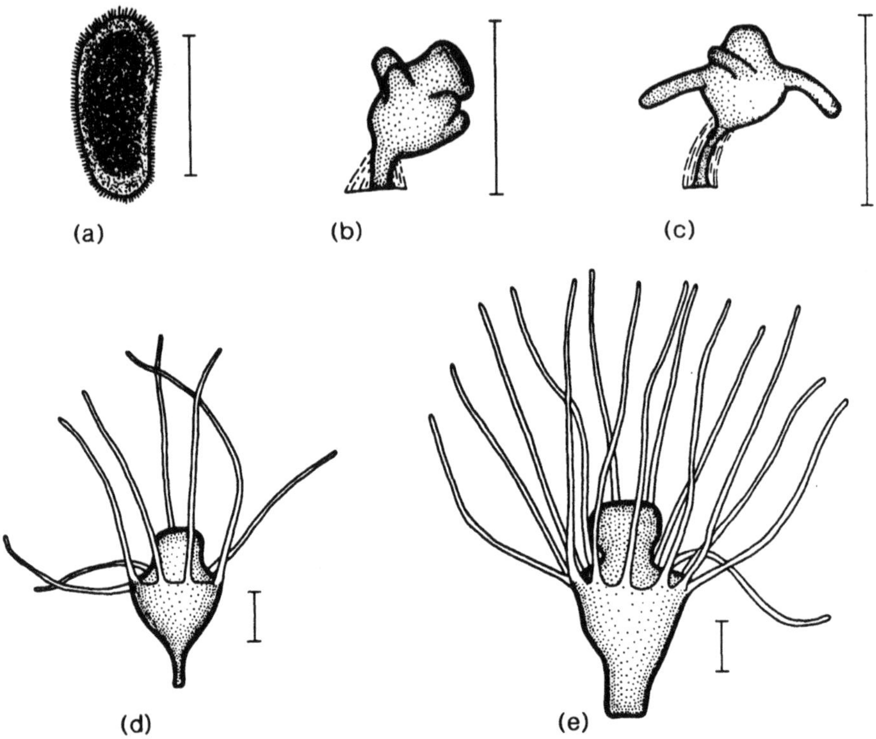

Figure 6.12 Planula and scyphistoma stages of *Stomolophus meleagris*. (a) Planula; (b) newly metamorphosed scyphistoma; (c) young scyphistoma; (d) intermediate, eight-tentacled scyphistoma; (e) fully developed scyphistoma. Scale bars = 250 μm. (Source: Calder, 1982, with permission of D.R. Calder and *Biological Bulletin*.)

planula and thence a complete polyp (Neumann, 1979). The posterior portion of the planula is unable to develop into a normal polyp, forming an unattached calyx without stalk or attachment disc. Similar unattached calyxes are formed if planulae of *Cassiopea xamachana* are treated with hydroxyurea which is a temporary antagonist of DNA synthesis (Lesh-Laurie and Suchy, 1986).

External stimuli are required for metamorphosis of planulae of *Cassiopea andromeda*. Metamorphosis is decreased in sterile natural seawater, and absent in artificial sea water without any organic substances. It may, however, be induced by addition of a filtrate containing low molecular weight compounds secreted by the marine bacterium *Vibrio cholerae*, as well as by such non-natural compounds as thyrotropin, and pancreatic casein hydrolysate peptides (Neumann, 1979; Wolk *et al.*, 1985; Fitt *et al.*, 1987). Activity has been shown

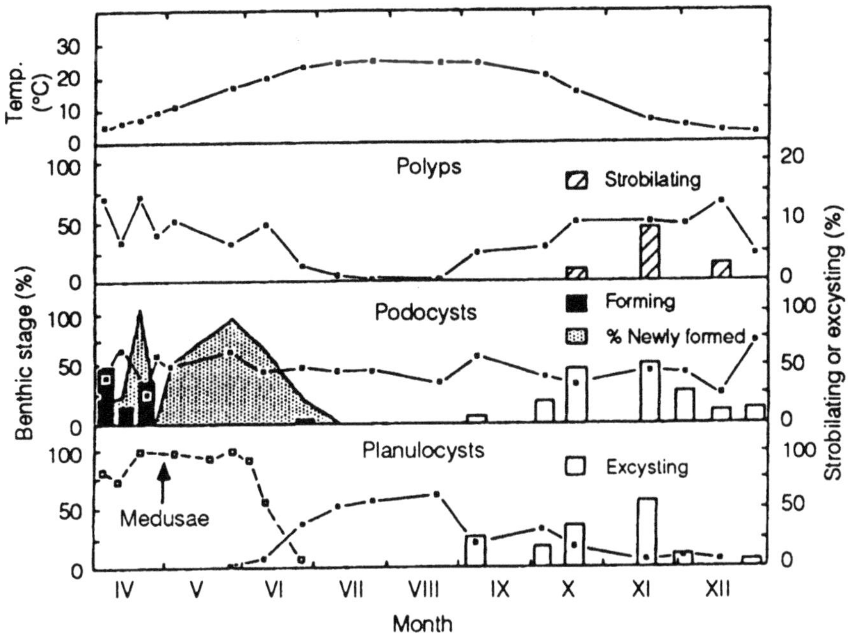

Figure 6.13 Populations of the benthic reproductive stages (solid lines) and of medusae (dashed line) of *Cyanea* in the Niantic River, Connecticut, and the monthly surface temperature. The diagram shows: when planulocysts appear; when polyps form podocysts (solid bars); proportion of podocysts recently formed (stippled polygon); when planulocysts and podocysts excyst (open bars); and when polyps strobilate (striped bars). Medusae are expressed as number of individuals observed/minute × 50. (Source: Brewer and Feingold, 1991, with permission of R.H. Brewer and Elsevier Science.)

for some peptides containing a proline residue next to the carboxyterminal amino acid (Fitt and Hoffmann, 1985; Rahat and Hofmann, 1987; Hofmann and Brand, 1987).

Planulae of *Cyanea capillata* form cysts (planulocysts) prior to forming a polyp (Brewer, 1976 and references therein). The cysts are shiny, circular, plano-convex, cuticle-covered and about 1 mm in diameter (Cargo, 1974). Similar cysts are formed by planulae of *Cyanea lamarcki* (Widersten, 1968). The planulae may remain on the oral folds until the medusae deteriorate, transporting the planulae to the benthos (Brewer, 1989). They then settle and form the cysts which excyst later in the same season (Figure 6.13) (Brewer, 1989, 1991; Brewer and Feingold, 1991).

Planulae of some coronate scyphozoa form a flattened film of protoplasm, the scyphorhiza, as they attach. The scyphorhiza of *Linuche unguiculata* then develops a group of polyps one by one, which remain attached to form a colony (Werner, 1979; Ortiz-Corp's, Cutress and Cutress, 1987).

As stauromedusan planulae settle, they attach to the substrate or to other planulae to form aggregates (Otto, 1978). The endodermal cells of each planula then rearrange to form a minute gastric cavity and a tiny polyp begins to form. In *Haliclystus octoradiatus* the aggregates may fuse, and then reform planuloid buds similar in form to the original planulae (Wietrzykowski, 1912). Each of the planuloid buds then forms a polyp. As in the semaeostome and rhizostome species, a minute polyp with four tentacles and four septa is formed.

6.3.4 Direct development

Although most of the well-known neritic scyphozoa have a life cycle that includes a polyp, direct development from planula to ephyra is also possible. For example, *Aurelia aurita* planulae may settle and develop into scyphistomae. Alternatively they may settle very briefly and, without developing the tentacles of the polyp, develop into ephyrae (Figure 6.14) (Hirai, 1958; Kakinuma, 1975). The ephyrae at first retain a peduncle from the attached stage, but that soon disappears. This process requires less time than scyphistoma formation and strobilation: 3–11 days depending on the water temperature (Yasuda, 1975b, 1979). The fast production of ephyrae after planulae development has been found in Wakasa Bay, Japan Sea.

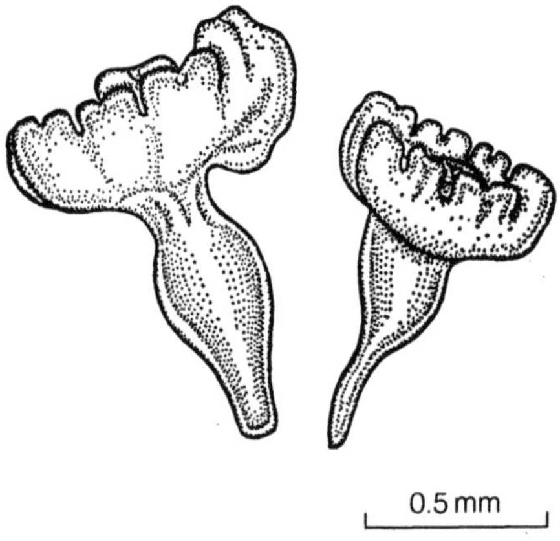

Figure 6.14 Stage in direct development of *Aurelia aurita* from planula to single ephyra. (Source: Yasuda, 1975b, with permission of T. Yasuda and Seto Marine Biological Laboratory.)

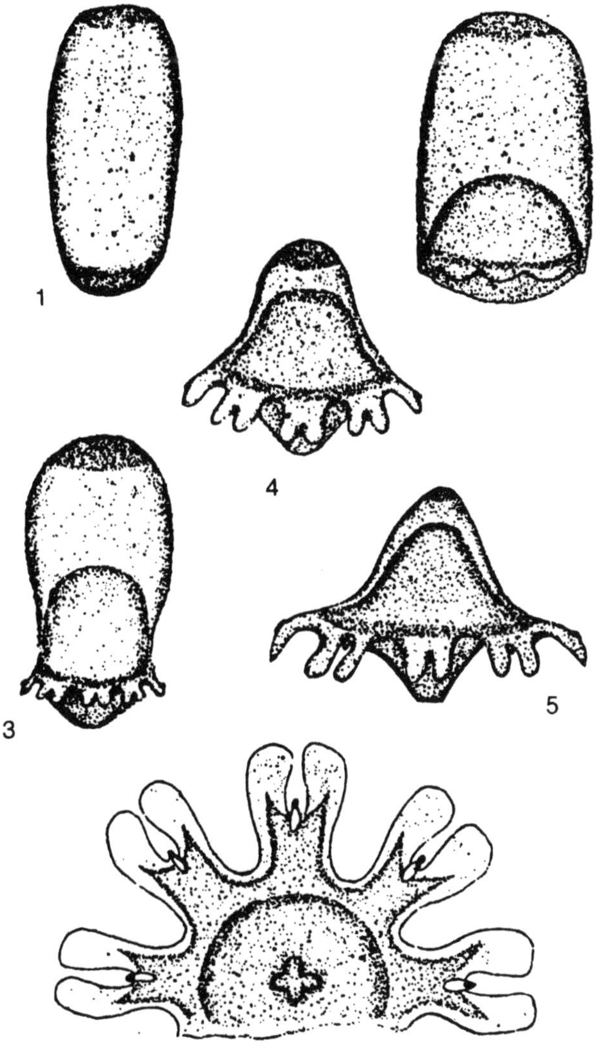

Figure 6.15 Direct development of *Pelagia noctiluca*, stages from planula to ephyra. Drawings 1–4, ×75; 5, ×45; 6, ×60. (After Delap, 1906; cilia omitted.)

Only direct development has been observed in *Pelagia noctiluca*. The planula of this species develops into an ephyra, without settling on the bottom (Figure 6.15) (Goette, 1893; Delap, 1906; Rottini-Sandrini and Avian, 1983; Avian, 1986b). The development rate is temperature-dependent. Development is prevented at 4.5°C, and faster at 19°C than at 13.5°C (Rottini-Sandrini, Avian and Zanelli, 1985; Avian, 1986b; Morand, Goy and Dallot, 1992).

It seems probable that direct development, without dependence on a benthic stage, would be advantageous for oceanic medusae. Berrill (1949) has suggested that direct development may be favoured by large eggs (which when fertilized form wide gastrulae with the blastocoel only partially obliterated). This may facilitate lappet formation. Larson (1986b) has compiled data showing that oceanic species such as the coronates *Atolla* spp. and *Periphylla periphylla* and the semaeostome *Poralia rufescens* have eggs an order of magnitude larger than those of related neritic forms. However, there is no evidence of the type of reproduction used by these oceanic species.

6.4 POLYP

Although direct development from planula to medusa can occur (section 6.3.4), the scyphozoan life cycle usually includes a sessile polyp stage between the planula and the adult. Whether it arises directly from the planula, a bud, a cyst or a scyphorhiza, it typically develops through a tetraradiate stage with four tentacles and four septa.

The scyphistoma polyps of the semaeostome and rhizostome species develop an oral calyx and a narrower stalk extending to the pedal disc attached to the substrate (Figure 6.12). The calyx includes an oral disc with a central mouth, and a ring of up to 24 tentacles around the margin. Internally four septa extend from the oral disc to the base of the calyx.

These scyphistomae may reproduce asexually by budding, by the formation of cysts, or by strobilation to form medusae. Each of these processes will be described in subsequent sections. They also rarely divide by longitudinal fission. In this last mode of reproduction *Aurelia aurita* may expand to each side, adding additional tentacles, and then divide into two polyps (Kakinuma, 1975).

The known coronate species differ from other orders in possessing a chitinous tube which surrounds the scyphopolyps. Scyphistomae of semaeostome and rhizostome species may have a delicate cuticle surrounding the base of the polyp (section 1.3), but only in the coronates is it highly developed, allowing tall thin vase-shaped solitary polyps or coloniality (Figure 6.16). The tubes have a two-layered construction, often with internal teeth (Jarms, 1991). The coronate polyps that have so far been identified all belong to species of the Nausithoidae or Linuchidae; other coronates from the deep sea may

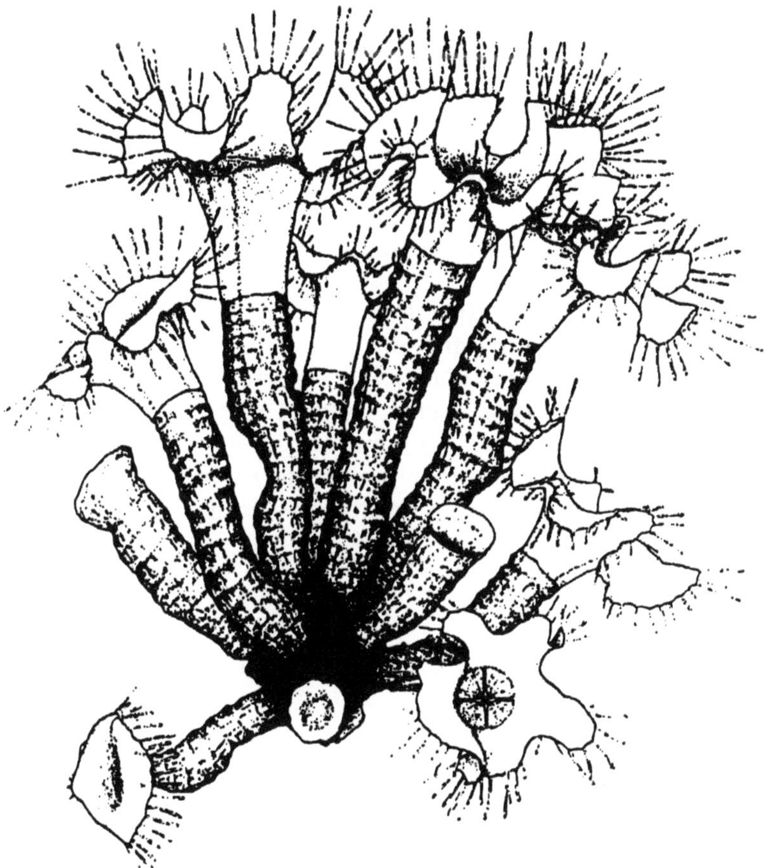

Figure 6.16 Part of a young colony of *Nausithoe racemosa* with fully expanded polyps. (Drawn by F. Heckmann. Source: Werner, 1970, with permission of Seto Marine Biological Laboratory.)

have direct development (section 6.3.4). These coronate polyps reproduce by strobilation, the modifications of which will be described in section 6.4.3.

Three species of coronates are known to form colonies. In *Nausithoe punctata* the polyps arise from a small basal disc and the polyps branch. In *Linuche unguiculata* polyps arise singly from an expanding scyphorhiza, and in *Nausithoe racemosa* the scyphorhiza forms a stolonal plate and there is also branching in older colonies (Figure 6.16) (Werner, 1979; Ortiz-Corp's, Cutress and Cutress, 1987).

The adult 'medusae' of the Stauromedusae remain sessile, somersaulting about or firmly attached, i.e. they are actually polyps (Figure

1.4). By a convention (which resulted from speculation on possible origins of the order from ancestors with medusae) the adults are often referred to as medusae (Uchida, 1929; Thiel, H., 1966). The resulting terminology is sometimes confusing. The stalked polyps develop from the planulae or the planuloid buds. In genera such as *Haliclystus* eight tentacles develop around the margin of the oral disc before the arms appear (Hirano, 1986a). These tentacles develop into the primary tentacles (anchors) of the adult. Eight clusters of secondary, capitate (knobbed) tentacles develop. In most species each of these clusters is borne on an arm, but in species such as *Kyopoda lamberti* they are located directly on the margin of the oral disc (Larson, 1988). The primary tentacles are secondarily lost in adult specimens of species such as *Kishinouyea corbini* (see Larson, 1980). The stauromedusan polyps gradually become sexually mature. The process of strobilation by which species in the other orders form medusae is therefore unnecessary. To date podocyst formation has not been described in this order.

6.4.1 Budding

Budding by polyps is very varied. Most possibilities may be classified as formation of buds similar in form to the parent polyp, of planuloids, or of elongated stolons.

In the first form of budding a bud forms on the calyx of the polyp, develops into a new polyp, and separates. In *Aurelia aurita* this form of budding is most common when there is an abundant food supply (Uchida and Nagao, 1963; Coyne, 1973; Keen and Gong, 1989). However even if the food supply per polyp is held approximately the same, population growth rate decreases as population density increases (Coyne, 1973), indicating that other factors are also involved.

Planuloid buds superficially resemble planulae. Unciliated planuloid buds are produced by the interstitial stauromedusa *Stylocoronella riedli* from the calyx (Figure 1.2) (Salvini-Plawen, 1966) and by *S. variabilis* from the tentacle tips (Kikinger and Salvini-Plawen, 1995).

Rhizostome scyphistomae of *Cassiopea andromeda*, *Cassiopea xamachana* and *Cotylorhiza tuberculata* produce ciliated planuloid buds (Curtis and Cowden, 1971; Hofmann, Neumann and Henne, 1978; Kikinger, 1992). Often a chain of three or four buds is formed on the underside of the calyx of the *Cassiopea* species, which releases the planuloids one by one. These ciliated planuloids differ from planulae in larger size and greater number of cell types (Hofmann and Honegger, 1990; van Lieshout and Martin, 1992). Marking experiments show that ectoderm of the parent polyp is incorporated into the emerging bud (Neumann, Schmahl and Hofmann, 1980; Hofmann and Gottlieb, 1991). Nevertheless the planuloids swim, settle and form scyphistomae much as the planulae do which were described in section 6.3.3. The distal end of the bud forms the anterior end of the planuloid and eventually forms the stalk of the new polyp (Hofmann, Manitz and Reckenfelderbäumer, 1993). In the laboratory metamorphosis can be initiated by peptides containing proline as the preterminal amino acid at the carboxyl terminus (Fleck and Hofmann, 1990, 1995). Induction of metamorphosis by phorbol esters, which activate protein kinase C (PKC), and blockage of metamorphosis by psychosine, which inhibits PKC, indicate that PKC may play a role in cellular regulation of metamorphosis (Fleck and Bischoff, 1993, quoted in Fleck and Hofmann, 1995).

Stolons are elongated tendrils extending from a scyphopolyp. They have been observed on *Aurelia aurita* (see Gilchrist, 1937; Kakinuma, 1975), *Chrysaora hysoscella* (see Chuin, 1930), *C. quinquecirrha* (see Cargo and Rabenold, 1980), *Cyanea capillata* (see Hargitt and Hargitt, 1910) and *Sanderia malayensis* (see Uchida and Sugiura, 1978). They are formed in the stalk region. Transplant experiments on *A. aurita* indicate that the presence of tentacles inhibits the formation of stolons on the upper portion of the column (Schmahl, 1986).

Some pedal stolons are involved in locomotion, attaching and pulling the main portion of the polyp toward the attachment point (section 2.6.3). Attachment of the stolons of *Aurelia aurita* can be triggered by contact with a bacterium of the family Micrococcaceae during its logarithmic growth phase (Schmahl, 1985a). The effective substances are the glycolipids monogalactosidyldiglyceride and acylgalactosidyldiglyceride (Schmahl, 1985b). In addition to locomotion, stolons can also be involved in formation of new polyps or cysts (Kakinuma, 1975; Uchida and Sugiura, 1978).

6.4.2 Cysts including podocysts

Several types of cysts are formed by scyphozoa. Planulocysts may be formed as planulae settle (section 6.3.3). Podocysts (pedal cysts) are formed at the base of some scyphistomae by a process which allows continuation of the parent scyphistoma. Cysts of species such as *Chrysaora quinquecirrha* may also be formed directly by whole scyphistomae or just at the tips of stolons (Littleford, 1939; Cargo and Schultz, 1966; Cargo and Rabenold, 1980). It is not known to what extent these last two types of cysts may differ from podocysts in structure or function. Planulocysts and podocysts of species such as *Cyanea capillata* clearly differ in size, shape and surface structure, as well as in physiological properties. Nevertheless ecological papers often do not distinguish adequately between the types of cysts and may use the terminology incorrectly, making it difficult to evaluate the importance of these stages.

Podocysts are cysts which form beneath the pedal discs of scyphistomae. They are surrounded by chitin. They have been described in most detail in the semaeostome species *Aurelia aurita* (see Chapman, D.M., 1966, 1968, 1970a) and *Chrysaora quinquecirrha* (see Black, Enright and Sung, 1976; Magnusen, 1980). Podocysts are also formed by *Chrysaora hysocella* (see Chuin, 1930) and *Cyanea capillata* (see Widersten, 1969; Gröndahl and Hernroth, 1987), and by the rhizostome species *Rhopilema esculenta* (see Ding and Chen, 1981; Guo, 1990), *Rhopilema nomadica* (see Lotan, Ben-Hillel and Loya, 1992), *Rhopilema verrilli* (see Cargo, 1971), *Rhizostoma pulmo* (see Paspaleff, 1938; Kühl, 1972) and *Stomolophus meleagris* (see Calder, 1982) (Figure 6.17).

Formation of podocysts by *Aurelia aurita* involves migration of amoebocytes through the mesoglea toward the base of the scyphistoma (Chapman, D.M., 1968, 1970a; Widersten, 1969). With epidermal cells the amoebocytes form a plano-convex aggregation, dome upward, and become encapsulated at the base of the scyphistoma. The extent of the amoebocyte migration and the extent of the breakdown or participation of the aboral epidermis of the parent scyphistoma are varied. Prior to podocyst formation the pedal disc contains desmocytes, cells which form protein 'rivets' binding the mesoglea through the epidermis to the substrate. These rivets remain in place as the epidermis disintegrates. As the cellular aggregation of amoebocytes and epidermal cells forms a chitinous cuticle around itself, it incorporates the rivets, and hence binds itself to the substrate. The epidermis of the parent scyphistoma then regrows over the surface of the dome. When the scyphistoma is again entire it can move off leaving the podocyst behind.

Podocyst formation by *Chrysaora quinquecirrha* differs in that a stolon is formed prior to the podocyst (Magnusen, 1980). The calyx of the scyphistoma forms the stolon which attaches and flattens at the tip. The calyx then moves so that the main axis of the polyp is over the new attachment, and a podocyst is formed through movement of mesoglea and epidermis. As in *A. aurita*, yolk-filled cells are surrounded by cuticle which incorporates rivets basally.

Podocysts can remain viable for some time. For example, the podocysts of *Chrysaora quinquecirrha* can remain dormant for at least 25 months at 25°C and 15‰ salinity (Black, Enright and Sung, 1976). They are protected by a series of concentric lamellae of chitin

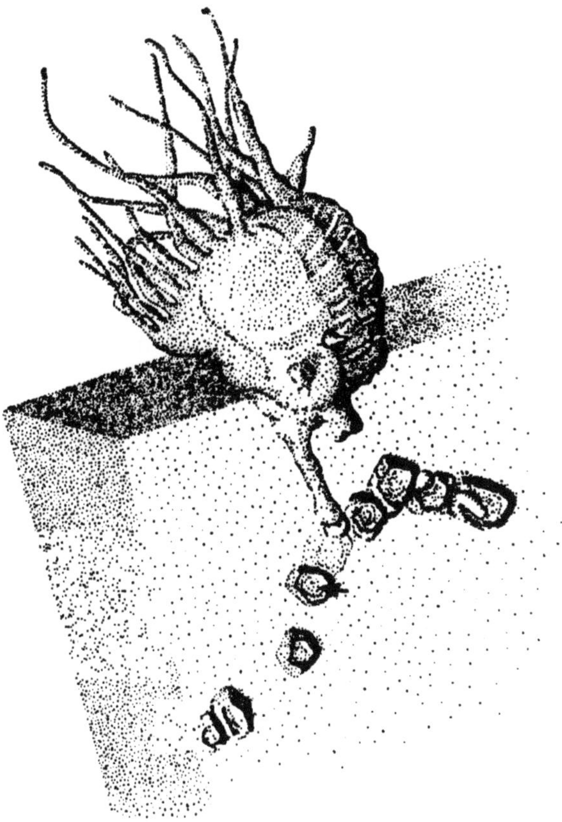

Figure 6.17 Scyphistoma of *Cyanea capillata* during podocyst formation: arrow indicates the typical crater in a *C. capillata* podocyst. (Source: Gröndahl and Hernroth, 1987, with permission of F. Gröndahl and Elsevier Science.)

(Blanquet, 1972a). The interior cell mass is originally arranged with a central mass of yolk-filled cells surrounded by a population of more active cells. Metabolism is low, as shown by low oxygen uptake, nevertheless over a year half of the DNA, one-third of the protein, and one-fifth of the lipid are lost (Black, 1981). Eventually a small, four-tentacled polyp is formed which emerges through the apex of the dome. Removal of the chitin coat results in an increase in metabolism and a faster formation of the polyp.

The timing of podocyst production and of excystment varies between species. It is dependent on adequate nutritional state of the scyphistoma (Guo, 1990). Rates are also loosely correlated with temperature changes. For example polyps of *Cyanea* sp. from the

Niantic River, Connecticut, form podocysts during rising temperatures in April–June, which excyst in falling temperatures in September–December (Brewer and Feingold, 1991) (Figure 6.13). Many of the podocysts remain encysted into the winter months, in contrast to planulocysts of the same species.

6.4.3 Strobilation

The term strobilation is used to describe the entire process by which scyphopolyps give rise to ephyrae (Spangenberg, 1965b). This requires disc formation ('segmentation') leading to fission, and also metamorphosis in which structures of the polyp are lost and replaced in each disc with those of the developing ephyra. A scyphopolyp in which developing discs can be seen due to constricting rings between them is referred to as a strobila (Figures 6.1 and 6.18). A single disc may develop at one time (monodisc strobilation) or a number may develop simultaneously (polydisc strobilation).

Earlier work on the metamorphic processes by which the scyphistoma develops into the ephyra was reviewed by Chuin (1930), Berrill (1949), Spangenberg (1968a) and Russell (1970). The tentacles of the scyphopolyp are lost, and later a new ring of tentacles is formed on the polyp below the developing discs (Figure 6.18). Lappets and rhopalia develop on the margin. Internally the septal muscles of the polyp are lost and replaced with lappet muscle. The first ephyra will use the mouth of the polyp from which it is developing, but each subsequent ephyra develops a mouth, as does the basal polyp. Separation of the epidermal layers of two discs begins from the mouth, moving outward. Gastrovascular canals and gastric cirri develop. The nematocysts of the polyp are replaced with other types characteristic of the ephyra.

When metamorphosis into an ephyra is complete, or nearly so, the ephyra begins to pulsate and separates from the polyp. The remaining scyphistoma resumes other activities such as budding.

Since Spangenberg's review the sequence of anatomical changes during strobilation have been confirmed in a number of semaeostome and rhizostome species. There has been further work on *Aurelia aurita* (see Kato, Aochi and Ozato, 1973; Spangenberg and Kuenning, 1976; Spangenberg, 1977; Kato, Tomioka and Sakagami, 1980). Workers have examined other semaeostome species including *Chrysaora melanaster*, *C. quinquecirrha* and *Sanderia malayensis* (see Kakinuma, 1967; Cones, 1969; Uchida and Sugiura, 1978). Among the rhizostomes, species studied include *Cassiopea andromeda*, *Cephea cephea*, *Rhopilema esculenta*, *R. nomadica*, *R. verrilli* and *Stomolophus meleagris*

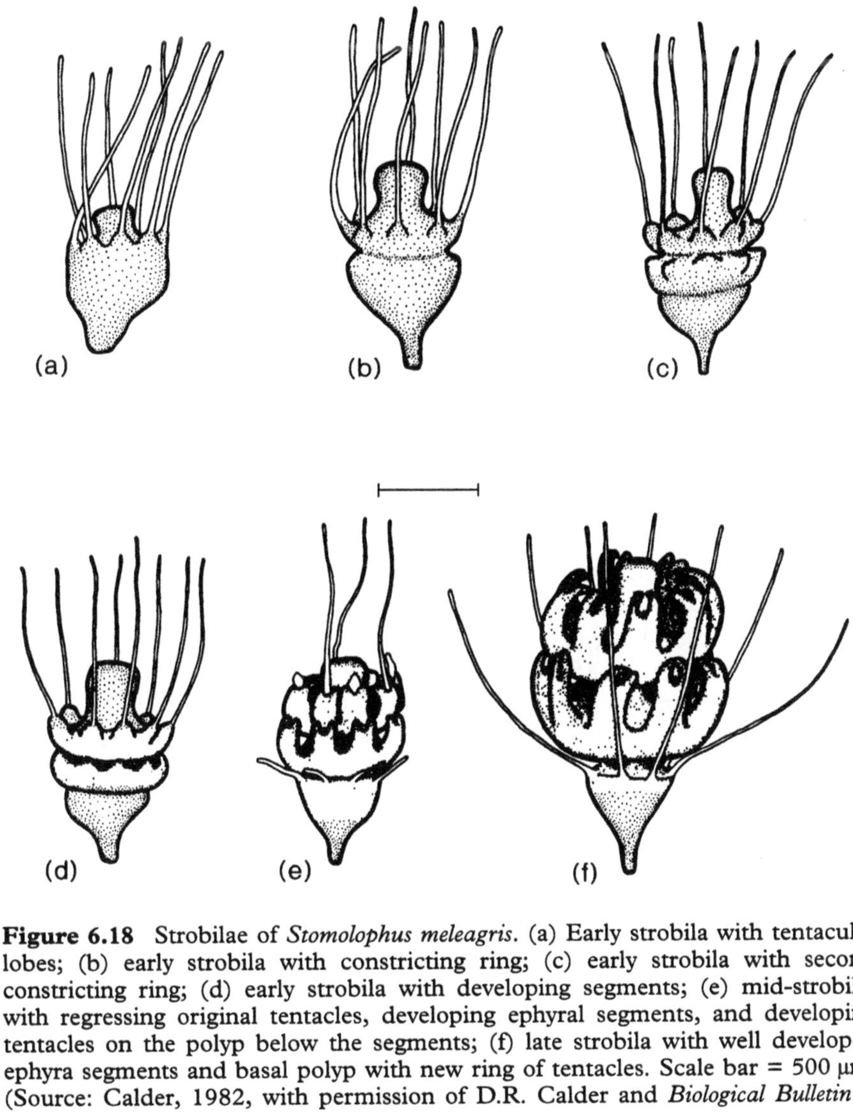

Figure 6.18 Strobilae of *Stomolophus meleagris*. (a) Early strobila with tentacular lobes; (b) early strobila with constricting ring; (c) early strobila with second constricting ring; (d) early strobila with developing segments; (e) mid-strobila, with regressing original tentacles, developing ephyral segments, and developing tentacles on the polyp below the segments; (f) late strobila with well developed ephyra segments and basal polyp with new ring of tentacles. Scale bar = 500 μm. (Source: Calder, 1982, with permission of D.R. Calder and *Biological Bulletin*.)

(see Sugiura, 1966; Ludwig, 1969; Calder, 1973, 1982; Ding and Chen, 1981; Lotan, Ben-Hillel and Loya, 1992).

The details of rhopalial development by *Aurelia aurita* have been examined by several workers. The structure of the adult rhopalium was described in section 2.4.1 (Figure 2.6). Even before polyp tentacle loss is complete, statoliths will begin to form at the bases of the tentacles. The development of the statoliths was described in section 2.4.3.

Strobilation patterns in coronate species are varied. Strobilation of some species such as *Atorella vanhoeffeni*, *Linuche unguiculata*, *Nausithoe punctata* and *N. werneri* is similar to that in the other two orders, except that many more ephyrae are produced (Werner, 1967; Ortiz-Corp's, Cutress and Cutress, 1987; Jarms, 1990). For example, a polyp of *L. unguiculata* can produce a string of up to 40 ephyrae. However, in *Nausithoe racemosa* eumedusoids (reduced medusae) are formed, and germ cells are present in the strobila (Komai, 1935; Komai and Tokuoka, 1939; Werner, 1973). *Nausithoe eumedusoides* produces a chain of hermaphroditic eumedusoids which remain attached (Werner, 1971a, 1974). The germ cells are fertilized and develop into planulae before being released. Finally, no sexual phase has been observed in cave-dwelling *Nausithoe planulophora* although it has been raised through several generations (Werner, 1971b, 1983; Werner and Hentschel, 1983). Ephyra-like discs are formed by strobilation and develop into free-swimming planuloid larvae while still in the tube. These larvae settle after 2–7 weeks and form young polyps.

Strobilation is controlled by endogenous factors. If the strobila is sectioned those discs which would have formed ephyrae continue to develop (Spangenberg, 1965b; Kato, Aochi and Ozato, 1973; Kato, Aochi and Sakaguchi, 1973; Kakinuma, 1975; Kakinuma and Sugiura, 1980; Kato, Tomioka and Sakagami, 1980; Schmahl, 1980). Neurons are present in the epidermis which release as yet unidentified neurosecretory material during segmentation (Crawford and Webb, 1972; Loeb and Hayes, 1981; Van der Linden and Decleir, 1982). However, the time and rate of strobilation are also influenced by exogenous physical and chemical factors including iodinated compounds, polypeptides, temperature, light and nutrition.

Iodinated compounds are important in the initiation of strobilation. Addition of potassium iodide to the sea water surrounding *Aurelia aurita*, *Rhizostoma pulmo* or *Chrysaora quinquecirrha* greatly increases the number of strobilating polyps, provided that they are in an adequate temperature regime (Paspaleff, 1938; Spangenberg, 1967; Black and Webb, 1973; Silverstone, Tosteson and Cutress, 1977). A similar response can be elicited by minute quantities of various iodine-containing compounds including thyroxine (T_4), triiodothyronine (T_3), diiodotyrosine (DIT), monoiodotyrosine (MIT) and thyroglobulin (Spangenberg, 1967, 1971; Silverstone, Tosteson and Cutress, 1977). Under experimental conditions radioactive iodine is accumulated by the strobila, particularly in the segmenting region (Spangenberg, 1971; Black and Webb, 1973; Olmon and Webb, 1974; Silverstone, Galton and Ingbar, 1978). The iodine is incorporated into several compounds, only some of which have been tentatively identified by chromatography. It is not known which is the active compound within the polyp in stimulating strobilation, or what the normal sources of iodine are in nature.

In addition to the iodinated high molecular weight compounds, there is at least one other, naturally produced, factor that stimulates strobilation: the neck-inducing factor (NIF). When NIF is released into the sea water by one polyp, it induces neck formation in neighbouring scyphistomae, i.e. formation of the first circular groove below the tentacles (Loeb, 1974a,b). NIF is a protein or large polypeptide. When strobilation of chilled *Chrysaora quinquecirrha* polyps is induced by temperature increase, there is a sharp peak of NIF release 2–5 hours after warming, whereas the peak concentration of iodinated compounds occurs 3 days after warming and falls gradually as strobilation proceeds (Loeb and Gordon, 1975).

Temperature effects on strobilation of a number of scyphozoa are indicated by the seasonal appearance of the ephyrae (if direct development has not been observed). In the laboratory strobilation from the scyphistomae of *Aurelia aurita*, *Cassiopea andromeda*, *Cephea cephea*, *Chrysaora quinquecirrha*, *Mastigias papua*, *Rhopilema esculenta* and *Rhopilema nomadica* can be induced by warming following a cooler period (Sugiura, 1965, 1966; Spangenberg, 1967; Loeb, 1972; Hofmann, Neumann and Henne, 1978; Rahat and Adar, 1980; Chen, J. and Ding, 1983; Lotan, Fine and Ben-Hillel, 1994). For example, *C. quinquecirrha* scyphistomae must be held for a minimum of 7 weeks at 20°C before they are able to strobilate in response to warming to 26°C (Loeb, 1972; Loeb and Gordon, 1975). In nature this species produces ephyrae in the spring or early summer (Cargo and Schultz, 1967; Calder, 1974b; Cargo and King, 1990). However, this is by no means a universal response to temperature change. Some polyps such as those of *Cyanea* sp. are stimulated by falling temperatures in the laboratory (Brewer and Feingold, 1991). The same species normally produces ephyrae in the fall in the Niantic River, Connecticut.

The rate of strobilation may also be affected by light or nutrition, factors which may be interrelated in symbiotic species. Light may be important for strobilation even of nonsymbiotic species. For example, absence of light delays the onset of strobilation of *Chrysaora quinquecirrha* (see Loeb, 1973). Poor nutrition may not prevent strobilation, but rather it reduces the number of ephyrae produced. For example, *Aurelia aurita* scyphistomae may strobilate even after two months' starvation, but they then usually produce only one ephyrae whereas the species normally has polydisc strobilation (Spangenberg, 1967). Similar effects of both these factors have been found for the rhizostome species *Rhopilema esculenta* (see Chen, J., Ding and Liu, 1984,

1985). The presence of symbionts was discussed in section 4.5.4, where it was concluded that they were important, but probably not indispensable, for the strobilation of any symbiotic species. This is probably also related to nutrition.

6.5 EPHYRA

The ephyrae of most semaeostome and rhizostome species resemble that of *Stomolophus meleagris* shown in Figure 6.19 (Calder, 1982). Most ephyrae have eight marginal lobes, with a rhopalium between a pair of lappets at the tip of each. Species such as *Cassiopea xamachana*, with greater numbers of rhopalia in the adult, also have more rhopalia and lappets in each ephyra (Bigelow, 1900). In coronate ephyrae such as *Atorella vanhoeffeni*, *Linuche unguiculata*, *Nausithoe punctata* and *N. werneri* the marginal lobes are absent, so that the 16 lappets arise directly from the margin.

Ephyra development is often imperfect. 'Monster' ephyrae may be formed with abnormal structures (Vannucci, 1957; Thiel, H., 1963a,b). Presumably these ephyrae do not survive. A single strobila may produce both normal and abnormal ephyrae (Low, 1921).

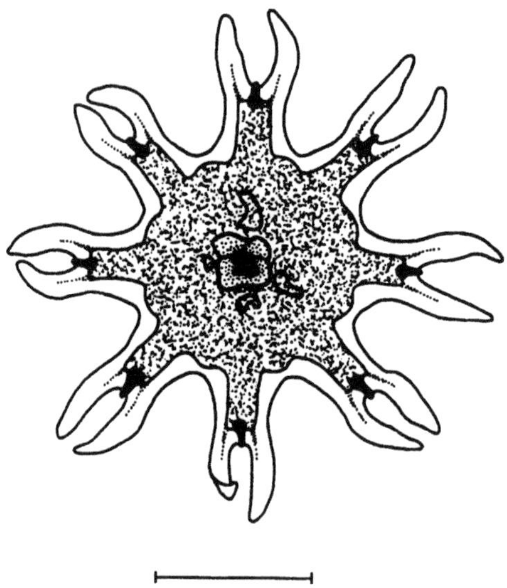

Figure 6.19 Newly liberated ephyra of *Stomolophus meleagris*. Note the eight marginal lobes with a rhopalium between a pair of lappets at the tip of each. Scale bar = 500 μm. (Source: Calder, 1982, with permission of D.R. Calder and *Biological Bulletin*.)

Development of an ephyra into an adult medusa involves growth of the bell margin between the rhopalia, tentacle growth (if the medusa of the species possesses tentacles), elaboration of the oral appendages and gastrovascular cavity, and sexual maturation. The developmental stages have been examined in a number of species including the semaeostome species *Aurelia aurita* (see Southward, 1955; Spangenberg, 1965a; Suckow, 1971; Yasuda, 1983), *A. limbata* (see Uchida and Nagao, 1963) and *Chrysaora quinquecirrha* (see Calder, 1972), and the rhizostome species *Cassiopea andromeda* (see Gohar and Eisawy, 1961b), *Cassiopea xamachana* (see Bigelow, 1900), *Cephea cephea* (see Sugiura, 1966), *Coltylorhiza tuberculata* (see Avian, 1986a), *Mastigias papua* (see Uchida, 1926), *Rhopilema nomadica* (see Lotan, Ben-Hillel and Loya, 1992), *R. verrilli* (see Calder, 1973) and *Stomolophus meleagris* (see Stiasny, 1922). Details vary among the different species with the anatomy of the mature medusae, but are otherwise similar.

Growth of the ephyrae is often very fast, as will be discussed in Chapter 7. However, in some situations ephyrae are produced which remain slow growing until better conditions are available. In the Gullmar Fjord, western Sweden, the peak abundance of *Aurelia aurita* ephyrae is in November. These ephyrae overwinter in deeper water, slowly maturing without increasing in size (Hernroth and Gröndahl, 1983, 1985a; Hernroth, 1986). They ascend to the warmer surface layers in April and then grow rapidly. A similar lag in growth occurs in Kiel Bight, Germany (Möller, 1980a). *Cyanea* sp. also show a similar time lag between production and growth in the Niantic River, Connecticut (Brewer and Feingold, 1991).

7 *Growth*

Quantitative measurements of growth of scyphozoa are largely based on tracing the growth of cohorts of medusae in the field, although a number of stages in the life cycle have been raised in the laboratory, as decribed in the previous chapter. The present chapter will first discuss the measurement of organic matter, which presents particular problems in gelatinous animals. It will then discuss growth, including the possibility of de-growth during starvation, and regrowth following refeeding.

Growth and regeneration are associated with DNA synthesis and cell division (Black, 1972; Balcer and Black, 1991; Lesh-Laurie, Hujer and Suchy, 1991; Napara and Chaga, 1992a,b). However, there is little knowledge of how these processes are controlled. Factors controlling strobilation were described in section 6.4.3. Another possibly active compound identified is testosterone in the gonads of *Aurelia aurita* (see Sadak, Hekim and Güzel, 1980).

7.1 MEASUREMENT OF GROWTH

7.1.1 Units

Growth can be measured in terms of dimensions such as diameter of the medusa, or height of a polyp. These units are useful as they do not require invasive procedures during an ongoing series of measurements. However, for comparison of growth with nutrient intake, and with alternative nutrient uses, it is necessary to measure growth in terms of change in content of organic material or energy. These measurements are invasive. Measurements are therefore made of similar sized animals, and equations are developed relating content to the dimension measured in the actual growth experiment.

In order to compare growth rates between different species or stages of the life cycle, it is useful to consider growth rate independently of the organism's size. The increase in content of a component over a unit of time may be compared with the initial content of the same component. This is the specific growth rate.

7.1.2 Methods

Measurement of the contents of gelatinous animals poses some particular problems. As mentioned in section 5.4.1, an unusually large amount of bound water of hydration remains after drying, unless the temperature is raised to a level which also causes oxidation of the organic compounds (Larson, 1986d). Routinely the methods used for drying are freeze-drying, or drying at 60°C, which largely prevents oxidation but retains some bound water. Elemental or organic compound measurements, expressed as percentage of dry weight, are decreased by this bound water. However, ash-free dry weight measurements are increased, as the bound water and organic compounds are driven off when the samples are ashed.

Ideally growth should be expressed as change in energy content. However, direct caloric measurements, usual for other animals, have not been very successful in scyphomedusae. Accurate combustion in a calorimeter, such as the Phillipson microbomb, is impossible in medusae with large amounts of mesoglea because of the high salt content compared with organic compounds. Combustion requires addition of such high amounts of benzoic acid that accuracy is decreased (Lutcavage and Lutz, 1986). A single measurement of the energy content of a scyphozoan has been made using bichromate oxidation (Shushkina and Musayeva, 1982). Oxidation with bichromate may not be complete (Ostopenya, 1965) and is not generally used by modern workers.

Measurement of the protein, carbohydrate and lipid content of the animals is possible, and the energy content may be calculated using the combustion equivalents of these compounds. Here, the combustion equivalents are assumed to be similar to those from other better investigated phyla, such as molluscs (Beukema and De Bruin, 1979). These measurements are time consuming. They are low if expressed as percentage of dry weight, as noted above. Also comparison of protein with the concentration of elemental nitrogen shows that there is a nitrogen content not measured by the usual methods for protein (Clarke, A., Holmes and Gore, 1992). This may be an amino-carbohydrate or glycoprotein not measured by conventional methods for protein such as Lowry/Hartree.

The elemental composition may be measured. These measurements are fairly routine but the interpretation is not. Carbon and hydrogen are present in carbohydrate, protein and lipid, whereas nitrogen is present primarily in protein, and phosphorus is present primarily in lipid. For many animals, such as fish, the calorific content can be routinely calculated from the CHN analysis (Gnaiger and Bitterlich, 1984). However, for gelatinous animals the values are lowered by the bound water if expressed as percentage of dry weight. The estimation of protein from the elemental nitrogen, as is routinely done for other animals, is questionable due to the unknown source of extra nitrogen noted above.

In summary Clarke, Holmes and Gore (1992) may be quoted: 'Unresolved difficulties over the residual water content and the nature of the unmeasured organic component mean that valid energy contents can be calculated neither from proximate composition nor carbon content.' In practice organic content, estimated as ash-free dry weight (AFDW) or carbon (C), is most often used in budgets for these animals, but in awareness of the limitations.

7.2 ORGANIC COMPOSITION OF SCYPHOZOA

A number of measurements of organic composition of scyphozoa have been made. Measurements of ash-free dry weight, carbon, nitrogen, and phosphorus of scyphomedusae are listed in Table 7.1. Measurements of the composition of the organic compounds are listed in Table 7.2. Data in Table 7.2 show that lipid and carbohydrate contents of scyphomedusae are very low compared with the amount of protein. Similar measurements of the composition of planulae larvae of *Chrysaora hysocella* and *Aurelia aurita* are given in Teissier (1929, 1932), Schneider and Weisse (1985), Schneider (1988b) and Lucas (1994).

The tables illustrate the difficulties in measurement of composition described in section 7.1.2. For example, the reported carbon content of *Aurelia aurita* varies by an order of magnitude. By comparison with other species it is unlikely that the measurements by Borodkin, Nalbandov and Stunzhas (1982) are correct.

There have been very few attempts to calculate energy content. The results are surprisingly consistent, given the problems discussed above.

Table 7.1 Chemical composition of scyphomedusae (data presented as percentage of dry weight)

Species	*Tissue*	*Ash-free dry weight*	*Carbon*	*Nitrogen*	*Phosphorus*	*Source*
Atolla wyvillei	W		15.6	4		Clarke, A., Holmes and Gore, 1992
Aurelia aurita	W		43.2	8.2	0.12	Borodkin, Nalbandov and Stunzhas, 1982 (see text)
Aurelia aurita	W	21	4.3	1.3		Larson, 1986d
Aurelia aurita	W	21.4–30.0				Ryndina and Polikarpov, 1988
Aurelia aurita	W		5.1–5.2	1.4	0.14	Schneider, 1988a
	G		17.3	5.0		
	OA		6.8	1.8		
	U		3.9	1.0		
Aurelia aurita	W		2.4–3.3	0.9–1.8		Matsakis and Conover, 1991
Aurelia aurita	W	25.6–46.0				Lucas, 1994
Aurelia aurita ephyra	W	33–67				Båmstedt, Martinussen and Matsakis, 1994
Cassiopea xamachana	W	35.8	11.9	2.2		Lutcavage and Lutz, 1986
Chrysaora fuscescens	W	23.6	6.5			Shenker, 1985
	U	21.6	4.4			
	OA	37.4	13.9			
	G	68.2	33.7			
Chrysaora fuscescens	W		7.0	2		Larson, 1986d
Chrysaora melanaster	U	20.6				Koizumi and Hosoi, 1936
Cyanea capillata	U	19.3				Koizumi and Hosoi, 1936
Cyanea capillata	W	37.0	11.6	0.4–1.4	0.0–0.4	Curl, 1962
Cyanea capillata	W	31.4	12.8	3.7		Larson, 1986d
	U	23	7.3	2.2		
	OA	29	3.5	2.9		
	T	35	17	5.1		
Cyanea Capillata	W	31.5	10.0	2.5		Bailey, T. G., Youngbluth and Owen, 1995

Table 7.1 (*contd*)

Species	*Tissue*	*Ash-free dry weight*	*Carbon*	*Nitrogen*	*Phosphorus*	*Source*
Pelagia noctiluca	W	31.0	8.2–9.9	1.4	0.14–0.16	Curl, 1962
Pelagia noctiluca	W	42.2–46.4				Ivleva and Litovchenko, 1978
Pelagia noctiluca	W	43.5				Davenport and Trueman, 1985
Pelagia noctiluca	W		11.4	4		Morand, Carré and Biggs, 1987
Pelagia noctiluca	W		11.4	3.3		Gorsky *et al.*, 1988
Pelagia noctiluca	W			3.6	0.2	Malej, 1989b
Pelagia noctiluca	W	26.3–29.7	7.8–9.6	2.3–2.8	0.2–0.3	Malej, 1991; Malej, Faganelli and Pezdic, 1993
Periphylla periphylla	W	29.2–46.0				Fosså, 1992
Phacellophora camtschatica	W	30				Larson, 1986d
	U	29	2.6	0.8		
	T	39	13.5	3.6		
Poralia rufescens	W	27	0.6	0.14		Bailey, T. G., Youngbluth and Owen, 1995
Rhizostoma pulmo	W			2.6		Gubareva *et al.*, 1983
Stomolophus meleagris	W	33–88				Kraeuter and Setzler, 1975
Manania atlantica	W		25	6.3		Larson, 1986d

G, gonad; OA, oral arm; T, tentacle; U, umbrella; W, whole specimen

Schneider (1988a) calculated the organic composition and hence the energy content of *Aurelia aurita* from the carbon, nitrogen and phosphorus content. He estimated the energy content of specimens greater than 2 cm in diameter to be 2.3 J/mg dry weight, and that of smaller animals to be 3.6 J/mg dry weight. Axiak and Civili (1991) quote calculations by Malej based on measurements of the protein, carbohydrate and lipid content of *Pelagia noctiluca*. The caloric value was calculated to be 3.1–4.1 J/mg dry weight. Clarke, Holmes and Gore

Table 7.2 Biochemical composition of nonsymbiotic scyphomedusae (data presented as percentage of dry weight)

Species	*Tissue*	*Protein*	*Carbohydrate*	*Lipid*	*Source*
Atolla wyvillei	W			1.1	Reinhardt and Van Vleet, 1986
Atolla wyvillei	W	16.9	1.7	4.2	Clarke, A., Holmes and Gore, 1992
Aurelia aurita	W			1.2	Joseph, 1979
Aurelia aurita	W	5.9	2.9	1.9	Schneider, 1988a
	G	23.7	14.6		
	OA	7.3	2.6		
	U	4.2	1.5		
Aurelia aurita	W			0.2	Holland, Davenport and East, 1990
Aurelia aurita	W	2.1–28.6	0.4–1.1	1.2–3.4	Lucas, 1994
	G	4.4–23.0	1.1–2.1	2.6–6.0	
	OA	4.0–15.3	0.6–1.5	1.3–4.0	
	U	2.3–8.3	0.3–0.9	0.9–2.9	
Chrysaora hysocella	W			2.7	Holland, Davenport and East, 1990
Cotylorhiza tuberculata	U	12.0		0.7	Carli *et al.*, 1991
	UM	7.6		0.5	
	OA	20.0		6.4	
	G	36.8		6.0	
Cyanea capillata	W	9.6	1.0	1.6	Bailey, T. G., Youngbluth and Owen, 1995
Cyanea lamarcki	W			0.7	Holland, Davenport and East, 1990
Pelagia noctiluca	W	10.9–19.8	0.1–0.7	1.3–2.9	Malej, 1991; Malej, Faganelli and Pezdic, 1993
Poralia rufescens	W	0.2	0.1	0.4	Bailey, T. G., Youngbluth and Owen, 1995
Rhizostoma pulmo	W			2.3	Gubareva *et al.*, 1983
Rhizostoma pulmo	U	8.7		0.7	Carli *et al.*, 1991
	UM	13.7		1.0	
	OA	27.0		0.8	
	G	18.0		1.2	

G, gonad; OA, oral arm; U, umbrella; UM, umbrella margin; W, whole specimen.

(1992) estimated the energy content of *Atolla wyvillei* as 6.0 J/mg dry weight based on either the proximate composition or the carbon content, assuming 10% residual bound water and conversion factors from Gnaiger and Bitterlich (1984).

7.3 GROWTH CURVES

7.3.1 Laboratory data

In order to trace the growth of an individual medusa, or known group of medusae, they must be raised in the laboratory. Reproductive modes of a number of species have been examined in the laboratory, as described in Chapter 6. Fragmentary quantitative data have been obtained on the growth from ephyra to young medusa of a few species based on measurements of the diameter of the umbrella. Raising of *Cephea cephea* to 35 mm (Sugiura, 1966) and of *Aurelia aurita* to 10 mm (Olesen, Frandsen and Riisgard, 1994) are examples of such experiments. Scyphozoa can now be reliably raised for public aquarium display. Although they usually do not grow as fast as in the field, there is now the potential for laboratory experimentation on the factors influencing growth.

7.3.2 Field data

Most of the data on growth of scyphomedusae comes from tracing cohorts in the field. Ephyrae, whether from strobilation or from direct development from planulae, may be formed in a restricted period of the year. The relatively synchronous growth of the cohort from ephyrae to mature medusae may be traced (Figure 7.1). Normally the bell diameter is measured. The diameter may be converted to ash-free dry weight or carbon using separately determined composition data (section 7.1), and specific growth rates may be calculated. The main disadvantages of this method are that the nutritional and metabolic history is unknown.

Field data, presented as increase of umbrella diameter over a growth season, are available for semaeostome medusae *Aurelia aurita* (see Mironov, 1967; Yasuda, 1969, 1971; Hamner and Jenssen, 1974; Möller, 1980a; Panayotidis, Anagnostaki and Siokou-Frangou, 1986; Schneider, 1989a; Lucas and Williams, 1994), *Cyanea* sp. (see Gröndahl and Hernroth, 1987; Brewer, 1989) and *Pelagia noctiluca* (see Malej and Malej, 1992). Similar data for rhizostome medusae include those for *Cotylorhiza tuberculata* (see Kikinger, 1986, 1992),

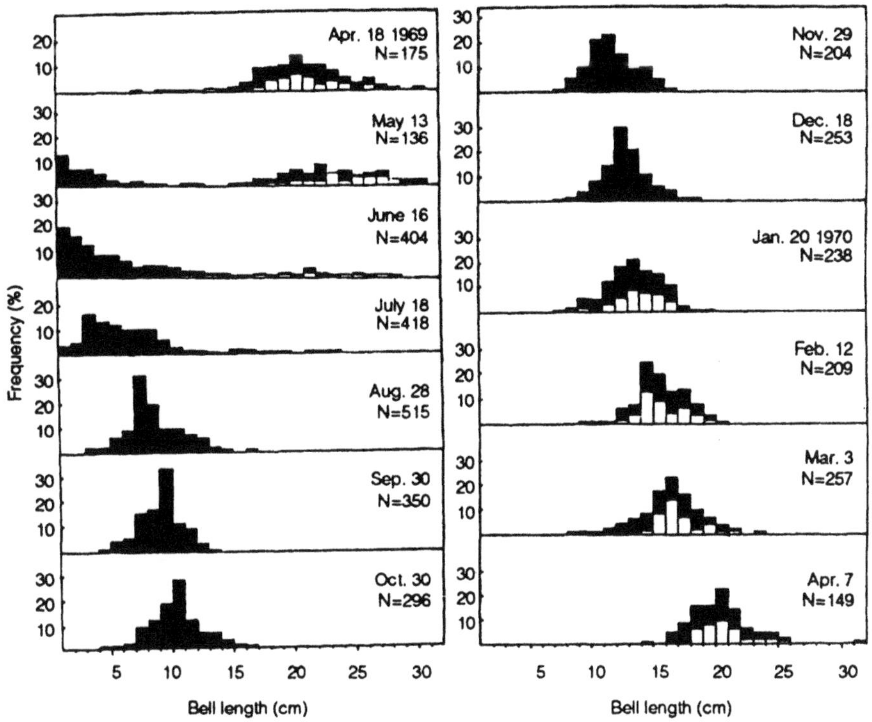

Figure 7.1 Growth of a cohort of *Aurelia aurita*. Monthly change in bell diameter of medusae collected from Urazoko Bay, April 1969 to April 1970. Unshaded area indicates the proportion of individuals brooding eggs or planulae. (Source: Yasuda, 1971, with permission of Japanese Society of Fisheries Science.)

Mastigias papua (see Sugiura, 1963), *Rhizostoma pulmo* (see Thiel, M.E., 1966; Husson and Fay, 1984) and *Rhopilema esculenta* (see Li, P., Tan and Ye, 1988).

The growth curve is sigmoid if it covers the entire period from appearance of ephyrae to production of mature medusae (Figure 7.2). The growth of the ephyrae is slow, that of the young medusa is rapid, and that of the mature medusa is again slowly approaching an upper asymptote, or even negative.

Conditions governing growth are more easily examined if growth rate is considered independently of the organism's size. The specific growth rate (u) is the rate of increase in a dimension or type of content per unit of the original dimension or content. This allows a comparison between various measurements as well as various ages, cohorts, or populations. In early work u was calculated directly from the change in diameter. In more recent work the diameter is usually converted

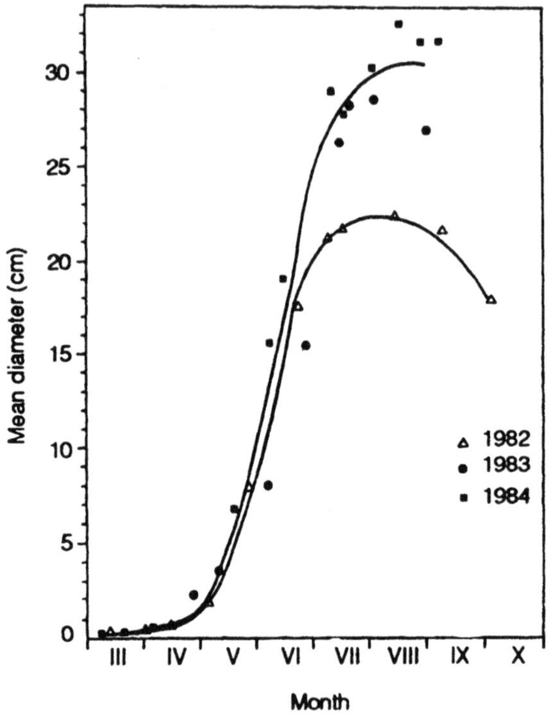

Figure 7.2 Sigmoid growth of *Aurelia aurita* medusae in Kiel Bight, 1982–1984. Data points show mean diameters for each cruise. Variability was ±30% in ephyrae and small medusae, and 15–20% in large specimens. (Source: Schneider, 1989a, with permission of Springer-Verlag.)

to ash-free dry weight or carbon. If growth is exponential, then the instantaneous specific growth rate may be calculated from the equation:

$$K = \frac{\ln W_T - \ln W_0}{T} = \frac{1}{T}\,[\ln W_T - \ln W_0] = \frac{1}{T} \times \left[\frac{W_T}{W_0}\right]$$

where T is time (days) and W_0 and W_T are the initial weight and the weight at time T, respectively.

Growth of scyphozoa is approximately exponential over the period of rapid growth of the young medusae, shown as a straight line on a semi-log plot. Figure 7.3 shows growth data for a population of *Aurelia aurita* plotted in this way (Möller, 1980a). Kruger (1968) plotted growth of 20–200 mm *Rhizostoma pulmo* in this way and found that the rate of growth increased with increasing body size. He proposed a new growth type based on this plot. However, subsequent workers

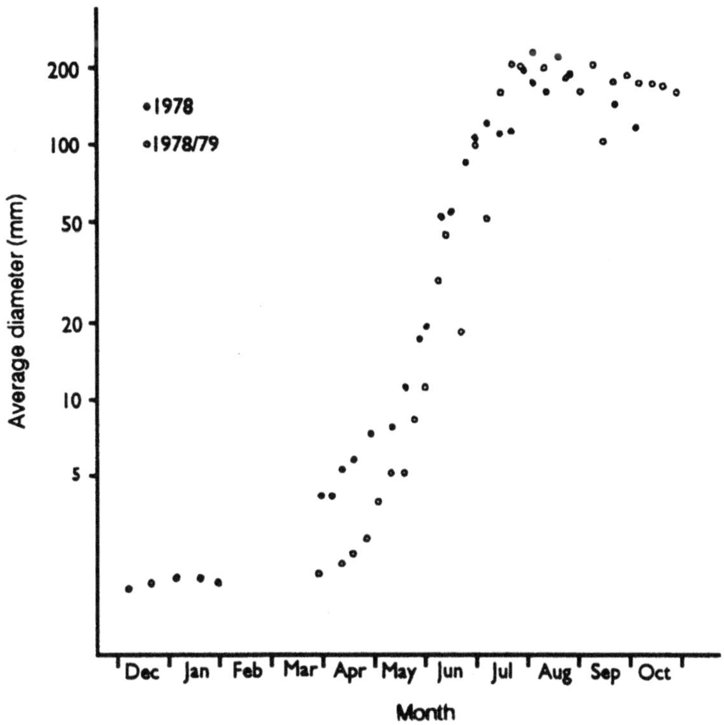

Figure 7.3 Growth of *Aurelia aurita* medusae in Kiel Fjord, 1978–1979, plotted against time on semi-log paper so that the exponential phase is shown as a straight line. (Source: Möller, 1980a, with permission of H. Möller and Springer-Verlag.)

have shown that scyphozoa show this type of growth when only the basal portion of the sigmoid curve is plotted, but normal sigmoid curves when the entire life cycle is plotted (Zaika, 1972).

The main interest in scyphozoa has been with this exponential growth rate of the young medusae, although it must always be remembered that this only represents a portion of the growth curve. The maximum specific growth rate of *Mastigias papua* is 0.3/day at less than 3 cm diameter, and it declines rapidly as diameter increases (Muscatine, Wilkerson and McCloskey, 1986). Similarly the specific growth rate of *Linuche unguiculata* declines from 0.13/day to 0.02/day over the growing season (Kremer *et al.*, 1990). For *Aurelia aurita* specific growth rates ranging from 0.1 to 0.3/day have been calculated from data in the literature from various locations (Larson, 1986c; Båmstedt, 1990). Olesen, Frandsen and Riisgard (1994) measured a maximum specific growth rate of 0.2/day for small *A. aurita* in the laboratory but observed a maximum of only 0.09/day in a shallow fjord. Similarly Brewer (1989) measured, or calculated from literature data, a range of instantaneous growth rates from 0.02 to 0.13/day for *Cyanea* sp. These correspond to doubling of the mass of these medusae during exponential growth in 5–34 days.

It is not known how the rate of growth is limited. In the field, it may be limited by the level or type of available food (section 7.6.2). In the laboratory, maximum specific growth rates of small *Aurelia aurita* medusae are obtained with moderate prey densities and do not increase at higher prey densities even though the ingestion rate continues to increase (Olesen, Frandsen and Riisgard, 1994). There may be limitations on assimilation or distribution of nutrients.

7.3.3 Life span

The life span of scyphomedusan individuals in the field is not known, although it can be assumed that the life span of individuals must be equal to or less than the period of occurrence of the population of medusae. In the field, populations of scyphomedusae often disappear after reproduction. For example, *Aurelia aurita* ephyrae produced in May in Urazoko Bay, Japan, develop into medusae which bear fertilized eggs and planulae by late January of the next year (Yasuda, 1971) (Figure 7.1). That generation of medusae disappear by late June, overlapping in time with the next cohort of ephyrae. The ephyrae and medusae stages of one cohort have therefore lasted for approximately 14 months. Other scyphomedusae may not survive through the winter, the species depending on the benthic stages for continuation of the life cycle. An example is *Cotylorhiza tuberculata* (see Kikinger, 1992) (Figure 6.6). Benthic stages may also be present for restricted periods of the year (see for example the *Cyanea* sp. populations shown in Figure 6.13).

It is not clear to what extent mortality is due to senescence, or to other causes, such as starvation, physical factors, predation, parasitism or disease, which will be discussed in this and the next two chapters. However, in the laboratory life spans may exceed those in the field, i.e. physiological longevity may be greater than ecological longevity. Individual polyps of *Cyanea capillata* may survive for two years or more (Brock and Strehler, 1963), those of *Aurelia aurita* for nearly three years (Spangenberg, 1965a). Medusae of *A. aurita* may live two years in captivity, and those of *Cassiopea* sp. for four years (Zahn, 1981).

For some medusae, mortality is correlated with reproduction and possibly caused by it. Gastric cirri of *Aurelia aurita* are extruded at the same time as sexual products (Spangenberg, 1965a). The medusa then undergoes morphological degradation, with reduction of tentacles, shortening of the oral arms and shrinkage of the central disc (Hamner and Jenssen, 1974; Möller, 1980a). Similarly a decreased number of *Cyanea* sp. feed during brooding, and there is a sequential deterioration of the tentacles, oral folds, gonads and sub- and exumbrellar epithelium (Brewer, 1989). The presence of gonads may also attract increased predation and parasite infestation. For example, the infestation of medusae by the amphipod *Hyperia galba* in the German Bight increases greatly during this period (Dittrich, 1988).

7.4 STARVATION AND REGENERATION

7.4.1 Degrowth and regrowth

There is little evidence of stored nutrients in scyphozoa. Lipid droplets are sometimes seen in the gastrovascular tracts, if lipid-rich prey are eaten (Larson and Harbison, 1989). They are not visible in other portions of the animal. There have been no tests reported for the presence of glycogen in these animals.

If scyphomedusae are starved they show 'degrowth', i.e decrease in the general structure of the body. The respiratory rate decreases as described in sections 5.2.3 and 5.3.1. They may be able to survive for some time without food, by slowly growing smaller. A starved *Cassiopea xamachana* can survive at least 42 days without food, shrinking to less than 1% of its original weight (Mayer, 1917). In this species the daily loss of weight is proportional to the weight of the animal throughout starvation (Mayer, 1914b; Cary, 1916; Hatai, 1917), whereas in *C. andromeda* the proportional rate of loss increases as starvation progresses (Gohar and Eisawy, 1961a).

During starvation of *Aurelia aurita* medusae the gonads are resorbed first. The spermatogonia that have already been formed continue to mature, but the remainder of the gonad decreases in size and shows dedifferentiation (de Beer and Huxley, 1924; Hamner and Jenssen, 1974). There may also be some decrease in bell diameter compared with that of the oral arms, but most of the degrowth is a reverse pattern to that of normal growth down to a diameter of approximately 2 cm. The polyps can also survive starvation for up to 30 days (Hiromi *et al.*, 1995).

When food is returned the medusa regrows toward and then may exceed the original size (Hamner and Jenssen, 1974) (Figure 7.4). If the medusa has lost the ability to produce gametes, it may again become fertile. In the laboratory this may be done repeatedly, showing

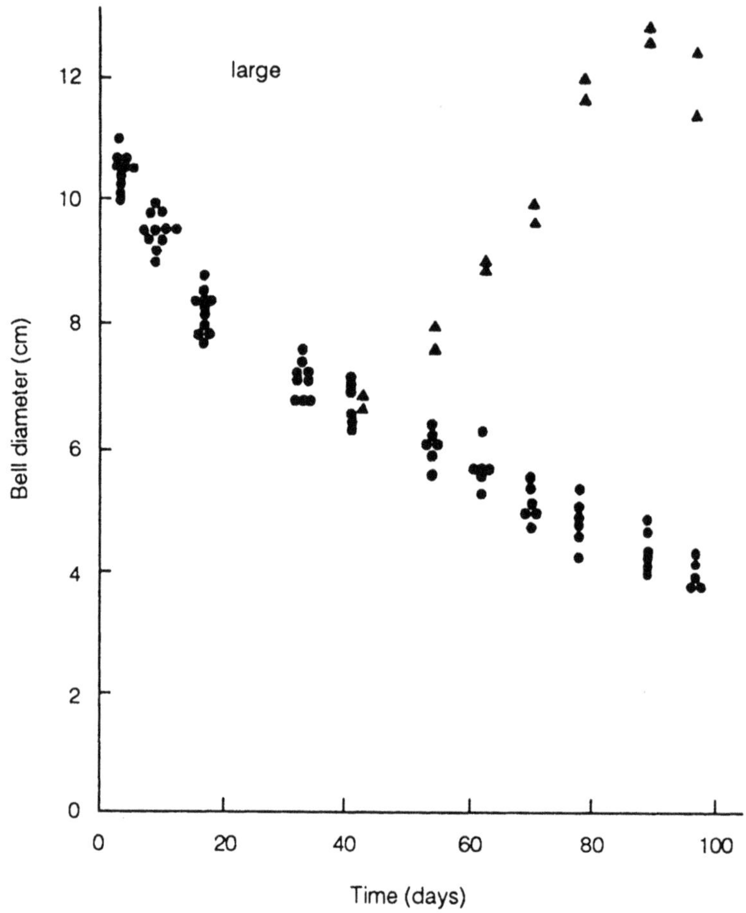

Figure 7.4 Degrowth and regrowth of *Aurelia aurita*. Sexually mature medusae of both sexes were starved until the gonads were completely regressed. On day 40 several were removed and fed separately until they again became sexually mature as determined by biopsy. Data points indicate selected individual animals. (Source: Hamner and Jensen, 1974, with permission of W.M. Hamner and American Society of Zoologists.)

that sexual maturation is a size or nutrient dependent phenomenon (section 6.2.2). Within the variation in these experiments, catch-up growth was not apparent, regrowth being at approximately the same rate as the original growth.

7.4.2 Regeneration

The ability to regenerate injured tissue is well developed in scyphozoa. Medusae reform lost marginal sense organs, as well as other appendages and internal structures (Hargitt, 1904; Stockard, 1908). Coronate polypoid colonies can regenerate from excised individual polyps (Werner, 1979).

There has been extensive investigation of regeneration of portions of scyphistomae. Portions as small as the distal one-third of the tentacles of *Aurelia aurita* can regenerate complete polyps (Lesh-Laurie and Corriel, 1973). The process requires DNA synthesis and cell division (Lesh-Laurie, Hujer and Suchy, 1991). This is an exception to the general rule that regeneration is prevented in proximal (basal) structures by the presence of more distal tissue. For example the isolated calyx of *Cassiopea* sp. does not regenerate the proximal parts but stem fragments will form a complete scyphopolyp (Curtis and Cowden, 1972, 1974; Polteva, Znidaric and Lui, 1985).

Isolated polyp epidermis, or even dissociated epithelia of scyphistomae, may also reform polyps. Dissociated cells have been formed from *Cassiopea* sp. epithelia by mechanical separation following temperature shock (Schmid *et al.*, 1981), and from *Chrysaora quinquecirrha* epithelia by trypsinization (Black and Riley, 1985). In the latter case reaggregates from the oral end of the polyp developed tentacles and mouths first and basal structures later, whereas cells from the lower gastric region formed basal structures first. It is not known what properties of the cells cause these differential reactions.

7.5 CONVERSION EFFICIENCIES

Growth efficiency is the efficiency with which ingested nutrients are converted into organic material of the consuming animal. Gross growth efficiency (GGE) is the percentage of ingested food converted, whereas net growth efficiency (NGE) is the percentage of assimilated food converted. As discussed in section 4.1.2 assimilation measurements are difficult in pelagic animals without discrete feces, so only gross growth efficiency has been directly measured for scyphomedusae.

The much quoted value of 37% for the GGE of *Cyanea capillata* and *Aurelia aurita* was based on only one specimen of each species fed a mixed diet (Fraser, 1969). More recent measurements are lower. Growth efficiencies of three *Chrysaora quinquecirrha* were 2–10% (Larson, 1986a); those of four *Pelagia noctiluca* were 7–12% (Larson,

1987d) and those of two *Drymonema dalmatinum* were 3–5% (Larson, 1987c) when fed other medusae. All of these measurements were on a wet weight basis. There is a need for further work using greater numbers of experimental animals and various types and levels of prey.

There have also been estimations of net growth efficiency based on the assumption that growth and respiration are equivalent to total assimilation. The extent of anaerobic metabolism is unknown for scyphomedusae (section 5.1), but for individuals growing in a well aerated environment prior to reproduction, the assumption may approximate reality. The estimated values of NGE for *Aurelia aurita* are 35% at maximum growth rate, but 18% and 24% under lower food conditions (Schneider, 1989b; Olesen, Frandsen and Riisgard, 1994).

It would be of interest to investigate the conversion efficiency based on content of organic matter in predator and prey. The low concentration of organic material in scyphomedusae compared with that of much of their prey makes it possible for medusae to attain large size with relatively low food intake. However, the large bulk may make organic conversion inefficient. Assuming that the assimilation efficiency was equal, Pepin, Shears and de Lafontaine (1992) estimated that the ratio of prey eaten to metabolic demands, reflected in oxygen consumption, was much higher for a stickleback than for *Aurelia aurita*. This would allow the fish more potential for growth.

7.6 DIETARY REQUIREMENTS

7.6.1 Energy budget

For growth to occur, there must be a surplus of nutrient intake beyond that needed for maintenance. However, the assumptions involved in prediction of feeding rates from metabolic and growth measurements presently make these predictions highly speculative (Arai, in press). The limitations of present measurements of rates of respiration, excretion of ammonia and somatic growth have been discussed (sections 5.2 and 5.3, and this chapter). The extent of anaerobic metabolism and of elimination of non-ammonium nitrogenous products and food wastes is unknown (sections 5.1, 5.3). Possible mucus production has not been examined. Reproduction does not usually occur during periods of maximum growth, but must be added to budgets for mature medusae (see section 6.2.2 re rates). On the intake side the extent of uptake of microplankton and dissolved organic material is also little known (sections 3.6 and 4.4).

7.6.2 Food supply

One question is to what extent growth in the field is limited by food supply. Specific growth rates in the field may be less than maximum rates measured in the laboratory (Olesen, Frandsen and Riisgard, 1994). For example, comparison between minimum rotifer concentrations required for maximum specific growth of *Aurelia aurita* in the laboratory and rotifer concentrations in a small shallow fjord showed that rotifers were at a lower concentration in the field conditions. If no other sources of nutrients were significant, *A. aurita* was food limited in the field. However, the medusae did grow faster in the field than in the laboratory at the same rotifer concentration, and continued to grow even when the zooplankton biomass became extremely low. This could be due to other sources of nutrients, such as ciliates or dissolved organic matter.

Other estimates have suggested food limitation in populations of *Aurelia aurita* (see Anninsky, 1988b; Båmstedt, 1990) and in *Pelagia noctiluca* (see Malej, 1989a, 1991). In the western Baltic adult *A. aurita* are smaller and lighter when abundant but larger and heavier in years of low abundance (Schneider and Behrends, 1994). It is probable that growth is limited by food in some situations. However, better information on alternative nutrient sources and predator and prey abundance and behaviour is needed before most field situations can be evaluated.

8 *Physical ecology*

8.1 BIOMASS

The biomass of scyphozoa is often very high. Swarms of *Chrysaora fuscescens* off Oregon reach concentrations of 18 litres of medusae per 10^3 m^3. This medusa density contains 50 mg C/m^3, at least 80% as much carbon as the densest concentrations of copepods along the same coast (Shenker, 1984).

There may be considerable interannual variations in biomass. Hay *et al.* (1990) provided a summary of 15 years of surveys by ICES member countries in the North Sea. They found a large variability in both total and relative abundances of *Aurelia aurita*, *Cyanea capillata* and *C. lamarcki* between years and areas of the sea. Schneider and Behrends (1994) found summer median abundances of *A. aurita* in Kiel Bight (western Baltic Sea) ranging from 0.2 to 16 individuals per 100 m^3.

Massive 'blooms' may occur when populations increase and decrease over periods of several years. For example, *Pelagia noctiluca* populations increased in the northwestern, central and Adriatic portions of the Mediterranean Sea from 1977 to a peak in 1981–1983, and declined again by 1986. The increased population exerted an impact on human activities such as fishing and tourism. A three-year regional programme of research, coordinated by the United Nations Environment Programme, produced a mass of new information on the species (Rottini-Sandrini *et al.*, 1991). The blooms were shown to be natural, recurrent phenomena. No cause was definitely identified. Causes hypothetically proposed included climatic and hydrological changes, hormesis (increase in growth with environmental stress such as reduced salinity) and reduced predation and competition (Legovic, 1987, 1991; Goy, Morand and Etienne, 1989; Axiak and Civili, 1991).

There has also been a massive increase and decrease in the *Aurelia aurita* populations of the Black Sea. Between surveys in 1949–1962 (Mironov, 1971) and surveys in 1978 the biomass of *A. aurita* increased to approximately 60 times the earlier level (Gomoiu, 1981; Zaitsev and Polishchuk, 1984). As a result, the wet weight of *A. aurita* exceeded by at least an order of magnitude the mass of all remaining plankton (Vinogradov, M.Y. and Grinberg, 1979; Shushkina and Musayeva, 1983). In estimates of caloric values, the medusae represented at least 45% of the plankton. It was presumed, although not proven, that the increase was due to eutrophication of the sea. In the 1980s the ctenophore *Mnemiopsis* was introduced, and radically affected the pelagic fauna of the sea. The population of *A. aurita* dropped abruptly (Shushkina and Musayeva, 1990; Shushkina and Vinogradov, 1991; Mutlu *et al.*, 1994; Lebedeva and Shushkina, 1994).

8.1.1 Measurement

The measurement of the biomass of scyphomedusae presents particular problems. As with most gelatinous material, most scyphozoa are delicate. The sheer bulk of larger scyphomedusae makes shipboard work a necessity, rather than allowing counts in the comfort of a shore-based laboratory. In addition many scyphomedusae are highly aggregated into swarms, or present very near the surface.

The biomass is often greatly underestimated by conventional methods. Comparison of data from net collections and observation from submersibles indicate that high proportions of medusae escape nets. Vinogradov and Shushkina (Vinogradov, M.Y. and Shushkina, 1982) estimated the number of *Aurelia aurita* present in the Black Sea using a steel wire cube observed from the *Argus* submersible. Simultaneous collections were made using a 50 cm conical net. Approximately three times the catch of *A. aurita* were observed from the submersible. Divers observed jellyfish being deflected by the 'bow wave' in front of the net, rather than being trapped in the net. Similarly comparison of numerical data from an underwater TV camera with that from a 1 m^2 plankton net indicated that approximately half of the medusae were caught by the net (Gomoiu, 1980).

The other problem with net collections is that they simply average the concentrations of medusae over the particular line traversed. When populations of medusae vary over several orders of magnitude, inside and outside swarms, the variance in net samples becomes very high (de Wolf, 1989). Reasonable accuracy in estimating the mean density of organisms in the sea requires large numbers of samples.

Measurements of biomass are often made for purposes of evaluating the importance of a species in a food web. The responses of the species may be very different inside and outside a swarm. Additional information on the size, distribution and biological features of the aggregations is necessary (Omori and Hamner, 1982). Near the surface observations may be made by scuba divers (Biggs, Bidigare and Smith, 1981; Biggs *et al.*, 1984). At depth observations may be made using echo-sounders (Inagaki and Toyokawa, 1991) or towed cameras. A remotely operated vehicle (ROV) may extend the use of a camera to areas away from the ship such as under pack ice, and may allow observation of individual medusae for prolonged periods (Bergström, B.I., Gustavsson and Strömberg, 1992). Manned submersibles are invaluable for direct observation but are unfortunately very expensive (Vinogradov, M.Y. and Shushkina, 1982; Mackie and Mills, 1983).

Biomass may be expressed as ash-free dry weight or carbon. The same methods, and limitations on accuracy, apply as were discussed for growth in section 7.1.

8.1.2 Production

The production of a population of zooplankton is the total amount of new biomass produced in a unit of time. Production includes growth, loss due to mortality, reproduction, and minor production such as mucus production in cnidaria. Unfortunately, there is no quantitative data on mucus production of scyphozoa, and very little on mortality or reproduction. Most estimates of production of scyphozoa have examined the period of rapid growth of a medusa cohort, neglecting mortality prior to maturity (Van der Veer and Oorthuysen, 1985; Schneider, 1989b; Lebedeva and Shushkina, 1991; Olesen, Frandsen and Riisgard, 1994).

In continuously reproducing populations it is possible to assume approximate steady state, and determine the integrated production over a time period from observations of the instantaneous production rate. With this method mortality need not be measured. García (1990) applied this method to production of the symbiotic rhizostome *Phyllorhiza punctata* in a Puerto Rico lagoon, although definite cohorts were present. He found maximum production rates of 1.6 mg AFDW/m^3 per day and 14.3 mg AFDW/m^3 per day in autumn and summer, respectively.

8.2 MORTALITY AND ADAPTATION TO PHYSICAL FACTORS

It is generally assumed that the distribution and survival of scyphozoa are affected by a number of physical factors such as temperature and salinity. However, there is surprisingly little known about survival limits even for the medusa, and hardly any data on effects on the other stages in the life history. Physical factors correlated with distributions may not be acting on the scyphozoan at all but rather on another associated animal such as its prey.

8.2.1 Temperature

Because temperature affects the rates of chemical reactions it influences most functions of poikilotherm animals. It has already been discussed with reference to swimming (section 2.6.2), feeding (section 3.6.2), digestion (section 4.2.3), uptake of dissolved organic matter (section 4.4), respiration (section 5.2.4), podocyst formation (section 6.4.2) and strobilation (section 6.4.3). The present section will concentrate on temperature effects on rates of survival and locomotion.

To investigate survival an animal is first held at a particular non-lethal temperature until it is acclimated or adjusted to that temperature, then tested by transfer to possibly lethal higher or lower temperatures. For example, *Chrysaora quinquecirrha* polyps were acclimated to 10.5°C and then tested at a series of different temperatures for 24 hours each (Mihursky and Kennedy, 1967). The LD-50, or upper lethal temperature dose killing 50% of the test animals, was 35°C. Such responses of an animal to thermal stress depend on its thermal history. Acclimating polyps or medusae by holding them at higher or lower non-lethal temperatures changes the LD-50 temperature. Higher acclimating temperatures aid survival at somewhat higher test temperatures (Gatz, Kennedy and Mihursky, 1973; Li, M. *et al.*, 1992).

It is not clear which adjustments to higher or lower temperatures affect the lethal temperatures. There are changes in enzymatic levels. For example, *Chrysaora quinquecirrha* polyps acclimated to cold temperatures show a rise in glucose-6-phosphate dehydrogenase (Blanquet, 1972b). If there is an abrupt change into a stressful but nonlethal higher temperature there is production of a suite of heat shock proteins (Black and Bloom, 1984).

There is also acclimation of pulsation rates of medusae. As early as 1914 Mayer showed that *Aurelia aurita* from Halifax or from the Tortugas were acclimated to the local temperatures. The pulsation rates for the Halifax population at a summer temperature of 14°C

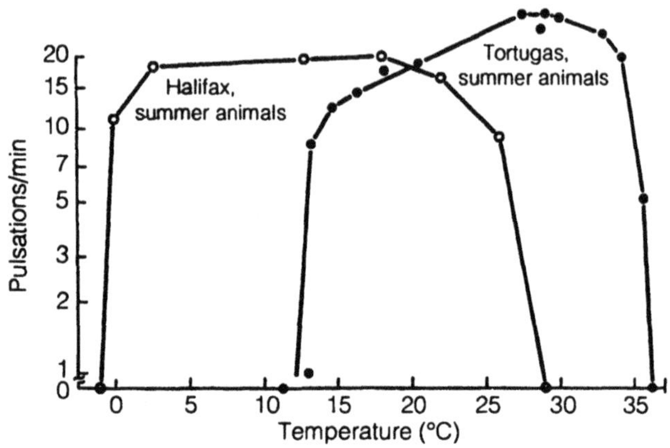

Figure 8.1 Rate–temperature curves for pulsation rates of northern and southern populations of *Aurelia aurita*. The Halifax population was acclimated to 14°C and the Tortugas population to 29°C. Acclimation is not perfect so that the southern animals still have a slightly higher pulsation rate than the northern ones at their respective acclimation temperatures. When observed following transfer to higher temperatures, the Tortugas population remained active at higher temperatures than the Halifax population. (Source: Bullock, 1955, redrawn from Mayer, 1914a, with permission of Cambridge University Press.)

were only slightly lower than those of the Tortugas population at a summer temperature of 29°C (Mayer, 1914a) (Figure 8.1). It should be noted that this data refers to two different populations of the same species. Similar acclimation has been demonstrated within single populations of *Chrysaora quinquecirrha* and *Cyanea capillata* (see Gatz, Kennedy and Mihursky, 1973; Mangum, Oakes and Shick, 1972).

8.2.2 Salinity

Survival in low salinity depends on the ability to survive dilution of the body fluids as was described in section 5.4. *Rhopilema esculenta* medusae can survive down to 8‰, the scyphistomae to 10‰ and the planulae to 12‰ (Lu, Liu and Guo, 1989). At the other extreme *Phyllorhiza peronlesueuri* has been collected from hypersaline water in Shark Bay, Australia where stromatolites were forming (Goy, 1990).

Recent increases in salinity levels in the Baltic and Azov Seas have influenced the distribution of some scyphozoa. In the Baltic Sea this has allowed expansion of *Aurelia aurita* and *Cyanea capillata* northwards in the sea (Hela, 1952; Segerstråle, 1951, 1953; Palmén, 1953; Hernroth and Ackefors, 1979; Haahtela and Lassig, 1967; Schulz, 1989). Following salinization of the Azov Sea *Rhizostoma pulmo* and

A. aurita have expanded into that sea from the Black Sea (Moshina, 1974; Zakutskiy, Kuropatkin and Gargopa, 1988). It is not known whether the geographic limits are set by direct effects of salinity on the medusa or some other stage of the life cycle.

8.2.3 Pollution

In modern seas, scyphozoa are exposed to a variety of sources of pollution in addition to the natural variables. There may be eutrophication caused by nutrients derived from agricultural lands and domestic sewage. There are other types of pollution including hydrocarbons (both oil and chlorinated hydrocarbons such as DDT), and heavy metals such as cadmium, copper, lead, mercury and zinc.

Eutrophication may be associated with an increase in numbers of scyphozoa. In a Mexican lagoon subject to eutrophication from tourist activity, the density of *Cassiopea frondosa* and *C. xamachana* was 42 medusae/m^2, whereas in nearby undisturbed lagoons it was 15 medusae/m^2 (Collado-Vides, Segura-Puertas and Merino-Ibarra, 1988). Populations of more than 1500 *Aurelia aurita* medusae/10^3 m^3 have been observed in Elefsis Bay, one of the most eutrophic areas of Greece (Papathanassiou, Panayotidis and Anagnostaki, 1986, 1987; Panayotidis, Anagnostaki and Siokou-Frangou, 1986). The large populations associated with eutrophication in the Black Sea were described in section 8.1. It is not known to what extent these increases are due to the ability of the medusa or polyp to utilize the increased nutrient, or to an another effect such as lack of predation (Wilkerson and Dugdale, 1984).

A variety of types of oil are released into the sea and cause acute mortality as well as chronic physiological and carcinogenic effects (Suchanek, 1993). Scyphozoa may be relatively hardy. For example, *Rhizostoma* sp. were among the species surviving in diesel oil polluted zones of Madras harbour (Fernandez, Daniel and Nicket, 1977). However Alaska crude petroleum, as well as a number of particular petroleum hydrocarbons, causes reduction or cessation of strobilation of *Aurelia aurita* polyps, and production of ephyrae and polyps with morphological and behavioural abnormalities (Spangenberg, Ives and Patten, 1980; Spangenberg, 1984, 1987).

Some pollutants become concentrated in scyphozoan tissues. *Pelagia noctiluca* in the Mediterranean Sea contain high concentrations of several elements including cadmium, lead, mercury and zinc (Cimino, Alfa and La Spada, 1983; Romeo, Gnassia-Barelli and Carré, 1987). *Aurelia* sp. from the coasts of Pakistan near Karachi contain high levels of copper and zinc (Siddiqui, Akbar and Qasim, 1988). *Chrysaora quinquecirrha* medusae concentrate the herbicide pendimethalin in the tentacles and show no change in behaviour at concentrations lethal to fish like white perch (Calton and Burnett, 1981).

8.2.4 Oxygen

Uptake of oxygen for metabolism depends on the partial pressure of oxygen in the surrounding sea water (section 5.2.5). Nevertheless scyphozoa may be present in areas of decreased oxygen.

The oxygen profile of the open ocean results in coronates being subjected to reduced oxygen. In the mixed layer near the surface of the sea the Po_2 is close to the pressure in the overlying air. There may be reduced O_2 at depths below 50–100 m where respiration exceeds photosynthesis due to decreased illumination. In most of the ocean the minimum oxygen layers are at about 400–1500 m and below this oxygen content rises again. The coronates *Periphylla periphylla* and *Nausithoe rubra* show high levels of the anaerobic enzyme lactate dehydrogenase (Thuesen and Childress, 1994; section 5.1.2), presumably as an adaptation to movement through the minimum oxygen layer.

In some restricted areas the bottom layers become stagnant and oxygen is depleted. In the Black Sea a layer of oxygen-depleted water lies over water containing hydrogen sulphide. Among the few animals surviving in this layer, with less than 0.5 ml 0_2 per litre, are planulae larvae of *Aurelia aurita* (see Vinogradov, M.E., Flint and Shushkina, 1985).

Hyperoxia is rarely encountered by nonsymbiotic species. The protections of symbiotic species against molecular oxygen produced by their symbionts was discussed in section 4.5.4. In the laboratory polyps of *Aurelia aurita* survived exposure to oxygen levels of approximately three times normal for seven days (Torres and Mangum, 1974).

8.3 DEPTH

8.3.1 Vertical distribution

Although the better known species of scyphozoa are those in the surface layers, coronate scyphozoa are present to at least 5000 m depth (Vinogradov, M.E., 1968). The epipelagic zone extends from the surface to 200 m, the mesopelagic from 200 to 1000 m, the bathypelagic from 1000 to 4000 m and the abyssopelagic from 4000 to 6000 m. Even epipelagic species of scyphomedusae can survive at least briefly at the high pressures present in deeper water. In a pressure chamber 50% of a sample of *Pelagia noctiluca* survived an hour at a pressure approximately corresponding to 5000 m (George and Marum, 1974). Nevertheless most species live in fairly restricted and characteristic depth ranges.

The depth ranges occupied are affected by buoyancy, light, pressure, presence of prey, and temperature, salinity and oxygen gradients. Buoyancy is adjusted, by exclusion of sulphate ions, so that medusae are identical in density to the surrounding sea water or slightly heavier (section 5.4.2). Light will be discussed in the next section on diel migration.

It is not known to what extent scyphozoa actively respond to pressure changes in nature, where the pressure changes slowly as a medusa swims vertically or sinks passively through the water. Pressure increases by 1 atmosphere for each 10 m increase in depth. Medusae do respond to small rapid changes in pressure in the laboratory. For example, in a pressure chamber ephyrae of *Aurelia aurita* respond to a pressure increase of an atmosphere or less by moving upward, irrespective of darkness or various orientations of light sources (Rice, 1964; Knight-Jones and Morgan, 1966; Digby, 1967). This response would have the effect of tending to keep a medusa at a constant depth.

Some medusae, present in arctic or temperate epipelagic or mesopelagic waters, show subtropical or tropical submergence. One example is *Periphylla periphylla* which lives at shallow depths north of 42°N in the eastern Atlantic (van der Spoel, 1987; Bleeker and van der Spoel, 1988). Submergence has been presumed to be due to increasing temperatures of the near surface layers towards the equator. However, although most *P. periphylla* are collected between 4°C and 11°C, they can tolerate temperatures up to 19.8°C (Larson, 1986b; Bleeker and van der Spoel, 1988). Van der Spoel and Shalk (1988) suggest that the ability to survive at higher temperatures depends on the balance between increased temperature-dependent metabolic rates and the food supply. Where vertical mixing leads to a highly productive area such as the Banda Sea, mesopelagic and bathypelagic organisms can survive in the upper layers at lower latitudes (van der Spoel and Bleeker, 1988; van der Spoel and Schalk, 1988).

8.3.2 Diel migration

Some species of medusae in the epipelagic and mesopelagic zones show diel vertical migration. Migrations may be for only a few metres or for hundreds of metres. In the northeast Atlantic, at sunset, *Atolla vanhoeffeni* migrates at least 200 m upward at a rate of at least 50 m

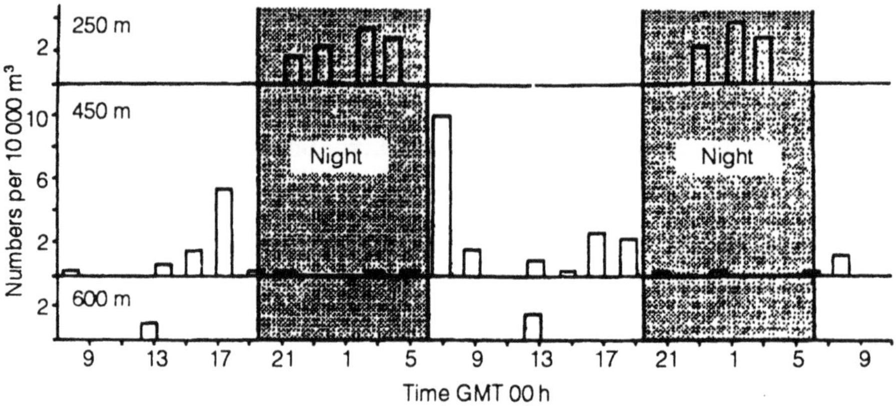

Figure 8.2 Diel migration of *Atolla vanhoeffeni* in the Northeast Atlantic. Numbers of individuals/haul per 10 000 m^3 of water filtered. (Source: Roe *et al.*, 1984. Reprinted with kind permission from Elsevier Science Ltd, The Boulevard, Langford House, Kidlington OX5 1GB, UK.)

per hour (Roe, James and Thurston, 1984) (Figure 8.2). Among the epipelagic species vertical migrations have been best described for *Pelagia noctiluca* (Franqueville, 1971; Maso and Castellon, 1985). *Mastigias* sp. in a saline lake in Palau migrate horizontally by day but make repeated vertical excursions between the surface and the chemocline at night (Hamner, Gilmer and Hamner, 1982).

The response of *Aurelia aurita* is varied. In a shallow bay in Japan it usually approaches the surface during the day and becomes scattered throughout the water column at night (Yasuda, 1972, 1973a,b, 1974, 1975a, 1982). The reverse is true in Eil Malk Jellyfish Lake in Palau (Hamner, Gilmer and Hamner, 1982). In Kiel Fjord in Germany it approaches the surface both at midday and at midnight, as do the copepods (Möller, 1984a). The reverse is true in Elefsis Bay, Greece (Papathanassiou, Panayotidis and Anagnostaki, 1987). In Saanich Inlet, British Columbia, Canada, it floats to the surface on still nights and migrates horizontally during the sunlit days (Hamner, Hamner and Strand, 1994).

Diel migrations of epipelagic medusae are at least in part active responses to light levels. *Aurelia aurita* maintained in a tank 10 m deep did not migrate in continuous darkness (Mackie *et al.*, 1981). If simulated day–night photic cycles were present they migrated, even if the cycle was 12 hours out of phase with the natural cycle. If the light intensity applied to *Pelagia noctiluca* in an aquarium was decreased from 3240 lux to 14 lux, the mean frequency of pulsation increased from 84 to 106 per minute (Axiak, 1984).

Migrations probably do not occur in the bathypelagic zone. Angel *et al.* (1982) found no evidence for migration by a number of invertebrates, including two species of *Atolla*, at 1000 m in the north Atlantic. However, migrations do occur at mesopelagic depths (Thurston, 1977; Roe, James and Thurston, 1984). In the clearest ocean water of the tropics, light, particularly in the blue range, may penetrate to more than 1000 m. It is possible therefore that migration in the mesopelagic zone is also dependent on light levels.

The rates of vertical migrations may be modified by the presence of thermoclines or haloclines. For instance most *Aurelia aurita* migrating toward the surface in Saanich Inlet remained below the thermocline (Hamner, Hamner and Strand, 1994).

8.3.3 Changes with life cycle

In some species the depth of occurrence changes with stage of the life cycle. Ephyrae may remain in deeper water during the winter when at the surface food levels are low and wave action high. They return to near-surface layers in the spring. In western Sweden the maximum release of ephyrae of *Aurelia aurita* occurs in early November (Hernroth and Gröndahl, 1985a). These ephyrae remain below the halocline through the winter and then appear in the surface layers in May. At this time there is an increase in temperature in the surface layers but neither temperature nor salinity changes occur below the halocline. A possible trigger of the ontogenetic migration may be increase in day length.

8.4 AGGREGATION AND HORIZONTAL MIGRATION

Scyphomedusae frequently occur in swarms, sometimes of immense size and density. The word swarm (rather than patch) is used accurately here because formation of these aggregations involves the activity of the medusa. In some cases this may include quite spectacular horizontal migration.

Aggregations of plankton are due to a combination of physical factors and active responses. At meso-scale (100–1000 km) or coarse-scale (1–100 km) levels, distributions are largely related to physical factors (Haury, McGowan and Wiebe, 1978). These factors may include gyres, eddies, currents, rings, upwelling, river plumes, island wakes, tides and oceanic fronts. At fine-scale (1–1000 m) level, further physical factors such as Langmuir circulation cells are added, as well as individual behavioural factors such as migration and interactions with other animals.

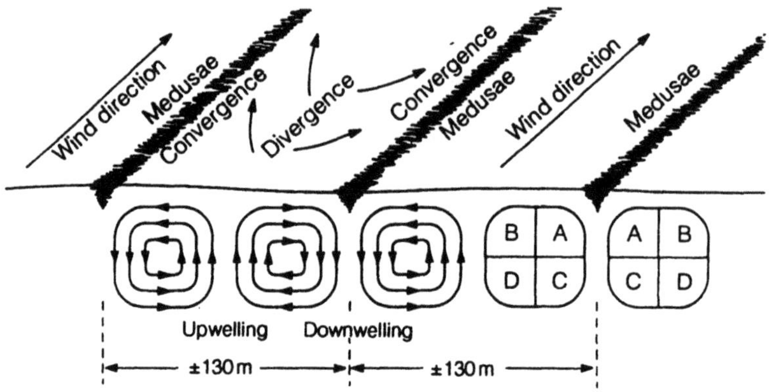

Figure 8.3 Three-dimensional diagram of Langmuir circulation cells and rows of medusae which aggregate between 'A' quadrats of adjacent circulation cyclinders. (Source: Hamner and Schneider, 1986, with permission of American Society of Limnology and Oceanography.)

Examples of meso-scale and coarse-scale effects are scattered through the literature, usually without much detail. Among better documented examples, the distribution of adult *Chrysaora hysocella* off Namibia is related to the current patterns of the Benguela system (Pagès, 1992). Juvenile *C. hysocella* off Namibia and *C. fuscescens* off Oregon are distributed near shore during coastal upwelling (Shenker, 1984). *Pelagia noctiluca* is distributed around the Maltese Islands by the eddy sea-water currents resulting from the prevalent northwesterly winds (Axiak, Galea and Schembri, 1991).

On a fine-scale, wind-driven Langmuir circulation cells may concentrate medusae. The cells form in alignment with the wind. Adjacent cells roll in opposite directions producing linear, parallel convergences and divergences (Figure 8.3). Medusae such as *Aurelia aurita*, *Chrysaora melanaster*, *Cyanea capillata*, *Linuche unguiculata* and *Stomolophus meleagris* become aggregated in the convergences swimming upward against the downwelling water (Hamner and Schneider, 1986; Shanks and Graham, 1987; Kingsford, Wolanski and Choat, 1991; Larson, 1992). *L. unguiculata* swims predominately in clockwise, circular, horizontal paths. This has the effect of maintaining the swarms for months even when the wind speed drops periodically (Larson, 1992).

If medusae are moved close to shore by wind or tide the swimming is modified. *Pelagia noctiluca* avoid both the sea surface and the bottom. They become concentrated in a wedge to form aggregations of up to 600 individuals/m^3 (Zavodnik, 1987). If *Stomolophus meleagris* bump the bottom or are tumbled by a breaking wave, they turn and swim at 180° to their initial heading (Shanks and Graham, 1987). Offshore *Aurelia aurita* and *Cyanea capillata* may 'tumble' as they swim, i.e. cyclically change direction (Seravin, 1987a,b). The cycle is modified or decreased in restricted spaces or against the shore.

In bays the populations of medusae such as *Aurelia aurita* may decrease at low tide, presumably swept out by the tidal currents (Feigenbaum and Kelly, 1984). There has been speculation by various authors that some medusae manage to remain in estuaries by selective vertical migration. However, although there is a tendency for medusae such as *Rhizostoma pulmo* to move up in the water column during higher current velocities, there is no differential effect of ebb and flood tides (Verwey, 1966).

Oriented horizontal migration has been described for only three species: *Mastigias* sp., *Stomolophus meleagris* and *Aurelia aurita*. Juvenile *Mastigias* migrate twice a day across three marine lakes in Palau, Western Caroline Islands (Hamner and Hauri, 1981; Hamner, Gilmer and Hamner, 1982). They form swarms with densities reaching $1000/m^2$ at either end of the migration routes in early morning and late afternoon. They are more dispersed during the migrations of up to 0.5 km each way. During the migrations individuals in a particular lake are oriented in the same directions (Figure 8.4), but the direction differs from lake to lake. Migration occurs according to a diel pattern and is unrelated to tidal currents. The medusae maximize their exposure to light by the migrations, and show avoidance reactions to shadows. However, the migrations are not a direct response to light, often preceding the actual light changes.

Mastigias (Figure 8.5) and *Stomolophus meleagris* also show oriented swimming in the open sea (Hamner and Hauri, 1981; Shanks and Graham, 1987). Within local populations of *S. meleagris* most individuals swim in approximately the same direction (Shanks and Graham, 1987), but populations 1 km apart may differ. If deflected by turbulence they reorient, although they may at first swim for up to 20 m in the opposite direction. The direction is not related to the sun's bearing, but may be either with or against the direction of the wind and surface waves, or of local currents.

The most complex orientation response known is that of *Aurelia aurita* in Saanich Inlet, British Columbia, Canada (Hamner, Hamner and Strand, 1994). This species has long been known to form large swarms in some localities (Kuwabara, Sato and Noguchi, 1969; Möller, 1980b; Roden *et al.*, 1990) but not in others (Hamner and Hauri, 1981). In Saanich Inlet Hamner and his co-workers examined the swimming orientation of approximately 2500 specimens under various light conditions (Hamner, Hamner and Strand, 1994;

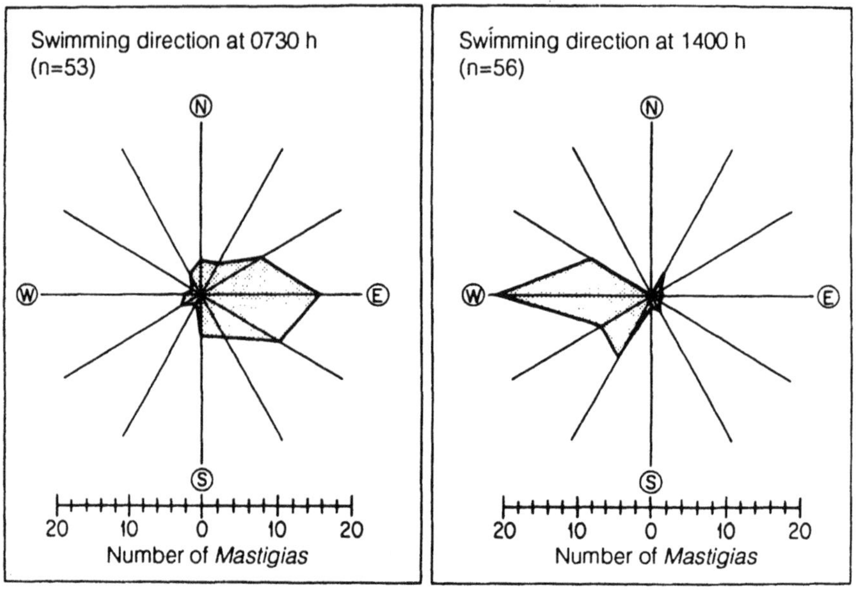

Figure 8.4 Compass orientation of *Mastigias* sp. medusae in the centre of Jellyfish Lake, Eil Malk, Palau, during diel horizontal migration back and forth across the lake. (Source: Hamner and Hauri, 1981, with permission of American Society of Limnology and Oceanography.)

Hamner, 1995). The adult medusae are oriented randomly and become passively dispersed by tidal currents when the sky is overcast or at night. When the sun is present they not only migrate, but migrate to the southeast regardless of the position of the sun!

The physiological basis for oriented swimming is not known. The muscular and nervous basis of locomotion was discussed in Chapter 2. The known neural responses explain the control of locomotion at the level of single beats, but do not even fully explain turning. The complex integration needed for horizontal migration is simply in the realm of speculation at this time.

One advantage of aggregation is that it facilitates spawning by the adults. In the swarm *Aurelia aurita* swim vertically up and down, rather than horizontally, with frequent collision and turning. Males release sperm strings (Hamner, Hamner and Strand, 1994). However adult activity is not restricted to spawning, and juvenile medusae also form swarms, so there may be other advantages to aggregations. Within aggregations medusae such as *Pelagia noctiluca* may continue to swim and fish actively (Malej, 1989a).

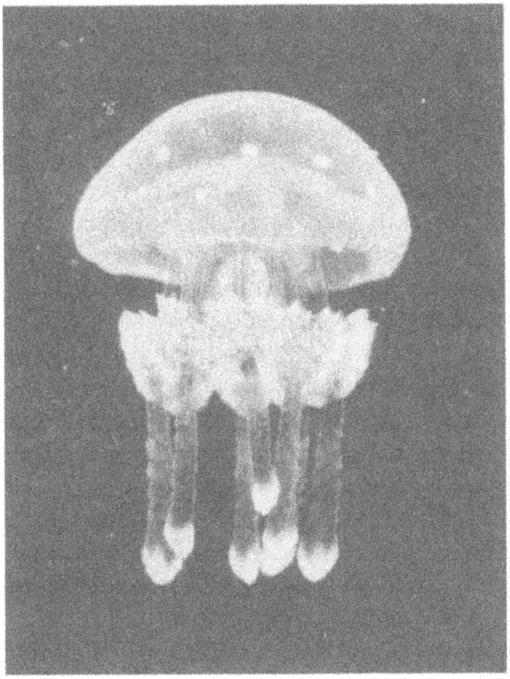

Figure 8.5 *Mastigias papua* medusa from Papua New Guinea, diameter 35 mm. (Courtesy of P.F.S. Cornelius.)

8.5 ZOOGEOGRAPHY

Much of the information on the distribution of scyphozoa is scattered through taxonomic papers and will not be included here. Information for the North Atlantic is summarized by Kramp (1947), and for the Antarctic Seas by Larson (Larson, 1986b).

Many epipelagic scyphomedusae show latitudinal distribution patterns, being found either in the waters of the tropics and subtropics, or in areas north or south of the tropics. These types of patterns may be due to temperature effects. However, the effect may not be on the medusa *per se* but rather on benthic stages, or indirectly through other animals with which there is interaction. Deeper water meso-and bathypelagic species usually have relatively wide distributions.

The most widely distributed species is *Aurelia aurita* with its many varieties. Although epipelagic it is eurythermal, and cosmopolitan in neritic areas other than the polar regions (Kramp, 1965).

Some tropical and subtropical species are present throughout the warm waters of the Atlantic, Pacific and Indian Oceans. These warm areas were connected until the late Tertiary. Other species may be restricted by recent barriers (such as the Isthmus of Panama) to one or two of the main oceans. The Order Rhizostomeae provides examples of each distribution pattern. The order includes epipelagic and neritic forms. They are mostly restricted to tropical waters, although a few ranges extend into temperate latitudes (Kramp, 1970). Four of the eight families are restricted to the Indo–West-Pacific Region.

The situation at the two poles differs. In the south an unbroken circumglobal ocean lies between Antarctica and the other continents. This allows continuous, concentric patterns of distribution around the continent (Larson, 1986b). Hence there is no barrier to movement of colder water species between the South Atlantic, the South Pacific and the Indian Ocean. In the north the North Atlantic and North Pacific Oceans are connected only through the Arctic Ocean. This connection has been intermittently closed during the late Tertiary and Pleistocene periods due to cooling of the Arctic and closure of the Bering Strait. The closures have allowed evolution of some distinct species in the two oceans, although many species are common to both (Larson, 1990).

Human activities influence the distributions of cnidaria. In addition to the effects of pollution and eutrophication discussed in section 8.2.3, many cnidaria have been transported in the sea-water ballast tanks of ocean-going vessels or as fouling organisms on the ship's hulls (Carlton, 1985). Although no scyphozoa have actually been observed attached to ships, the distributions of some scyphozoa suggest this kind of transport. *Phyllorhiza punctata*, originally described from the Indo-Pacific, is appearing near harbours used by ocean-going ships, such as San Diego Bay, California (Galil, Spanier and Ferguson, 1990; Larson and Arneson, 1990).

The building of canals may link biogeographical provinces. When the Suez Canal was opened in 1869 it linked the Red Sea with the Mediterranean. This allowed those Red Sea species hardy enough to survive the temperature and salinity variations of the canal, to spread into the Mediterranean (Spanier and Galil, 1991). The first recorded scyphozoan was *Cassiopea andromeda*, which was found in the canal in 1886 and reached Cyprus by 1903 (Galil, Spanier and Ferguson, 1990; Spanier and Galil, 1991). More recently *Rhopilema nomadica* has also appeared (Galil, Spanier and Ferguson, 1990; Spanier and Galil, 1991; Lotan, Ben-Hillel and Loya, 1992; Lotan, Fine and Ben-Hillel, 1994). It was first observed off Israel in 1977 and has been forming large aggregations there since the summer of 1986. It has been moving north, in the direction of the prevailing currents. It was first reported in Lebanese waters in 1988 and formed large aggregations there in 1991.

9 *Biological interactions*

In addition to physical factors, animals are affected by interaction with other animals. Feeding was described in Chapter 3, but the effects on the prey populations will be discussed below. Scyphozoa are also affected by predators (including humans), parasites and other associates. Transparency, pigmentation and bioluminescence have been included in this chapter because of their possible roles in affecting such interactions.

9.1 PREDATION

9.1.1 Natural predators: planktonic

Scyphozoa are eaten by a wide variety of predators. For nonquantitative purposes the presence of nematocysts or larvae of parasitic amphipods may indicate a cnidarian diet. Other species, such as the large ocean sunfish, *Mola mola*, have been directly observed feeding on scyphozoa. However, fast digestion of medusae, without resistant skeletons, has been an obstacle to quantitative measurements of predation on scyphomedusae. In order to observe portions of medusae amongst stomach contents, the guts of possible predators must be fixed immediately after collection.

Scyphomedusae are eaten by other pelagic coelenterates, including other scyphozoa. In the laboratory *Aurelia aurita* are eaten by the hydromedusae *Aequorea victoria*, *Eutonina indicans* and *Stomotoca atra* (see Arai and Jacobs, 1980) and by the scyphozoa *Chrysaora hysocella* and *Cyanea capillata* (see Lebour, 1923; Plotnikova, 1961). Field observations have confirmed predation by scyphozoa including *C. capillata*, *Drymonema dalmatinum* and *Phacellophora camtschatica* (see Loginova and Perzova, 1967; Larson, 1987c; Strand and Hamner, 1988). Apolemid siphonophores consume gelatinous zooplankters including coronate scyphozoa (Larson, Mills and Harbison, 1991).

Invertebrates feeding on scyphomedusae also include the parasites described in section 9.2, and possibly some of the associates described in section 9.3. Other predators include mesopelagic arthropods. The shrimp *Notostomus robustus* (see Larson, Mills and Harbison, 1991; Moore, P.G., Rainbow and Larson, 1993) and the gammaridean amphipod *Parandania boecki* (see Moore, P.G. and Rainbow, 1989; Coleman, 1990), feed on coronates such as *Atolla*.

Neritic species may also be captured by benthic predators. Some sea anemones eat scyphomedusae that stray within reach (Cargo and Schultz, 1967; Hamner, Gilmer and Hamner, 1982; Berryman, 1984; Fautin and Fitt, 1991). Barnacles may prey on ephyrae (Cones and Haven, 1969).

A large number of fish species eat cnidaria and ctenophora (see reviews by Arai, 1988; Ates, 1988). However, most of the studies of the contents of fish stomachs have not distinguished between scyphozoa and other gelatinous animals. The two authors list only 24 species known to eat scyphozoa *per se*. Several deep-water species of fish are included among those which eat scyphozoa. For example, below 1000 m in the North Atlantic *Atolla* sp. and *Periphylla periphylla* form a major portion of the diets of the smoothhead *Alepocephalus bairdii* and the roundnose grenadier *Coryphaenoides rupestris* (see Mauchline and Gordon, 1983; Gushchin and Podrazhanskaya, 1984; Gordon and Mauchline, 1990). Section 9.3.1 will discuss associations between larval fish and scyphozoa. Some of these species of larval fish become predators as they grow.

There has been speculation that the recent increase in populations of *Pelagia noctiluca* in the Mediterranean Sea is partly due to over-fishing of the fish (Legovic, 1987; Avian and Rottini-Sandrini, 1988). Species such as mackerel, bogue and saddle bream are predatory; others such as sardines might compete for the same food supply. However, if predators are controlling prey numbers, a negative correlation is probable between the populations of predator and prey. In the Mediterranean Sea analysis of high vs low '*Pelagia* years' showed a positive correlation between the common fish species and the populations of *P. noctiluca* (see Vucetic and Alegria Hernandez, 1988).

Sea turtles, especially leathery turtles, *Dermochelys coriacea*, feed on scyphomedusae and other pelagic cnidaria. Identifiable pieces of

Cyanea capillata and *Rhizostoma pulmo* have been found in *D. coriacea* guts (Bleakney, 1965; Duron, Quero and Duron, 1983). Tentative identification of the prey from the nematocyst types also indicates that scyphozoa are eaten by *D. coriacea* (see Den Hartog, 1980; Den Hartog and Van Nierop, 1984), as well as by loggerhead turtles, *Caretta caretta* (see Van Nierop and Den Hartog, 1984). The resemblance of the fatty acid compositions of the scyphozoa and turtles also supports the presence of predation (Sipos and Ackman, 1968; Hooper and Ackman, 1972). A leatherback turtle has been observed feeding on *Aurelia aurita* at the surface (Eisenberg and Frazier, 1983). However, *D. coriacea* are also able to dive to at least 1200 m, which may allow them to exploit deep-water medusae during seasons when surface supplies are poor (Davenport, 1988). There is no quantitative data on the amount eaten in the wild. In the laboratory hatchling turtles have been reared to six months on a diet of *Cassiopea xamachana* (see Witham and Futch, 1977; Lutcavage and Lutz, 1986).

Birds also may utilize scyphozoa in their diets. Of 17 bird species collected from the Bering Sea, 11 (including shearwaters, petrels, gulls, murres, auklets and puffins) had portions of scyphozoa in their guts (Harrison, 1984, 1990). Birds such as sandpipers may also feed on stranded medusae, particularly the gonads (Ates, 1991; Grimm, 1984). No non-human mammalian predation has been observed, although dolphins may play with medusae, throwing them into the air (dos Santos and Lacerda, 1987).

Scyphozoa have several mechanisms which reduce predation. Transparency will be discussed in section 9.1.4. It is assumed that cnidae form a defence against predation by fish, although there is little direct proof. If a *Stomolophus meleagris* is subjected to a simulated small bite, it releases clouds of nematocysts which drive off associated small fish but not crabs (Shanks and Graham, 1988). The nematocysts may be accompanied by mucus which sticks to fish, especially the gills, and increases the toxicity. For slowly swimming predators such as other scyphozoa, increased swimming may lead to escape. If *Aurelia aurita* contacts a tentacle of the predators *Cyanea capillata* or *Phacellophora camtschatica* it begins to swim rapidly and often escapes (Strand and Hamner, 1988; Hansson and Kultima, 1995). The ability to escape depends on the relative size of predator and prey, so one defence is simply to grow larger.

9.1.2 Natural predators: benthic

Scyphistomae such as those of *Cyanea capillata* and *Aurelia aurita* are subject to predation by nudibranchs. In Gullmar Fjord, Western

Sweden, the main predator is *Coryphella verrucosa* (see Hernroth and Gröndahl, 1985a,b; Gröndahl and Hernroth, 1987). A single nudibranch can consume up to 200 polyps per day. The nudibranchs mature in approximately six weeks during the period of maximum fall abundance of the *A. aurita* scyphistomae. They are a major cause of decline of scyphistomae populations but they have not been observed eating podocysts, which may therefore serve as a protective stage against this type of predation (Gröndahl, 1988a; Brewer and Feingold, 1991).

Similar predation by nudibranchs has been observed in other parts of the world. In Chesapeake Bay *Cratena pilata* (originally identified as *Coryphella* sp.) preys on polyps of *Chrysaora quinquecirrha* and other cnidaria (Oakes and Haven, 1971; Cargo and Schultz, 1967; Cargo and Burnett, 1982). The nudibranch stores the cnidae of the scyphozoan in cnidosacs at the tips of cerata, finger-like projections on the dorsal surface. If disturbed *C. pilata* curls the body ventrally, bristles the cerata, and releases nematocysts. *Dondice paraguensis* stores cnidae from the oral arms of *Cassiopea xamachana* in a similar manner (Brandon and Cutress, 1985). However, not all nudibranchs are able to eat all scyphozoa. For example, *Aplysia dactylometra* actively withdraws from any contact with *C. xamachana* (see Lederhendler, Bell and Tobach, 1975).

Other predators of polyps include caprellid amphipods, pycnogonids and decapods (Uchida and Hanaoka, 1933; Oakes and Haven, 1971; Hutton *et al.*, 1986). As in the pelagic environment, benthic stages of scyphozoa may eat one another. For example, scyphistomae of *Aurelia aurita* eat planulae larvae of *Cyanea capillata* and of their own species (Gröndahl, 1988a, b).

9.1.3 Fisheries

As well as natural predation, scyphozoa are subject to fisheries for human consumption. The most important markets are in China and Japan. In 1981 the value of scyphozoa exported from other Asian countries to Japan was US$40 million (Omori, 1981).

The fisheries are for large rhizostome medusae. The largest fishery is for *Rhopilema esculenta*. Other species utilized include *Lobonema smithi*, *Lobonemoides gracilis*, *Rhopilema hispidum* and *Stomolophus meleagris* (see Omori, 1981). Although fishing occurs on a small scale as far north as southern Korea and the west coast of Japan, and as far west as India, the main sources are China, the Philippines, Thailand, Malaysia and Indonesia (Omori, 1981; James, Vivekanandan and Srinivasarengan, 1985; Sloan, 1986). Other countries are now exploring the possibility of processing and marketing their species (Huang, Y.-W., 1988). The potential is mainly in countries with warm coastal waters and populations of rhizostome medusae. Semaeostome medusae, such as *Aurelia aurita*, yield a poor quality product after processing (Sloan and Gunn, 1985).

The medusae are fixed and preserved with a mixture of table salt and alum, and the semi-dried material is marketed (Wootton, Buckle and Martin, 1982; Huang, Y.-W., 1988). The main component of the resulting product is a collagen-like protein (Kimura, Miura and Park, 1983). To prepare the preserved medusae for eating it is soaked in water, cut into thin strips, and flavoured. It is often shredded into salads with vegetables and meat or fish (Ma, 1960; Chen, P.K., Chen and Tseng, 1983).

9.1.4 Transparency and pigmentation

The large number of transparent animals in the fauna of the upper layers suggests that transparency confers a selective advantage such as decreased predation, but there is no direct proof that transparency does affect predation. Chapman (1976a) has pointed out that the ability of a predator to discriminate depends on the ratio of the radiance transmitted by the prey to that of the background, and on the visual acuity of the predator. The background radiance near the surface depends on the angle of observation relative to the surface. The visual acuity of some fish is better than that of humans, but apart from cephalopods the visual acuity of most invertebrate predators is much less. Many, such as other scyphozoa, do not use visual information to feed. If transparency is an advantage it is probably relative to vertebrate predation.

Transparency depends on transmission of incident light rather than absorption, scattering or reflection. Transparency of scyphomedusae is largely due to the properties of the mesoglea. Even a thin layer of cells causes greater light scattering, i.e. less transmittance. For *Chrysaora* and *Aurelia* the transmittance of the isolated mesoglea is high in the visual spectrum, but falls, relative to sea water, in the UV range (Figure 9.1) (Chapman, G., 1976b). Even in the relatively transparent *Aurelia*, transmittance is greatly decreased by retention of the subumbrellar cell layer with the mesoglea.

In contrast to the transparent forms, many scyphomedusae, particularly in deep water, are heavily pigmented, i.e. there is absorption of some or all wavelengths of light. These pigments are involved in the external coloration of scyphozoa, often including elaborate radial patterning. The extent to which these patterns influence the degree

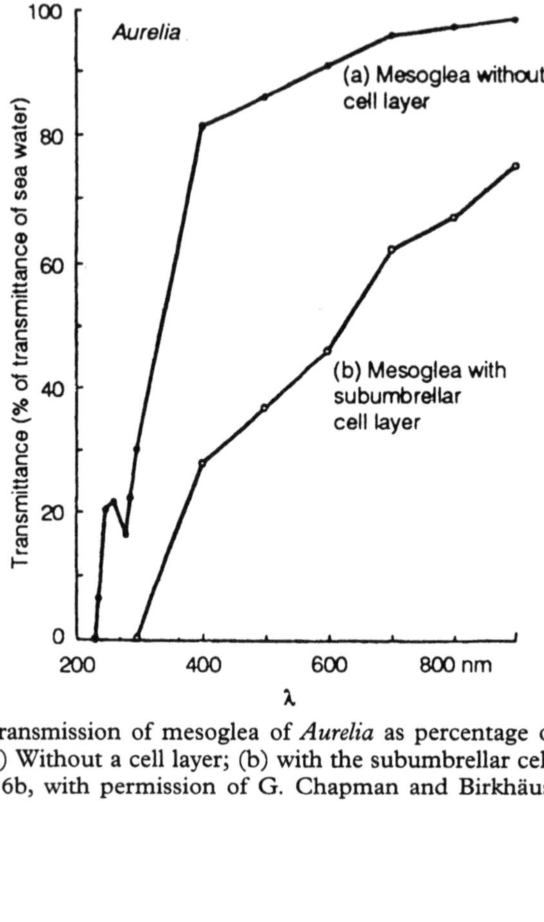

Figure 9.1 Transmission of mesoglea of *Aurelia* as percentage of transmittance of sea water. (a) Without a cell layer; (b) with the subumbrellar cell layer. (Source: Chapman, 1976b, with permission of G. Chapman and Birkhäuser Verlag.)

of predation is not known. Another function of pigments in relation to symbiosis with algae is discussed in section 4.5.4.

The pigments fall into different chemical classes. Carotenoids are widespread in cnidaria (Fox and Pantin, 1944; Kennedy, 1979). They can possess a wide variety of colours, but most often range from yellow and orange to rich red. They have been extracted from *Aurelia aurita* (see Czeczuga, 1970). Melanins are indoles, which are usually dark or black. An example is the magenta pigment of *Pelagia noctiluca* (see Fox and Millott, 1954; Millott and Fox, 1954). Protoporphyrin, related to chlorophyll and haemoglobin, is purple-brown. It is the characteristic pigment of bathypelagic species such as *Atolla wyvillei* and *Periphylla periphylla*, especially in the stomach wall and manubrium (Herring, 1972; Bonnett, Head and Herring, 1979). It causes a photosynthesizing effect leading to tissue damage, and so is not accumulated in shallow-living medusae.

9.2 PARASITES

A wide range of possible metazoan parasites has been recorded from scyphomedusae, although in many cases it is not clear whether they are actually feeding on the medusa, i.e. are parasitic, or are simply using the medusa as a substrate. The most extensively investigated are larval trematodes and cestodes and hyperiid amphipods, described separately below. Other parasites include larval actinian anemones, and a number of non-amphipod arthropods, such as isopods, and decapods. For reviews including records of associations with scyphomedusae, consult Theil (1976b), Lauckner (1980) and Théodoridès (1989). Recent papers have added a cirripid and a pycnogonid to the list of arthropods shown to feed on their hosts (Child and Harbison, 1986; Tabachnik, 1986). There are no data on parasites of the benthic stages of the scyphozoa.

The microbial flora has been described for *Aurelia aurita* from the Black Sea and *Chrysaora quinquecirrha* from Chesapeake Bay (Doores and Cook, 1976; Nizhegorodova and Nidzvetskaya, 1983). There is no evidence that these bacteria are pathogenic. However, the flora differs from that of the surrounding sea water, suggesting that antimicrobial substances are present, as is better documented for anthozoa.

It is not known how much stress the metazoan parasites cause their hosts. Scyphomedusae are able to regenerate injured tissue, including marginal sense organs (section 7.4.2). It is therefore probable that they can recover from lesions produced by parasites, provided they are able to continue feeding actively. The question is how much nutrient is diverted to maintenance of the parasite burden, both for regeneration and for extra costs of locomotion.

9.2.1 Larval trematodes and cestodes

Trematode and cestode larvae are widely distributed in pelagic cnidaria (Dollfus, 1963; Thiel, M.E., 1976b; Lauckner, 1980). In most cases the life cycle is unknown.

Two life cycles that have been completed in the laboratory are those of the digenetic trematodes *Neopechona pyriforme* (Figure 9.2) and *Lepocreadium setiferoides* (Stunkard, 1969, 1972). The cercariae of these parasites develop in rediae in snails, the unencysted metacercariae develop in various invertebrates including scyphomedusae, and fish are the definitive hosts. There is some evidence of specificity. *N. pyriforme* cercariae from infected molluscs will readily penetrate *Chrysaora quinquecirrha* and *Pelagia noctiluca*, but make no attempt to penetrate *Aurelia aurita* (see Stunkard, 1969).

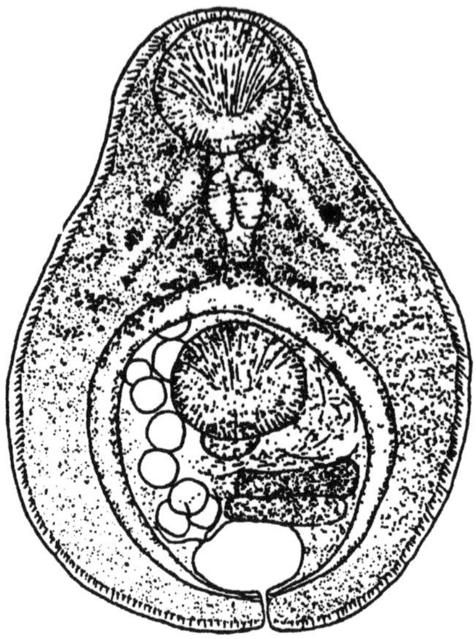

Figure 9.2 Metacercaria of the trematode *Neopechona pyriforme*, ventral view, specimen 0.18 mm long. (Source: Stunkard, 1969, with permission of *Biological Bulletin*.)

There is also evidence of specificity of cestode larvae. Over 500 specimens of *Stomolophus meleagris* taken in Mississippi, Louisiana and Texas coastal waters in 1967–1971 contained plerocercoids of an unidentified cestode (Phillips and Levin, 1973). Three other species of rhizostome medusae (*Rhopilema verrilli*, *Cassiopea xamachana* and *C. frondosa*) found in the same waters were uninfected. Possibly identical cestode larvae were described from the rhizostome *Catostylus ouwensi* from Indonesian waters (Moestafa and McConnaughey, 1966). They have not been collected from non-rhizostome scyphomedusae.

9.2.2 Hyperiid amphipods

Hyperiids are pelagic amphipods with large heads and eyes. They have been observed on or in a large number of different scyphomedusae and other gelatinous animals. Associations of the genus *Hyperia* with zooplankton were reviewed by Thurston (1977), and those of other genera of the order were reviewed by Laval (1980). More recent records include *Hyperia curticephala* with *Chrysaora plocamia* (see Vinogradov, M.E. and Semenova, 1985), *Hyperia medusarum* with an undescribed *Chrysaora* species (Martin, J.W. and Kuck, 1991), and

Figure 9.3 Adult *Hyperia galba* in characteristic resting position on a medusa. (Source: White and Bone, 1972, redrawn from Bowman *et al.*, 1963, with permission of British Antarctic Survey.)

Hyperiella dilatata with *Diplulmaris antarctica* (see Larson and Harbison, 1990).

Specificity is variable. Of 16 species of oceanic medusae examined by Thurston (1977) in the North Atlantic, only *Periphylla periphylla* contained *Hyperia*. On the other hand, *Hyperia galba* utilizes the five scyphozoan species available in the German Bight (Dittrich, 1988).

Only a few species have been thoroughly examined as to the location and type of association with the medusae. In the best known species, such as *Hyperia galba* and *H. spinigera*, the juveniles and adults are capable of swimming. They may be present on the exterior of the medusae or vertically migrate independently (Schriever, 1975). They often rest in a characteristic position with their back against the medusa, holding on with the recurved pereopods (Figure 9.3) (Bowman, Meyers and Hicks, 1963). However, the early larvae have reduced abdominal appendages and the eyes are not completely developed (Figure 9.4) (Dittrich, 1987). Being unable to swim, they are dependent on the medusa substrate. They are found in the gastrovascular cavities of the hosts or embedded in the mesoglea or in the gonads. The females hatch the larvae in brood pouches, then deposit them on the medusae near the gonads and canals of the gastrovascular system.

Whether the hyperiids are in fact parasitic (that is, whether they are eating the host tissue) has been a matter of debate even with reference to the internal larvae. Adult amphipods observed with

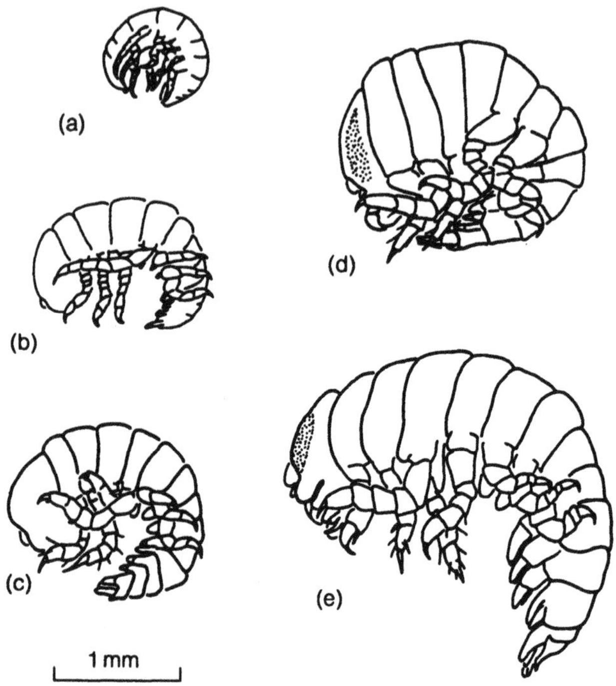

Figure 9.4 Series of instars of *Hyperia macrocephela* to show the incomplete development on release from the female brood pouch. (a) Pre-release instar; (b)–(e) instars from the gastrovascular system of the medusa *Desmonema gaudichaudi*. (Source: White and Bone, 1972, with permission of British Antarctic Survey.)

nematocysts in their stomachs include *Hyperia galba* feeding on *Cyanea capillata* (see Dahl, 1959a,b), *H. macrocephela* feeding on *Desmonema gaudichaudi* (see White and Bone, 1972) and *H. spinigera* feeding on *Periphylla periphylla* (see Thurston, 1977). Larval forms from within the gastrovascular system contain cell debris which might be either partially digested food of the medusa or the host's tissues (White and Bone, 1972). Those in the mesoglea or gonads are presumed to ingest host tissue.

The prevalence of the infestation of the medusae by *Hyperia galba* increases very rapidly after the maturation of the medusan gonads, i.e. in late summer in northern European waters (Metz, 1967; Rasmussen, 1973; Möller, 1984a; Dittrich, 1988). It is not known what triggers this reproduction of the amphipod. It is also not known to what extent the hyperiids decrease the fecundity of the medusae or contribute to the mortality of the host.

9.3 ASSOCIATIONS

In addition to being subject to whole animal predation and to parasite grazing, scyphozoa interact in other ways with a number of organisms. Feeding was described in Chapter 3. A number of associations are of mutual benefit, or do not fit clearly into one of the above categories. Symbiotic associations with algae were discussed in section 4.5, and associations with non-predatory fish will be discussed in section 9.3.1.

The medusae are associated with a number of arthropods which are not known to be parasitic. These arthropods use the medusae as a substrate. They presumably require the medusa to expend extra energy for locomotion, but are not necessarily otherwise harmful. One example is the phyllosoma larvae of the scyllarid decapods (Herrnkind, Halusky and Kanciruk, 1976). The sand lobsters hold on to the exumbrella apex of the medusae.

Associations between three species of papernautili and scyphomedusae have been observed (Thiel, M.E., 1971; Heeger, Piatkowski and Möller, 1992). The pelagic octopods clasp the exumbrella with their lateral and ventral arms. There are holes in the mesoglea but it is not clear whether the molluscs are eating the medusae, gaining access to nutrient in the gastric cavities, or simply using the medusae for transport since the medusae continue to pulsate.

The polyps of coronate scyphozoa such as *Nausithoe punctata* and *N. racemosa* may develop colonies within horny sponges (Werner, 1970, 1979). This is a mutualistic association (Uriz, Rosell and Maldonado, 1992). The polyps gain protection against physical disturbance and predation, and access to the organic particles in the inhalent current of the sponge. The sponge fuses its fibres with the theca of the polyp and saves on the metabolic costs of building its own skeleton. The polyp also cleans the water surrounding the sponge of large particles likely to foul the ostia.

9.3.1 Associations with fish

Scyphomedusae are often accompanied by teleost fish. Particularly common are young stages of the Gadidae and Carangidae, and the stromateoid families Centrolophidae, Nomidae and Stromateidae. The literature on associations with rhizostome and semaeostome medusae has been reviewed by Mansueti (1963) and Thiel (1970a, 1978). These papers contain lists of associations. Other species of medusae observed with fish include *Stygiomedusa gigantea* (see Harbison, Smith and Backus, 1973), *Desmonema gaudichaudi* (see Southcott and Glover,

1987), *Drymonema dalmatinum* (see Larson, 1987c) and *Rhopilema nomadica* (see Spanier and Galil, 1991).

The associations between the fish and medusa vary from simple opportunistic relationships, through commensalism, to ectoparasitism and predation. They vary both between different fish and medusa species, and with stage of development of the animals. The relationship may be commensal if the fish merely use the medusa as shelter. An example is the association of the Atlantic bumper, *Chloroscombrus chrysurus*, with *Aurelia aurita* where the fish eats only free-living animals like small crustacea (Tolley, 1987). The relationship may be mutualistic if the fish gain shelter and clean the medusae of associated arthropods. An example is whiting, *Merlangius merlangus*, associated with *Cyanea capillata* and *Rhizostoma pulmo*, and eating parasitic hyperiids as well as free-living copepods (Dahl, 1961; Nagabhushanam,

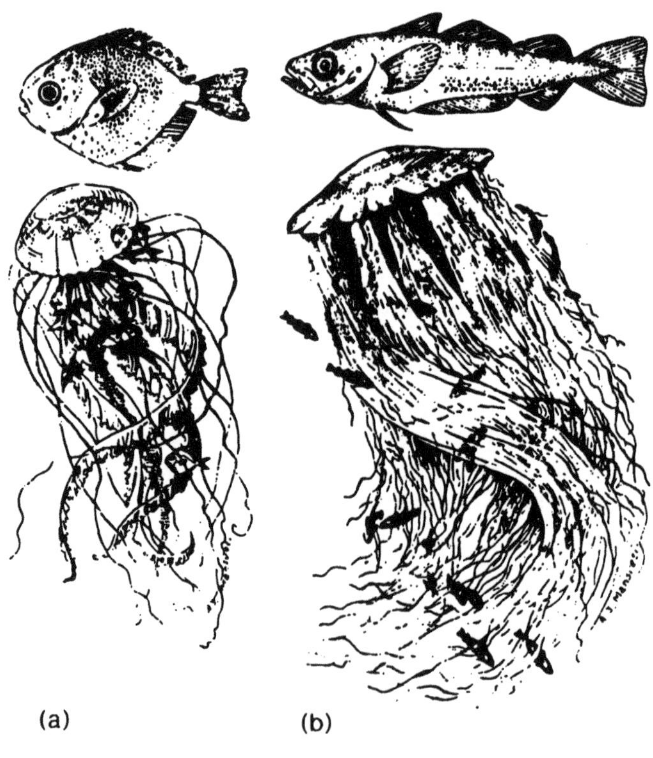

Figure 9.5 Symbiotic fish and scyphomedusae. (a) Harvest fish, *Peprilus alepidotus*, and *Chrysaora quinquecirrha*; (b) whiting, *Merlangius merlangus*, and *Cyanea capillata*. (Source: Mansueti, 1963, with permission of American Society of Ichthyologists and Herpetologists.)

1964, 1965) (Figure 9.5). Thiel (1970a) proposed that many associations with rhizostomes are also cleaning symbioses.

Some species of fish are ectoparasitic for only a portion of their life history. The developmentally variable diets have led to controversies as to the types of association present in various species. For example, Mansueti (1963) described the relationship of many individuals of the harvest fish *Peprilus alepidotus* with the scyphomedusa *Chrysaora quinquecirrha* as initially commensal, becoming ectoparasitic and finally nonsymbiotic but predatory as the fish grows (Figure 9.5). Other individuals, even of the larvae, did not associate with jellyfish, i.e. they were facultative symbionts. However, Phillips, Burke and Keener (1969), who examined a smaller number of *P. alepidotus*, did not find them feeding on medusae, although they found the related gulf butterfish *Peprilus burti* were eating *Cyanea capillata.*

It is not known to what extent fish larvae that form these associations are dependent on the presence of the medusae for survival. The distributions of the two species may coincide. For example the distributions of pelagic juveniles of haddock, *Melanogrammus aeglefinus*, and whiting extensively overlap those of *Cyanea* sp. on the Grand Banks and in the North Sea (Colton and Temple, 1961; Hay, Hislop and Shanks, 1990). However, the distributions may just be determined by common external forces, as suggested by Hay *et al.* (1990). Fish larvae often associate with other cover as well.

Fish gain protection from other predators, in some cases. In a clear example of protection Duffy (1988) observed common terns, *Sterno hirundo*, feeding on butterfish, *Peprilus tricanthus*, in Long Island Sound. Upstream from a dock no independent *P. tricanthus* were seen, whereas *Cyanea capillata* medusae each harboured three to 30 fish. When the tidal currents tumbled the medusae against the dock, the fish were displaced and caught by the waiting terns. However, in another case walleye pollock, *Theragra chalcogramma*, deserted *Cyanea* and darted into deeper water when threatened from above (Van Hyning and Cooney, 1974). Some fish move as a vanguard in advance of their associated medusa, which is therefore unlikely to afford them any protection. An example is larvae of the carangid *Seleroides leptolepis* associated with *Acromitus flagellatus* (see Jones, 1960). The fish probably respond to the pressure fields in front of the medusae (Thiel, M.E., 1970b).

There has been speculation on the mechanisms that allow fish to approach the medusa and form associations. Possible mechanisms include avoidance of contact, resistance to nematocyst toxin, or reduced nematocyst discharge because of mucus, etc. on the fish skin. All of these mechanisms are used in the association of the man-of-war fish *Nomeus gronovii* with the siphonophore *Physalia physalis*, but they have not been carefully investigated in any association with scyphomedusae. The stromateoid fish *Schedophilus medusophagus* associated with *Phacellophora camtschatica* and *Pelagia noctiluca* has a resistant integument with a keratinized external layer (Bone and Brook, 1973).

9.4 BIOLUMINESCENCE

Bioluminescence, the emission of visible light, has only been observed in five species of scyphozoa. Luminescence of the neritic *Pelagia noctiluca* was recognized at least as early as the first century AD by Pliny the Elder (Harvey, 1952; 1957). More recently it has been examined in another semaeostome medusa, *Poralia rufescens*, and in three coronate medusae, *Atolla parva*, *Atolla wyvillei* and *Periphylla periphylla*, all oceanic (Herring, 1990).

9.4.1 Anatomy of luminescent structures

Pelagia noctiluca medusae swimming normally through calm water do not spontaneously luminesce, but if they are stimulated they emit an intense blue-green light (Dahlgren, 1916). A slight mechanical contact with the exumbrella causes a spot of light at the point touched, and this spreads out in lines, streaks or patches. Stronger mechanical, chemical or electrical stimuli cause a general glow of the exumbrella, especially the marginal lobes, and of the outer surfaces of the oral arms and tentacles. Damaged epithelium may produce luminous mucus which sticks to objects, such as fingers or glass rods, that have made contact. Freshly captured specimens also show a flickering response from the subumbrellar surface (Herring, 1990).

The blue luminescence of *Atolla wyvillei* following stimulation is widely distributed over the exumbrella and subumbrella (Nicol, 1958; Herring, 1990). It is particularly concentrated at the bases of the marginal lappets and tentacles and adjacent to the coronal groove. In some specimens the ovaries are also luminous. *A. parva* and *Periphylla periphylla* have similar responses with greater release of luminous mucus.

In most animals luminescence is correlated with the presence of granule-filled cells, the photocytes. Early workers showed that the epithelium of *Pelagia noctiluca* has granule-containing cells (Panceri, 1872; Dahlgren, 1916). However, free granules in the mucus can luminesce and tissue luminescence is probably extracellular (Harvey,

1926; Morin and Reynolds, 1972). Similarly the distribution of luminescence is not correlated with the distribution of the granule-filled cells in *Atolla wyvillei* (see Herring, 1990).

9.4.2 Chemical basis of luminescence

Bioluminescence is a catalysed chemiluminescence in which chemical energy is converted to light energy. Emission does not depend either on the temperature of the excited molecule as in incandescence, or on prior absorption of light as in fluorescence and phosphorescence.

The enzymes (luciferases) and substrates (luciferins) involved in bioluminescence differ among taxa, as do various cofactors (Morin, 1974; Cormier, 1978). The luciferases are oxygenases which catalyse the formation of an intermediate peroxy compound. Its return to ground state results in emission of a photon in the visible range. In most taxa, the oxidation occurs at the time of light production so bioluminescence requires the presence of oxygen. This is true of the anthozoan *Renilla*. However, in various hydrozoa such as *Aequorea*, the luciferase apoaequorin reacts with the luciferin coelenterazine and oxygen to form a stable peroxide, the photoprotein aequorin. Aequorin then emits light upon the addition of calcium:

$$\text{apoaequorin} + \text{coelenterazine} + O_2 \rightarrow \text{aequorin} \xrightarrow{Ca^{++}} h\nu$$
$$+ \text{coelenteramide} + \text{apoaequorin}$$

The bioluminescent compounds of scyphozoa have not been extracted and identified. In *Pelagia noctiluca* molecular oxygen is not necessary at the time of bioluminescence (Harvey, 1926; Morin and Hastings, 1971a) and calcium activates light production (Morin and Reynolds, 1972). The *Pelagia* system may be similar to *Aequorea*. The luminescence of homogenized ovaries of *Atolla wyvillei* and *Periphylla periphylla* is not activated by calcium (Herring, 1990). It is probable that another activating ion or compound or an extra cofactor is needed.

Light emissions resulting from bioluminescent reactions span a range of wave lengths. Emission spectra of scyphozoa are shown in Figure 9.6. The majority of pelagic animals, including scyphomedusae, have blue or blue-green emissions with maxima centred in the 450–500 nm range (Herring, 1983) (Table 9.1). In many other coelenterates the luminescent system has an associated green-fluorescent protein. No spectra with green maxima have been observed in scyphomedusae. In *Pelagia noctiluca* the emission spectrum is the

Table 9.1 Emission maxima of scyphozoa

Species	*Emission maxima (nm)*	*Reference*
Atolla parva	465	Widder *et al.*, 1989
Atolla wyvillei	470	Nicol, 1958
	462	Widder, Latz and Case, 1983
Pelagia noctiluca	~475	Morin and Reynolds, 1972
Periphylla periphylla	463	Widder, Latz and Case, 1983
ovaries only	470–480	Herring, 1983

same *in vivo* or following extraction and separation by column chromatography, supporting the absence of any fluorescent protein (Morin and Hastings, 1971b).

9.4.3 Control of luminescence

Bioluminescence in scyphozoa has been triggered by application of mechanical and electrical stimuli as well as by various chemicals such as magnesium sulphate and sodium hydroxide (Heymans and Moore, 1924; Moore, A.R., 1926). It is unclear how many of these stimuli directly affect the luminescent reactions and how many may trigger responses in a control system. In many cases, waves of luminescence travel over the bodies of the medusae. These may involve nervous control since no epithelial conduction has been found in scyphozoan medusae.

The main response of *Pelagia noctiluca* to a series of electrical stimuli delivered at intervals of at least 0.5 seconds is a series of exumbrellar flashes each corresponding to a single stimulus (Morin and Reynolds, 1972). Responses may also be propagated on the subumbrellar surface (Herring, 1990). In *Atolla wyvillei*, the main pathways for propagated responses are the coronal groove and the exumbrellar margin (Herring, 1990). A single electrical stimulus may induce one or several waves propagated at velocities of 70–490 mm per second, most frequently at 150–260 mm per second. *Periphylla periphylla* responds to a short pulse of alternating current with a series of flashes (Clarke, G.L. *et al.*, 1962). Responses have not been correlated with specific nerve nets.

9.4.4 Ecological significance

Although bioluminescence is spectacular, its functional importance to gelatinous plankton interacting with other animals is not known (Galt,

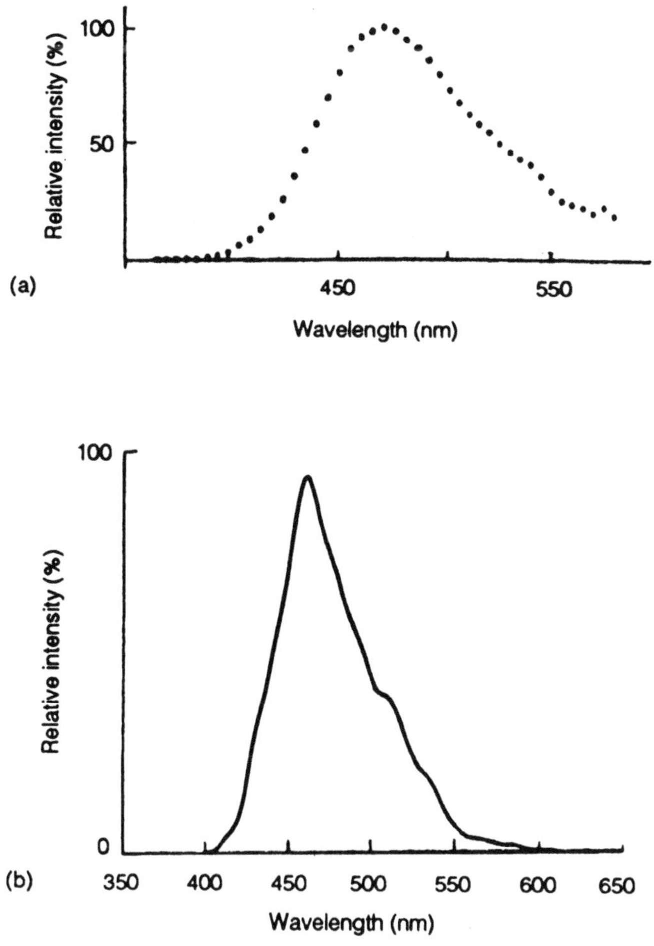

Figure 9.6 Emission spectra of scyphozoa. (a) *Periphylla periphylla*; (b) *Atolla parva*. (Sources: (a) Herring, 1983, with permission of P.J. Herring and The Royal Society; (b) Widder *et al.*, 1989, with permission of Springer-Verlag.)

1989). It could be of use in counter-illumination, or ventral camouflage. It may also attract photosensitive prey. However, the short duration of light production in scyphomedusae makes both of these functions unlikely. A more likely possibility is that production of a luminous cloud or of flashes following mechanical stimulation may be useful in defence. Even in this case, light may attract as well as repel possible fish or turtle predators (Davenport, 1988).

9.5 TROPHIC RELATIONSHIPS

9.5.1 Impact on prey populations

Scyphozoa utilize a wide selection of zooplankton prey (section 3.3.1). However, their predation on larval fish, or on species utilized by fish such as copepods, could impact commercial fish stocks. Recent reviews of coelenterate predation on fish include Purcell (1985), Arai (1988) and Bailey and Houde (1989).

Predation rates (number of food animals eaten per predator per day) are calculated from stomach contents and digestion rates (section 3.6). In order to calculate the impact on a prey population, measurements of the abundances of predator and prey are also needed. The necessary data for predation by scyphozoa has been gathered only in a few cases. Möller (1980b, 1984a,b) calculated that at least 2–5% of the herring larvae in Kiel Fjord were consumed daily in May by *Aurelia aurita*. Purcell (1992, 1994) found that in July and August *Chrysaora quinquecirrha* ate 0–3% daily of the standing stock of copepods in Chesapeake Bay. However, in the tributaries to the bay, predation was generally 20–50% of the standing stock, and in some areas was as high as 94%.

In some studies field data on predator and prey abundance have been combined with clearance rates measured in the laboratory. Laboratory measurements of feeding are less accurate due to the effects of confinement in containers and the absence of alternative prey (section 3.6). This method was applied by Fancett and Jenkins (1988) to predation by *Pseudorhiza haeckeli* and *Cyanea capillata* on fish eggs and larvae, and on copepods, in Port Phillip Bay, Australia. The impact of *P. haeckeli* ranged from 0.1% to 3.8% of the fish eggs and larvae per day, and that of *C. capillata* from 0.1% to 2.4% per day. Predation on copepods ranged from 0.1% to 4.8% per day for *P. haeckeli* and 0.1% to 1.6% per day for *C. capillata*. Cowan and Houde (1993) estimated that *Chrysaora quinquecirrha* have the potential to consume 20–40% per day of the eggs and larvae of bay anchovy in Chesapeake Bay. Garcia and Durbin (1993), examining predation by *Phyllorhiza punctata* in a Puerto Rico lagoon, found a maximum August clearance of copepods from 35% of the lagoon's volume per day.

Calculations of predation have also been based on the energy requirements of the predator, as measured in growth and metabolism, combined with field data on population size. However, there is little or no data on other terms in the energy equation such as assimilation, intake of microorganisms or DOM, non-ammonia excretion, mortality, mucus production, anaerobic metabolism and reproduction. Hence these calculations are highly speculative (Arai, in press). An example of this kind of speculation where most of the assumptions are clearly stated is that of Schneider (1989). He calculated that the daily food uptake of *Aurelia aurita* in Kiel Fjord was 3–15% of the mesozooplankton, or 1–6% of both the micro- and mesozooplankton stocks. Other calculations of this type have been made by Mironov (1967), Shushkina and Musayeva (1983) and Schneider and Behrends (1994) on *A. aurita* in the Black and Baltic Seas, and by Malej (1989a) on *Pelagia noctiluca* in the Mediterranean Sea.

The above figures compare predation with the standing stock of prey. The actual effect of predation on the prey population depends also on the rate of production of the prey.

One advantage for fish and some other prey is that they may remain vulnerable to predation by scyphozoa for only a small portion of their life cycle (section 3.6.2). If the peak fish egg and larvae production precedes that of the medusae, the effect on the fish will be minimized. For example, the seasonal immigration of plaice larvae into the Wadden Sea precedes peak abundance of *Aurelia aurita*. Flounder larvae immigration coincides with a period of abundance of the medusae so that flounder larvae are more vulnerable to predation than plaice larvae (Van der Veer, 1985; Van der Veer and Oorthuysen, 1985). Year to year variability in temporal succession is an important topic for future research in areas like the Wadden Sea and Chesapeake Bay (Bailey, K.M. and Houde, 1989; Cowan and Houde, 1993).

If prey production is low, a negative relationship between population abundances of the scyphozoa and their prey may occur. An example is the inverse relationship between *Aurelia aurita* and herring larvae abundances in Kiel Fjord (Moller 1984a,b,c). Conversely some authors have used a negative relationship as an indicator of predation. However, as noted by Hunter (1984), Frank and Leggett (1985) and Purcell (1985), there can be other explanations for such a negative relationship. For example, in coastal Newfoundland macroinvertebrate predators and coastal ichthyoplankton occupy discrete water masses (Frank and Leggett, 1985). The alternative presence of these masses inshore depends on oscillatory wind conditions.

Most controlled experiments involving scyphomedusae in large containers have related to rates of feeding on fish larvae and were discussed in section 3.6.2. Olsson *et al.* (1992) added *Aurelia aurita* to cylinders containing natural phytoplankton and microzooplankton with varying additions of copepods. The medusae caused a small reduction in copepod grazing pressure, allowing stronger growth of diatoms.

9.5.2 Competition

Many scyphozoa eat herbivorous zooplankton such as copepods (section 3.3.1). When the same prey are utilized by other predators, such as fish, it is tempting to assume that competition is occurring. However, competition is difficult to prove as it needs measurements of rates of predation of the two predators, and also examination of the prey population to show that the population is limited by predation rather than other factors such as food or the environment. The requisite data is rarely available in marine or estuarine environments. For example, in Chesapeake Bay the populations of the copepod *Acartia tonsa* are not limited by predation of the abundant medusae and ctenophores and only rarely by food (Purcell, White and Roman, 1994). If competition between predators is occurring, it must be due to the presence also of non-gelatinous predators such as fish.

If competition is occurring, then intraguild predation may occur. Intraguild predation refers to consumption of species that are potential competitors for a resource. Purcell (1991) has recently reviewed the scyphozoa feeding on other gelatinous predators. However, the extent of dietary overlap and the occurrence of competition is unknown.

9.5.3 Trophic levels

In the past the ctenophores or medusae were often considered as trophic dead ends. For example, Greve and Parsons (1977) suggested that:

> . . . two principal pathways exist for the transfer of energy up the food web of the sea. These are: Nanophytoplankton (e.g. small flagellates) → small zooplankton → ctenophores or medusae, or, alternately, Microphytoplankton (e.g. large diatoms) → large zooplankton → young fish.

This hypothesis may be criticized on the basis that there is no special feeding relationship between coelenterates and small zooplankton (Longhurst, 1985), but also on the basis that coelenterates are eaten by a wide variety of other carnivores (section 9.1.1). Until rates of predation on various sizes of scyphozoa are known, it is premature to assume that they are always at the highest trophic levels in the complex marine food webs.

Recently, stable isotopes have been used as indicators of animal trophic position. Heavy isotopes are enriched along the food web for reasons that are imperfectly understood (Fry and Sherr, 1984; Mills,

Pittman and Tan, 1984). The only study of this type including scyphozoa is that of Malej, Faganelli and Pezdic (1993) on *Pelagia noctiluca* . This showed enrichment (i.e. a higher trophic position) of the medusa relative to general zooplankton collected with a 250 μm mesh net.

Appendix: Classification of extant scyphozoa

This classification is modified from Dunn (1982), Stepanjants and Sheiko (1989), Franc (1993) and Cornelius (in press). Although there are approximately 200 species of scyphozoa, in this appendix only families and species of scyphozoa mentioned in the text or figures are included.

PHYLUM CNIDARIA, CLASS SCYPHOZOA

Order Stauromedusae

Family Cleistocarpidae
Craterolophus convolvulus (Johnston, 1835)
Manania atlantica (Berrill, 1962)
Manania distincta (Kishinouye, 1899)
Manania gwilliami Larson and Fautin, 1989
Family Eleutherocarpidae
Haliclystus auricula (Rathke, 1806)
Haliclystus octoradiatus (Lamarck, 1816)
Haliclystus salpinx Clark, 1863
Haliclystus stejnegeri Kishinouye, 1899
Kishinouyea corbini Larson, 1980
Kyopoda lamberti Larson, 1988
Lucernaria quadricornis O.F.Müller, 1776
Lucernariopsis campanulata (Lamouroux, 1815)
Stylocoronella riedli Salvini-Plawen, 1966

Stylocoronella variabilis Salvini-Plawen, 1987

Order Coronatae

Family Atollidae (=Collaspidae)
Atolla parva Russell, 1958
Atolla vanhoeffeni Russell, 1957
Atolla wyvillei Haeckel, 1880
Family Linuchidae
Linuche unguiculata (Schwartz, 1788)
Family Nausithoidae
Atorella japonica Kawaguti and Matsuno, 1981
Atorella vanhoeffeni Bigelow, 1909
Nausithoe eumedusoides (Werner, 1971)
Nausithoe planulophora (Werner,1971)
Nausithoe punctata Kölliker, 1853
Nausithoe racemosa (Komai, 1936)
Nausithoe rubra Vanhöffen, 1902
Nausithoe werneri Jarms, 1990
Family Paraphyllinidae
Paraphyllina intermedia Maas, 1903
Paraphyllina ransoni Russell, 1956
Family Periphyllidae
Periphylla periphylla (Péron and Lesueur, 1810)

Order Semaeostomeae

Family Cyaneidae
Cyanea capillata (Linnaeus, 1758)
Cyanea lamarcki Péron and Lesueur, 1810
Desmonema gaudichaudi (Lesson, 1830)
Drymonema dalmatinum Haeckel, 1880
Family Pelagiidae
Chrysaora fuscescens Brandt, 1835
Chrysaora hysoscella (Linnaeus, 1766)
Chrysaora melanaster Brandt, 1835
Chrysaora plocamia (Lesson, 1830)
Chrysaora quinquecirrha (Desor, 1848)
Pelagia colorata Russell, 1964
Pelagia noctiluca (Forskål, 1775)
Sanderia malayensis Goette, 1886

Family Ulmaridae
Aurelia aurita (Linnaeus, 1758)
Aurelia limbata (Brandt, 1838)
Deepstaria enigmatica Russell, 1967
Deepstaria reticulum Larson, Madin and Harbison, 1988
Diplulmaris antarctica Maas, 1908
Discomedusa lobata Claus, 1877
Phacellophora camtschatica Brandt, 1838
Poralia rufescens Vanhöffen, 1902
Stygiomedusa gigantea (Browne, 1910)

Order Rhizostomeae

Family Cassiopeidae
Cassiopea andromeda (Forskål, 1775)
Cassiopea frondosa (Pallas, 1774)
Cassiopea ornata Haeckel, 1880
Cassiopea xamachana R.P. Bigelow, 1892
Family Catostylidae
Catostylus ouwensi Moestafa and McConnaughey, 1966
Acromitus flagellatus (Maas, 1903)
Family Cepheidae
Cephea cephea (Forskål, 1775)
Cotylorhiza tuberculata (Macri, 1778)
Family Lobonematidae
Lobonema smithi Mayer, 1910
Lobonemoides gracilis Light, 1914
Family Lychnorhizidae
Pseudorhiza haeckeli Haacke, 1884
Family Mastigiidae
Mastigias albipunctatus Stiasny, 1920
Mastigias papua (Lesson, 1830)
Phyllorhiza peronlesueuri Goy, 1990
Phyllorhiza punctata von Lendenfeld, 1884
Family Rhizostomatidae
Rhizostoma pulmo (Macri, 1778)
Rhopilema esculenta Kishinouye, 1891
Rhopilema hispidum (Vanhöffen, 1888)
Rhopilema nomadica Galil, Spanier and Ferguson, 1990
Rhopilema verrilli (Fewkes, 1887)
Family Stomolophidae
Stomolophus meleagris L. Agassiz, 1862
Stomolophus nomurai (Kishinouye)

Scyphozoa *incertae sedis*

Tetraplatia chuni Carlgren, 1909
Tetraplatia volitans Busch, 1851

References

Afzelius, B.A. and Franzén, Å. (1971) The spermatozoon of the jellyfish *Nausithoë*. *Journal of Ultrastructure Research*, **37**, 186–199.

Alexander, R.McN. (1964) Visco-elastic properties of the mesogloea of jellyfish. *Journal of Experimental Biology*, **41**, 363–369.

Aleyev, Y.G. and Khvorov, S.A. (1980) Age-related changes in the mean body density and buoyancy of *Daphnia magna* and *Cyclops strenuus*. *Hydrobiological Journal (English Translation of Gidrobiologicheskii Zhurnal)*, **16**, 27–30.

Alvariño, A. (1985) Predation in the plankton realm; mainly with reference to fish larvae. *Investigaciones Marinas CICIMAR (Centro Interdisciplinario de Ciencias Marinas)*, **2**, 1–122.

Anderson, P.A.V. (1985) Physiology of a bidirectional, excitatory, chemical synapse. *Journal of Neurophysiology (Bethesda)*, **53**, 821–833.

Anderson, P.A.V. (1987) Properties and pharmacology of a TTX-insensitive Na^+ current in neurones of the jellyfish *Cyanea capillata*. *Journal of Experimental Biology*, **133**, 231–248.

Anderson, P.A.V. (1989) Ionic currents of the Scyphozoa, in *Evolution of the First Nervous Systems*, (ed. P.A.V. Anderson), Plenum Press, New York, pp. 267–280.

Anderson, P.A.V. and Grünert, U. (1988) Three-dimensional structure of bidirectional, excitatory chemical synapses in the jellyfish *Cyanea capillata*. *Synapse*, **2**, 606–613.

Anderson, P.A.V. and McKay, M.C. (1987) The electrophysiology of cnidocytes. *Journal of Experimental Biology*, **133**, 215–230.

Anderson, P.A.V. and Schwab, W.E. (1981) The organization and structure of nerve and muscle in the jellyfish *Cyanea capillata* (Coelenterata: Scyphozoa). *Journal of Morphology*, **170**, 383–399.

Anderson, P.A.V. and Schwab, W.E. (1982) Recent advances and model systems in coelenterate neurobiology. *Progress in Neurobiology (Oxford)*, **19**, 213–236.

Anderson, P.A.V. and Schwab, W.E. (1983) Action potentials in neurons of motor nerve net of *Cyanea* (Coelenterata). *Journal of Neurophysiology (Bethesda)*, **50**, 671–683.

Anderson, P.A.V. and Schwab, W.E. (1984) An epithelial cell-free preparation of the motor nerve net of *Cyanea* (Coelenterata; Scyphozoa). *Biological Bulletin (Woods Hole)*, **166**, 396–408.

Anderson, P.A.V. and Spencer, A.N. (1989) The importance of cnidarian synapses for neurobiology. *Journal of Neurobiology*, **20**, 435–457.

Anderson, P.A.V., Holman, M.A. and Greenberg, R.M. (1993) Deduced amino acid sequence of a putative sodium channel from the scyphozoan jellyfish *Cyanea capillata*. *Proceedings of the National Academy of Sciences of the United States of America*, **90**, 7419–7423.

Anderson, P.A.V., Moosler, A. and Grimmelikhuijzen, C.J.P. (1992) The presence and distribution of Antho-RFamide-like material in scyphomedusae. *Cell and Tissue Research*, **267**, 67–74.

Anderson, P.A.V., Schwab, W.E., Gilbert, S. *et al.* (1986) Fast axonal transport by neurons from the jellyfish *Cyanea capillata*. *Journal of Neurobiology*, **17**, 29–37.

Angel, M.V., Hargreaves, P., Kirkpatrick, P. *et al.* (1982) Low variability in planktonic and micronektonic populations at 1,000 m depth in the vicinity of 42°N, 17°W; Evidence against diel migratory behavior in the majority of species. *Biological Oceanography*, **1**, 287–319.

Anninsky, B.E. (1988a) The effect of food concentration on the diet of *Aurelia aurita* (L.) at different temperatures. *Ekologiya Morya (Kiev)*, **29**, 31–38. In Russian; English abstract.

Anninsky, B.E. (1988b) Physiological requirements and real possibilities of *Aurelia aurita* (L.) nutrition under conditions of the Black Sea. *Ekologiya Morya (Kiev)*, **29**, 38–45. In Russian; English abstract.

Anninsky, B.E. (1988c) The rate and efficiency of copepod assimilation by scyphoid medusa *Aurelia aurita* L. *Ekologiya Morya (Kiev)*, **28**, 58–64. In Russian; English abstract.

Arai, M.N. (1965) Contractile properties of a preparation of the column of *Pachycerianthus torreyi* (Anthozoa). *Comparative Biochemistry and Physiology*, **14**, 323–337.

Arai, M.N. (1988) Interactions of fish and pelagic coelenterates. *Canadian Journal of Zoology*, **66**, 1913–1927.

Arai, M.N. (1991) Attraction of *Aurelia* and *Aequorea* to prey. *Hydrobiologia*, **216/217**, 363–366.

Arai, M.N. (in press) Coelenterates in pelagic food webs, in *Proceedings of the 6th International Conference on Coelenterate Biology 1995*, (ed. J.C. den Hartog), Nationaal Natuurhistorisch Museum, Leiden.

Arai, M.N. and Jacobs, J.R. (1980) Interspecific predation of common Strait of Georgia planktonic coelenterates: laboratory evidence. *Canadian Journal of Fisheries and Aquatic Sciences*, **37**, 120–123.

Aronova, M.Z., Kharkeevich, T.A. and Tsirulis, T.P. (1986) Role of carbonic anhydrase in the formation of the otolithic apparatus of certain marine invertebrates. *Journal of Evolutionary Biochemistry and Physiology (English Translation of Zhurnal Evolyutsionnoi Biokhimii i Fiziologii)*, **22**, 385–392.

Aronova, M.Z., Titova, L.K., Kharkeevitch, T.A. *et al.* (1979) Ultrastructural peculiarities of the chemical transmission of the nerve impulse in the gravity receptors of some invertebrates. *Zeitschrift fuer Mikroskopisch-Anatomische Forschung (Leipzig)*, **93**, 91–103.

Ates, R.M.L. (1988) Medusivorous fishes, a review. *Zoologische Mededelingen (Leiden)*, **62**, 29–42.

Ates, R.M.L. (1991) Predation on Cnidaria by vertebrates other than fishes. *Hydrobiologia*, **216/217**, 305–307.

Avian, M. (1986a) Biological cycle of *Cotylorhiza tubercolata* (Macri 1778): Morphological aspects of the development from ephyra to young medusa. *Nova Thalassia*, **8 Suppl. 2**, 47–58.

Avian, M. (1986b) Temperature influence on *in vitro* reproduction and development of *Pelagia noctiluca* (Forskål). *Bollettino di Zoologia*, **53**, 385–391.

Avian, M. and Rottini-Sandrini, L. (1988) Fishery and swarmings of *Pelagia noctiluca* in the Central and Northern Adriatic Sea: middle term analysis. *Rapports et Proces-Verbaux des Reunions Commission Internationale pour l'Exploration Scientifique de la Mer Mediterranee Monaco*, **31**, 231.

Avian, M. and Rottini-Sandrini, L. (1991) Oocyte development in four species of scyphomedusa in the northern Adriatic Sea. *Hydrobiologia*, **216/217**, 189–195.

Avian, M., Del Negro, P. and Rottini-Sandrini, L. (1991) A comparative analysis of nematocysts in *Pelagia noctiluca* and *Rhizostoma pulmo* from the North Adriatic Sea. *Hydrobiologia*, **216/217**, 615–621.

Avian, M., Giorgi, R. and Rottini-Sandrini, L. (1991) Seasonal influence on the vitellogenesis of *Pelagia noctiluca* in the northern Adriatic Sea, in *UNEP: Jellyfish Blooms in the Mediterranean. Proceedings of the II Workshop on Jellyfish in the Mediterranean Sea*, MAP Technical Reports Series No. 47, UNEP, Athens, pp. 22–31.

Avian, M., Rottini-Sandrini, L. and Bratina, F. (1991) Six types of cnidocysts in *Pelagia noctiluca* identified by means of light-scanning electron microscopy and videoanalyzer, in *UNEP: Jellyfish Blooms in the Mediterranean. Proceedings of the II Workshop on Jellyfish in the Mediterranean Sea*, MAP Technical Reports Series No. 47, UNEP, Athens, pp. 217–226.

Avian, M., Rottini-Sandrini, L. and Stravisi, F. (1991) The effect of seawater temperature on the swimming activity of *Pelagia noctiluca* (Forsskål). *Bollettino di Zoologia*, **58**, 135–141.

Avian, M., Spanier, E. and Galil, R. (1995) Nematocysts of *Rhopilema nomadica* (Scyphozoa: Rhizostomeae), an immigrant jellyfish in the eastern Mediterranean. *Journal of Morphology*, **224**, 221–231.

Avian, M., Micali, F., Sciancalepore, M. *et al.* (1987) Vitellogenesi in *Pelagia noctiluca* primi risultati di tecniche autoradiografiche. *Atti della Accademia delle Scienze dell'Istituto di Bologna Classe di Scienze Fisiche Rendiconti*, **3**, 139–146. In Italian; English abstract.

Axiak, V. (1984) Effect of decreasing light intensity on the activity of the scyphomedusa, *Pelagia noctiluca* (Forskal), in *Workshop on Jellyfish Blooms in the Mediterranean*, UNEP, Athens, pp. 121–127.

Axiak, V. and Civili, F.S. (1991) Jellyfish blooms in the Mediterranean: causes, mechanisms, impact on man and the environment. A programme review, in *UNEP: Jellyfish Blooms in the Mediterranean. Proceedings of the II Workshop on Jellyfish in the Mediterranean Sea*, UNEP, Athens, pp. 1–21.

Axiak, V., Galea, C. and Schembri, P.J. (1991) Coastal aggregations of the jellyfish *Pelagia noctiluca* (Scyphozoa) in Maltese coastal waters during 1980–1986, in *UNEP: Jellyfish Blooms in the Mediterranean. Proceedings of the II Workshop on Jellyfish in the Mediterranean Sea*, MAP Technical Reports Series No. 47, UNEP, Athens, pp. 32–40.

Babcock, L.E. (1991) The enigma of conularid affinities, in *The Early Evolution of Metazoa and the Significance of Problematic Taxa*, (eds A.M. Simonetta and S. Conway Morris), Cambridge University Press, Cambridge, pp. 133–143.

Bailey, K.M. (1984) Comparison of laboratory rates of predation on five species of marine fish larvae by three planktonic invertebrates: effects of larval size on vulnerability. *Marine Biology (Berlin)*, **79**, 303–309.

Bailey, K.M. and Batty, R.S. (1983) A laboratory study of predation by *Aurelia aurita* on larval herring (*Clupea harengus*): experimental observations compared with model predictions. *Marine Biology (Berlin)*, **72**, 295–301.

Bailey, K.M. and Batty, R.S. (1984) Laboratory study of predation by *Aurelia aurita* on larvae of cod, flounder, plaice and herring: development and vulnerability to capture. *Marine Biology (Berlin)*, **83**, 287–291.

Bailey, K.M. and Houde, E.D. (1989) Predation on eggs and larvae of marine fishes and the recruitment problem. *Advances in Marine Biology*, **25**, 1–83.

Bailey, T.G., Youngbluth, M.J. and Owen, G.P. (1995) Chemical composition and metabolic rates of gelatinous zooplankton from midwater and benthic boundary layer environments off Cape Hatteras, North Carolina, USA. *Marine Ecology Progress Series*, **122**, 121–134.

Balcer, L.J. and Black, R.E. (1991) Budding and strobilation in *Aurelia* (Scyphozoa, Cnidaria): functional requirement and spatial patterns of nucleic acid synthesis. *Roux's Archives of Developmental Biology*, **200**, 45–50.

Balderston, W.L. and Claus, G. (1969) A study of the symbiotic relationship between *Symbiodinium microadriaticum* Freudenthal, a zooxanthella and the upside down jellyfish, *Cassiopea* sp. *Nova Hedwigia*, **17**, 373–382.

Båmstedt, U. (1988) Interspecific, seasonal and diel variations in zooplankton trypsin and amylase activities in Kosterfjorden, western Sweden. *Marine Ecology Progress Series*, **44**, 15–24.

Båmstedt, U. (1990) Trophodynamics of the scyphomedusae *Aurelia aurita*. Predation rate in relation to abundance, size and type of prey organism. *Journal of Plankton Research*, **12**, 215–229.

Båmstedt, U., Martinussen, M.B. and Matsakis, S. (1994) Trophodynamics of the two scyphozoan jellyfishes, *Aurelia aurita* and *Cyanea capillata*, in western Norway. *ICES (International Council for the Exploration of the Sea) Journal of Marine Science*, **51**, 369–382.

Barnes, R.D. and Harrison, F.W. (1991) Introduction, in *Microscopic Anatomy of Invertebrates Volume 2 Placozoa, Porifera, Cnidaria, and Ctenophora*, (eds F.W. Harrison and J.A. Westfall), Wiley-Liss, New York, pp. 1–12.

Barnes, W.J.P. and Horridge, G.A. (1965) A neuropharmacologically active substance from jellyfish ganglia. *Journal of Experimental Biology*, **42**, 257–267.

Bateman, J.B. (1932) The osmotic properties of medusae. *Journal of Experimental Biology*, **9**, 124–127.

Berger, V.Y. (1977) Effect of dibazole and vitamin B12 on the processes of adaptations of the jellyfish *Aurelia aurita* (L.) larvae to salinity fluctuations. *Biologiya Morya (Vladivostok)*, **5**, 48–54. In Russian; English abstract.

Berger, V.Y., Lukanin, V.V. and Khlebovich, V.V. (1970) The effect of actinomycin D on salinity acclimation in larvae of the jellyfish *Aurelia aurita* and the mollusc *Epheria vinata. Zhurnal Evolyutsionnoi Biokhimii i Fiziologii*, **6**, 636–638. In Russian; English abstract.

Bergström, B.I., Gustavsson, A. and Strömberg, J.-O. (1992) Determination of abundance of gelatinous plankton with a remotely operated vehicle (ROV). *Archiv fuer Hydrobiologie Beihefte. Ergebnisse der Limnologie*, **36**, 59–65.

Bergström, R.M. (1971) Unit spike activity in the coelenteran neural network. *Naturwissenschaften*, **58**, 153.

Berrill, M. (1962) The biology of three New England stauromedusae, with a description of a new species. *Canadian Journal of Zoology*, **40**, 1249–1262.

Berrill, N.J. (1949) Developmental Analysis of Scyphomedusae. *Biological Reviews of the Cambridge Philosophical Society*, **24**, 393–410.

Berryman, J. (1984) Predation of *Sagartiogeton laceratus* upon *Aurelia aurita* in shallow water. *Journal of the Marine Biological Association of the United Kingdom*, **64**, 725.

Beukema, J.J. and De Bruin, W. (1979) Calorific values of the soft parts of the tellinid bivalve *Macoma balthica* (L.) as determined by two methods. *Journal of Experimental Marine Biology and Ecology*, **37**, 19–30.

Bidigare, R.R. and Biggs, D.C. (1980) The role of sulfate exclusion in buoyancy maintenance by siphonophores and other oceanic gelatinous zooplankton. *Comparative Biochemistry and Physiology*, **66A**, 467–471.

Bigelow, R.P. (1900) The anatomy and development of *Cassiopea xamachana*. *Memoirs of the Boston Society of Natural History*, **5**, 191–236, pls 31–38.

Bigelow, R.P. (1910) A comparison of the sense-organs in medusae of the family Pelagidae. *Journal of Experimental Zoology*, **9**, 751–785.

Biggs, D.C. (1977) Respiration and ammonium excretion by open ocean gelatinous zooplankton. *Limnology and Oceanography*, **22**, 108–117.

Biggs, D.C., Bidigare, R.R. and Smith, D.E. (1981) Population density of gelatinous plankton: *In situ* estimation in oceanic surface waters. *Biological Oceanography*, **1**, 157–173.

Biggs, D.C., Smith, D.E., Bidigare, R.R. *et al.* (1984) *In situ* estimation of the population density of gelatinous planktivores in Gulf of Mexico surface waters. *Memorial University of Newfoundland Occasional Papers in Biology*, **9**, 17–34.

Bischoff, G.C.O. (1989) Bryoniida new order from early Paleozoic strata of eastern Australia (Cnidaria, thecate scyphopolyps). *Senckenbergiana Lethaea*, **69**, 467–521.

Black, R.E. (1972) Nucleic acid and protein levels in strobilating polyps of *Chrysaora quinquecirrha* and *Aurelia aurita*. *Marine Biology (Berlin)*, **16**, 134–137.

Black, R.E. (1981) Metabolism and ultrastructure of dormant podocysts of *Chrysaora quinquecirrha* (Scyphozoa). *Journal of Experimental Zoology*, **218**, 175–182.

Black, R.E. and Bloom, L. (1984) Heat shock proteins in *Aurelia* (Cnidaria, Scyphozoa). *Journal of Experimental Zoology*, **230**, 303–307.

Black, R.E. and Riley, G.K. (1985) Dissociation and reaggregation of cells of *Chrysaora quinquecirrha* (Cnidaria, Scyphozoa). *Journal of Experimental Zoology*, **233**, 369–375.

Black, R.E. and Webb, K.L. (1973) Metabolism of 131I in relation to strobilation of *Chrysaora quinquecirrha* (Scyphozoa). *Comparative Biochemistry and Physiology*, **45A**, 1023–1029.

Black, R.E., Enright, R.T. and Sung, L.-P. (1976) Activation of the dormant podocyst of *Chrysaora quinquecirrha* (Scyphozoa) by removal of the cyst covering. *Journal of Experimental Zoology*, **197**, 403–414.

Blank, R.J. (1986) Unusual chloroplast structures in endosymbiotic dinoflagellates: a clue to evolutionary differentiation within the genus *Symbiodinium (Dinophyceae)*. *Plant Systematics and Evolution*, **151**, 271–280.

Blank, R.J. and Huss, V.A.R. (1989) DNA divergency and speciation in *Symbiodinium (Dinophyceae)*. *Plant Systematics and Evolution*, **163**, 153–163.

Blank, R.J. and Trench, R.K. (1985) Speciation and symbiotic dinoflagellates. *Science (Washington D. C.)*, **229**, 656–658.

Blank, R.J., Huss, V.A.R. and Kersten, W. (1988) Base composition of DNA from symbiotic dinoflagellates: a tool for phylogenetic classification. *Archives of Microbiology*, **149**, 515–520.

Blanquet, R.S. (1972a) Structural and chemical aspects of the podocyst cuticle of the scyphozoan medusa, *Chrysaora quinquecirrha*. *Biological Bulletin (Woods Hole)*, **142**, 1–10.

Blanquet, R.S. (1972b) Temperature acclimation in the medusa, *Chrysaora quinquecirrha*. *Comparative Biochemistry and Physiology*, **43B**, 717–723.

Blanquet, R.S. and Phelan, M.A. (1987) An unusual blue mesogleal protein from the mangrove jellyfish *Cassiopea xamachana*. *Marine Biology (Berlin)*, **94**, 423–430.

Blanquet, R.S. and Riordan, G.P. (1981) An ultrastructural study of the subumbrellar musculature and desmosomal complexes of *Cassiopea xamachana* (Cnidaria: Scyphozoa). *Transactions of the American Microscopical Society*, **100**, 109–119.

Blanquet, R.S. and Wetzel, B. (1975) Surface ultrastructure of the scyphopolyp, *Chrysaora quinquecirrha. Biological Bulletin (Woods Hole)*, **148**, 181–192.

Blanquet, R.S., Emanuel, D. and Murphy, T.A. (1988) Suppression of exogenous alanine uptake in isolated zooxanthellae by cnidarian host homogenate fractions: species and symbiosis specificity. *Journal of Experimental Marine Biology and Ecology*, **117**, 1–8.

Bleakney, J.S. (1965) Reports of marine turtles from New England and eastern Canada. *Canadian Field-Naturalist*, **79**, 120–128.

Bleeker, J. and van der Spoel, S. (1988) Medusae of the Amsterdam mid North Atlantic plankton expeditions (1980–1983) with descriptions of two new species. *Bijdragen tot de Dierkunde*, **58**, 227–258.

Bocquet, J., Pujol, J.-P., Rolland, M. *et al.* (1972) Absence de mucopolysaccharides acides dans la mésoglée de *Rhizostoma pulmo* L. (Scyphoméduse). *Comptes Rendus Hebdomadaires des Seances de l'Academie des Sciences*, **274**, 101–104. In French.

Bodansky, M. and Rose, W.C. (1922) Comparative studies of digestion I The digestive enzymes of coelenterates. *American Journal of Physiology*, **62**, 473–481.

Bone, Q. and Brook, C.E.R. (1973) On *Schedophilus medusophagus* (Pisces: Stromateoidei). *Journal of the Marine Biological Association of the United Kingdom*, **53**, 753–761.

Bonnett, R., Head, E.J. and Herring, P.J. (1979) Porphyrin pigments of some deep-sea medusae. *Journal of the Marine Biological Association of the United Kingdom*, **59**, 565–573.

Borodkin, S.O., Nalbandov, Y.R. and Stunzhas, P.A. (1982) Elemental chemical composition of *Aurelia aurita* marine jellyfish and their role in the cycle of chemical elements in the Black Sea, in *Sezonnye Izmeneniya Chernomorskogo Planktona (Seasonal Changes in the Black Sea Plankton)*, NAUKA, Moscow, pp. 133–139. In Russian.

Bouillon, J. (1981) Origine et phylogenèse des cnidaires et des hydropolypes-hydroméduses. *Annales de la Société Royale Zoologique de Belgique*, **111**, 44–56. In French; English abstract.

Bouillon, J. and Coppois, G. (1977) Étude comparative de la mésoglée des Cnidaires. *Cahiers de Biologie Marine*, **18**, 339–368. In French; English abstract.

Bouillon, J. and Nielsen, M. (1974) Étude de quelques organes sensoriels de cnidaires. *Archives de Biologie*, **85**, 307–328. In French.

Bouillon, J. and Vandermeerssche, G. (1956) Structure et nature de la mésoglée des hydro- et scyphoméduses. *Annales de la Société Royale Zoologique de Belgique*, **87**, 7–25. In French.

Bowman, T.E., Meyers, C.D. and Hicks, S.D. (1963) Notes on associations between hyperiid amphipods and medusae in Chesapeake and Narragansett Bays and the Niantic River. *Chesapeake Science*, **4**, 141–146.

Bozhenova, O.V. (1988) Present views on the classification of the nematocysts of Cnidaria, in *Porifera and Cnidaria: Modern and Perspective Investigations*, (eds V.M. Koltun and S.D. Stepanjants), USSR Academy of Sciences, Moscow, pp. 57–71. In Russian; English abstract.

Bozhenova, O.V., Grebel'nyi, S.D. and Stepanjants, S.D. (1988) The possible ways of the cnidaria nematocysts evolution, in *Porifera and Cnidaria: Modern and Perspective Investigations*, (eds V.M. Koltun and S.D. Stepanjants), USSR Academy of Sciences, Moscow, pp. 71–74. In Russian; English abstract.

Bozler, E. (1926a) Sinnes- und Nervenphysiologische Untersuchungen an Scyphomedusen. *Zeitschrift fuer vergleichende Physiologie*, **4**, 37–80. In German.

Bozler, E. (1926b) Weitere Untersuchungen zur Sinnes- und Nervenphysiologie der Medusen: Erregungsleitung, Funktion der Randkörper, Nahrungsaufnahme. *Zeitschrift fuer vergleichende Physiologie*, **4**, 797–817. In German.

Bozler, E. (1927a) Untersuchungen über das Nervensystem der Coelenteraten I Kontinuität oder Kontakt zwischen den Nervenzellen? *Zeitschrift fuer Zellforschung und Mikroskopische Anatomie Abteilung Histochemie*, **5**, 244–262. In German.

Bozler, E. (1927b) Untersuchungen über das Nervensystem der Coelenteraten II Über die Struktur der Ganglienzellen und die Funktion der Neurofibrillen nach Lebenduntersuchungen. *Zeitschrift fuer vergleichende Physiologie*, **6**, 255–263. In German.

Brandon, M. and Cutress, C.E. (1985) A new *Dondice* (Opisthobranchia: Favorinidae), predator of *Cassiopea* in southwest Puerto Rico. *Bulletin of Marine Science*, **36**, 139–144.

Bratina, F., Rottini-Sandrini, L. and Avian, M. (1981) Simultaneous presence of *Pelagia noctiluca* (Forskal) planulae, ephyrae, and adults, index of sexual reproduction in the North Adriatic sea. *Bollettino della Societa Adriatica de Scienze Trieste*, **65**, 61–68.

Brewer, R.H. (1976) Larval settling behavior in *Cyanea capillata* (Cnidaria: Scyphozoa). *Biological Bulletin (Woods Hole)*, **150**, 183–199.

Brewer, R.H. (1978) Larval settlement behavior in the jellyfish *Aurelia aurita* (Linnaeus) (Scyphozoa: Semaeostomeae). *Estuaries*, **1**, 120–122.

Brewer, R.H. (1984) The influence of the orientation, roughness, and wettability of solid surfaces on the behavior and attachment of planulae of *Cyanea* (Cnidaria: Scyphozoa). *Biological Bulletin (Woods Hole)*, **166**, 11–21.

Brewer, R.H. (1989) The annual pattern of feeding, growth, and sexual reproduction in *Cyanea* (Cnidaria: Scyphozoa) in the Niantic River estuary, Connecticut. *Biological Bulletin (Woods Hole)*, **176**, 272–281.

Brewer, R.H. (1991) Morphological differences between, and reproductive isolation of, two populations of the jellyfish *Cyanea* in Long Island Sound, USA. *Hydrobiologia*, **216/217**, 471–477.

Brewer, R.H. and Feingold, J.S. (1991) The effect of temperature on the benthic stages of *Cyanea* (Cnidaria: Scyphozoa), and their seasonal distribution in the Niantic River estuary, Connecticut. *Journal of Experimental Marine Biology and Ecology*, **152**, 49–60.

Bridge, D., Cunningham, C.W., Schierwater, B. *et al.* (1992) Class-level relationships in the phylum Cnidaria: Evidence from mitochondrial genome structure. *Proceedings of the National Academy of Sciences of the United States of America*, **89**, 8750–8753.

Bridge, D., Cunningham, C.W., Desalle, R. *et al.* (1995) Class-level relationships in the Phylum Cnidaria: molecular and morphological evidence. *Molecular Biology and Evolution*, **12**, 679–689.

Brock, M.A. and Strehler, B.L. (1963) Studies on the comparative physiology of aging IV Age and mortality of some marine cnidaria in the laboratory. *Journal of Gerontology*, **18**, 23–28.

Browder, L.W., Erickson, C.A. and Jeffery, W.R. (1991) *Developmental Biology*, Saunders College Publishing, Philadelphia.

Bruton, D.L. (1991) Beach and laboratory experiments with the jellyfish *Aurelia* and remarks on some fossil 'medusoid' traces, in *The Early Evolution of Metazoa and the Significance of Problematic Taxa*, (eds A.M. Simonetta and S. Conway Morris), Cambridge University Press, Cambridge, pp. 125–129.

Bullock, T.H. (1943) Neuromuscular facilitation in scyphomedusae. *Journal of Cellular and Comparative Physiology*, **22**, 251–272.

Bullock, T.H. (1955) Compensation for temperature in the metabolism and activity of poikilotherms. *Biological Reviews of the Cambridge Philosophical Society*, **30**, 311–342.

Burggren, W. and Roberts, J. (1991) Respiration and metabolism, in *Environmental and Metabolic Animal Physiology*, Comparative Animal Physiology, 4, (ed. C.L. Prosser), Wiley-Liss, New York, pp. 353–435.

Burnett, J.W. (1971) An electron microscopic study of two nematocytes in the tentacle of *Cyanea capillata*. *Chesapeake Science*, **12**, 67–71.

Burnett, J.W. (1991a) Clinical manifestations of jellyfish envenomation. *Hydrobiologia*, **216/217**, 629–635.

Burnett, J.W. (1991b) Jellyfish envenomation syndromes worldwide, in *UNEP: Jellyfish Blooms in the Mediterranean. Proceedings of the II Workshop on Jellyfish in the Mediterranean Sea*, MAP Technical Reports Series No. 47, UNEP, Athens, pp. 227–235.

Burnett, J.W. and Calton, G.J. (1987) Venomous pelagic coelenterates: chemistry, toxicology, immunology and treatment of their stings. *Toxicon*, **25**, 581–602.

Burnett, J.W. and Sutton, J.S. (1969) The fine structural organization of the sea nettle fishing tentacle. *Journal of Experimental Zoology*, **172**, 335–348.

Burnett, J.W., Calton, G.J. and Burnett, H.W. (1986) Jellyfish envenomation syndromes. *Journal of the American Academy of Dermatology*, **14**, 100–106.

Bynum, M.A. and Black, R.E. (1974) Ultrastructure of the mesoglea in strobilae of *Chrysaora quinquecirrha* (Scyphozoa). *Journal of Experimental Zoology*, **187**, 323–333.

Calder, D.R. (1971) Nematocysts of polyps of *Aurelia, Chrysaora* and *Cyanea*, and their utility in identification. *Transactions of the American Microscopical Society*, **90**, 269–274.

Calder, D.R. (1972a) Development of the sea nettle *Chrysaora quinquecirrha* (Scyphozoa, Semaeostomeae). *Chesapeake Science*, **13**, 40–44.

Calder, D.R. (1972b) Nematocysts of the medusa stage of *Rhopilema verrilli* (Scyphozoa, Rhizostomeae). *Transactions of the American Microscopical Society*, **91**, 213–216.

Calder, D.R. (1973) Laboratory observations on the life history of *Rhopilema verrilli* (Scyphozoa: Rhizostomeae). *Marine Biology (Berlin)*, **21**, 109–114.

Calder, D.R. (1974a) Nematocysts of the coronate scyphomedusa, *Linuche unguiculata*, with a brief reexamination of scyphozoan nematocyst classification. *Chesapeake Science*, **15**, 170–173.

Calder, D.R. (1974b) Strobilation of the sea nettle, *Chrysaora quinquecirrha*, under field conditions. *Biological Bulletin (Woods Hole)*, **146**, 326–334.

Calder, D.R. (1977) Nematocysts of the ephyra stages of *Aurelia*, *Chrysaora*, *Cyanea*, and *Rhopilema* (Cnidaria, Scyphozoa). *Transactions of the American Microscopical Society*, **96**, 13–19.

Calder, D.R. (1982) Life history of the cannonball jellyfish, *Stomolophus meleagris* L. Agassiz, 1860 (Scyphozoa, Rhizostomida). *Biological Bulletin (Woods Hole)*, **162**, 149–162.

Calder, D.R. (1983) Nematocysts of stages in the life cycle of *Stomolophus meleagris*, with keys to scyphistomae and ephyrae of some western Atlantic Scyphozoa. *Canadian Journal of Zoology*, **61**, 1185–1192.

Calder, D.R. and Peters, E.C. (1975) Nematocysts of *Chiropsalmus quadrumanus* with comments on the systematic status of the Cubomedusae. *Helgolaender Wissenschaftliche Meeresuntersuchungen*, **27**, 364–369.

Calton, G.J. and Burnett, J.W. (1981) Uptake of the herbicide pendimethalin by the sea nettle *Chrysaora quinquecirrha*. *Bulletin of Environmental Contamination and Toxicology*, **26**, 386–388.

Calton, G.J. and Burnett, J.W. (1982a) Partial purification and characterization of the acid protease of sea nettle (*Chrysaora quiquecirrha*) nematocyst venom. *Comparative Biochemistry and Physiology B Comparative Biochemistry*, **72**, 93–97.

Calton, G.J. and Burnett, J.W. (1982b) Partial purification and characterization of the alkaline protease of sea nettle (*Chrysaora quinquecirrha*) nematocyst venom. *Comparative Biochemistry and Physiology C Comparative Pharmacology*, **74**, 361–364.

Campbell, R.D. (1988) Migration of nematocysts in hydrozoans, in *The Biology of Nematocysts*, (eds D.A. Hessinger and H.M. Lenhoff), Academic Press, San Diego, pp. 123–142.

Cargo, D.G. (1971) The sessile stages of a scyphozoan identified as *Rhopilema verrilli*. *Tulane Studies in Zoology and Botany*, **17**, 31–34.

Cargo, D.G. (1974) Comments on the laboratory culture of Scyphozoa, in *Culture of Marine Invertebrate Animals*, (eds W.L. Smith and M.H. Chanley), Plenum Publishing Corporation, New York, pp. 145–154.

Cargo, D.G. (1979) Observations on the settling behavior of planular larvae of *Chrysaora quinquecirrha*. *International Journal of Invertebrate Reproduction*, **1**, 279–287.

Cargo, D.G. (1984) Some laboratory techniques for the culture of scyphozoa, in *Workshop on Jellyfish Blooms in the Mediterranean*, UNEP, Athens, pp. 129–138.

Cargo, D.G. and Burnett, J.W. (1982) Observations on the ultrastructure and defensive behavior of the cnidosac of *Cratena pilata*. *Veliger*, **24**, 325–327.

Cargo, D.G. and King, D.R. (1990) Forecasting the abundance of the sea nettle, *Chrysaora quinquecirrha*, in the Chesapeake Bay. *Estuaries*, **13**, 486–491.

Cargo, D.G. and Rabenold, G.E. (1980) Observations on the asexual reproductive activities of the sessile stages of the sea nettle *Chrysaora quinquecirrha* (Scyphozoa). *Estuaries*, **3**, 20–27.

Cargo, D.G. and Schultz, L.P. (1966) Notes on the biology of the sea nettle, *Chrysaora quinquecirrha*, in Chesapeake Bay. *Chesapeake Science*, **7**, 95–100.

Cargo, D.G. and Schultz, L.P. (1967) Further observations on the biology of the sea nettle and jellyfishes in Chesapeake Bay. *Chesapeake Science*, **8**, 209–220.

Cariello, L., Romano, G., Spagnuolo, A. *et al.* (1988) Isolation and partial characterization of rhizolysin, a high molecular weight protein with hemolytic activity, from the jellyfish *Rhizostoma pulmo*. *Toxicon*, **26**, 1057–1065.

Carlberg, M. and Rosengren, E. (1985) Biochemical basis for adrenergic neurotransmission in coelenterates. *Journal of Comparative Physiology B Biochemical Systematic and Environmental Physiology*, **155**, 251–255.

Carlberg, M., Alfredsson, K., Nielsen, S.-O. *et al.* (1995) Taurine-like immunoreactivity in the motor nerve net of the jellyfish *Cyanea capillata*. *Biological Bulletin (Woods Hole)*, **188**, 78–82.

Carli, A., Pane, L., Valente, T. *et al.* (1991) Lipid and protein content of jellyfish from the Ligurian Sea. First results, in *UNEP: Jellyfish Blooms in the Mediterranean. Proceedings of the II Workshop on Jellyfish in the Mediterranean Sea*, MAP Technical Reports Series No. 47, UNEP, Athens, pp. 236–240.

Carlton, J.T. (1985) Transoceanic and interoceanic dispersal of coastal marine organisms: The biology of ballast water. *Oceanography and Marine Biology; an Annual Review*, **23**, 313–374.

Carré, D. and Carré, C. (1980) On triggering and control of cnidocyst discharge. *Marine Behaviour and Physiology*, **7**, 109–117.

Carroll, S. and Blanquet, R.S. (1984a) Alanine uptake by isolated zooxanthellae of the mangrove jellyfish, *Cassiopea xamachana* II Inhibition by host homogenate fraction. *Biological Bulletin (Woods Hole)*, **166**, 419–426.

Carroll, S. and Blanquet, R.S. (1984b) Alanine uptake by isolated zooxanthellae of the mangrove jellyfish, *Cassiopea xamachana* I Transport mechanisms and utilization. *Biological Bulletin (Woods Hole)*, **166**, 409–418.

Cary, L.R. (1916) The influence of the marginal sense organs on metabolic activity in *Cassiopea xamachana*. *Proceedings of the National Academy of Sciences of the United States of America*, **2**, 709–712.

Cates, N. (1975) Productivity and organic consumption in *Cassiopea* and *Condylactus*. *Journal of Experimental Marine Biology and Ecology*, **18**, 55–59.

Cates, N. and McLaughlin, J.J.A. (1976) Differences of ammonia metabolism in symbiotic and aposymbiotic *Condylactus* and *Cassiopea* spp. *Journal of Experimental Marine Biology and Ecology*, **21**, 1–5.
Chang, S.S. and Trench, R.K. (1982) Peridinin-chlorophyll *a* proteins from the symbiotic dinoflagellate *Symbiodinium* (=*Gymnodinium*) *microadriaticum*, Freudenthal. *Proceedings of the Royal Society of London B Biological Sciences*, **215**, 191–210.
Chang, S.S. and Trench, R.K. (1984) The isoelectric forms, quaternary structure and amino acid composition of peridinin-chlorophyll *a* proteins from the symbiotic dinoflagellate *Symbiodinium microadriaticum* Freudenthal. *Proceedings of the Royal Society of London B Biological Sciences*, **222**, 259–271.
Chapman, D.M. (1965) Co-ordination in a scyphistoma. *American Zoologist*, **5**, 455–464.
Chapman, D.M. (1966) Evolution of the scyphistoma, in *The Cnidaria and their Evolution*, (ed. W.J. Rees), Academic Press, London, pp. 51–75.
Chapman, D.M. (1968) Structure, histochemistry and formation of the podocyst and cuticle of *Aurelia aurita*. *Journal of the Marine Biological Association of the United Kingdom*, **48**, 187–208.
Chapman, D.M. (1969) The nature of cnidarian desmocytes. *Tissue and Cell*, **1**, 619–632.
Chapman, D.M. (1970a) Further observations on podocyst formation. *Journal of the Marine Biological Association of the United Kingdom*, **50**, 107–111.
Chapman, D.M. (1970b) Reextension mechanism of a scyphistoma's tentacle. *Canadian Journal of Zoology*, **48**, 931–943.
Chapman, D.M. (1973) Behavior and flagellar currents in coronate polyps (Scyphozoa) and comparisons with semaeostome polyps. *Helgolaender Wissenschaftliche Meeresuntersuchungen*, **25**, 214–227.
Chapman, D.M. (1974) Cnidarian histology, in *Coelenterate Biology: Reviews and New Perspectives*, (eds L. Muscatine and H.M. Lenhoff), Academic Press, New York, pp. 1–92.
Chapman, D.M. (1985) X-ray microanalysis of selected coelenterate statoliths. *Journal of the Marine Biological Association of the United Kingdom*, **65**, 617–627.
Chapman, D.M. and James, R. (1973) Intraepithelial flagella in the medusa of *Aurelia aurita* (L.). *Publications of the Seto Marine Biological Laboratory*, **20**, 731–743. Proc. Second Int. Symp. Cnidaria.
Chapman, G. (1953) Studies of the mesogloea of coelenterates I Histology and chemical properties. *Quarterly Journal of Microscopical Science*, **94**, 155–176.
Chapman, G. (1959) The mesogloea of *Pelagia noctiluca*. *Quarterly Journal of Microscopical Science*, **100**, 599–610.
Chapman, G. (1966) The structure and functions of the mesogloea, in *The Cnidaria and their Evolution*, (ed. W.J. Rees), Academic Press, London, pp. 147–168.
Chapman, G. (1976a) Reflections on transparency, in *Coelenterate Ecology and Behavior*, (ed. G.O. Mackie), Plenum Press, New York, pp. 491–499.

Chapman, G. (1976b) Transparency in organisms. *Experientia (Basel)*, **32**, 123–125.
Chen, J. and Ding, G. (1983) Effect of temperature on the strobilation of jellyfish (*Rhopilema esculenta* Kishinouye – Scyphozoa, Rhizostomeae). *Acta Zoologica Sinica*, **29**, 195–206. In Chinese; English abstract.
Chen, J., Ding, G. and Liu, C. (1984) Effect of light on the strobilation of edible medusa, *Rhopilema esculenta* Kishinouye (Cnidaria, Scyphozoa). *Oceanologica et Limnologica Sinica*, **15**, 310–316. In Chinese; English abstract.
Chen, J., Ding, G. and Liu, C. (1985) Effect of nutritional conditions on the strobilation of edible medusa, *Rhopilema esculenta* Kishinouye. *Journal of Fisheries of China*, **9**, 321–329. In Chinese; English abstract.
Chen, P.K., Chen, T.C. and Tseng, R.Y.L. (1983) *Everything You Want to Know about Chinese Cooking*, Barron's, New York.
Chia, F.-S., Amerongen, H.M. and Peteya, D.J. (1984) Ultrastructure of the neuromuscular system of the polyp of *Aurelia aurita* L. 1758 (Cnidaria, Scyphozoa). *Journal of Morphology*, **180**, 69–79.
Chia, F.-S. and Bickell, L.R. (1978) Mechanisms of larval attachment and the induction of settlement and metamorphosis in coelenterates: A review, in *Settlement and Metamorphosis of Marine Invertebrate Larvae*, (eds F.-S. Chia and M.E. Rice), Elsevier, New York, pp. 1–12.
Chia, F.-S., Buckland-Nicks, J. and Young, C.M. (1984) Locomotion of marine invertebrate larvae: a review. *Canadian Journal of Zoology*, **62**, 1205–1222.
Child, C.A. and Harbison, G.R. (1986) A parasitic association between a pycnogonid and a scyphomedusa in midwater. *Journal of the Marine Biological Association of the United Kingdom*, **66**, 113–117.
Christomanos, A. (1954) A violet pigment from the Mediterranean medusa *Rhizostoma pulmo*. *Nature (London)*, **173**, 875–876.
Chuin, T.T. (1929a) La digestion chèz le scyphistome de *Chrysaora*. *Comptes Rendus des Séances de la Société de Biologie et de ses Filiales*, **102**, 520–522. In French.
Chuin, T.T. (1929b) Les phénomènes cytologiques au cours de la digestion intracellulaire chez le scyphistome de *Chrysaora*. *Comptes Rendus des Seances de la Societe de Biologie et de ses Filiales*, **102**, 557–558. In French.
Chuin, T.-T. (1930) Le cycle évolutif du scyphistome de *Chrysaora*. *Travaux de la Station Biologique de Roscoff*, **8**, 1–179, pls 1–5. In French.
Cimino, G., Alfa, M. and La Spada, G. (1983) Trace elements in tentacles from the jellyfish *Pelagia noctiluca*. *Marine Pollution Bulletin*, **14**, 197–198.
Clarke, A., Holmes, L.J. and Gore, D.J. (1992) Proximate and elemental composition of gelatinous zooplankton from the Southern Ocean. *Journal of Experimental Marine Biology and Ecology*, **155**, 55–68.
Clarke, G.L., Conover, R.J., David, C.N. *et al.* (1962) Comparative studies of luminescence in copepods and other pelagic marine animals. *Journal of the Marine Biological Association of the United Kingdom*, **42**, 541–564, pls 1–2.

Claus, C. (1877) Studien über Polypen und Quallen der Adria. *Denkschriften der Kaiserlichen Akademie der Wissenschaften Mathematisch-Naturwissenschaftliche Klasse* , **38**, 1–64, pls 1–11. In German.

Clifford, H.C. and Cargo, D.G. (1978) Feeding rates of the sea nettle, *Chrysaora quinquecirrha*, under laboratory conditions. *Estuaries*, **1**, 58–61.

Cobbs, C.S., Drzymala, R.E., Shamoo, A.E. *et al.* (1983) Sea nettle (*Chrysaora quinquecirrha*) lethal factor: effect on black lipid membranes. *Toxicon*, **21**, 558–561.

Coleman, C.O. (1990) Anatomy of the alimentary canal of *Parandania boecki* (Stebbing, 1888) (Crustacea, Amphipoda, Stegocephalidae) from the Antarctic Ocean. *Journal of Natural History*, **24**, 1573–1585.

Collado-Vides, L., Segura-Puertas, L. and Merino-Ibarra, M. (1988) Observaciones sobre dos escifomedusas del género *Cassiopea* en la Laguna de Bojórquez, Quintana Roo, México. *Revista de Investigaciones Marinas*, **9**, 21–27. In Spanish; English abstract.

Colley, N.J. and Trench, R.K. (1983) Selectivity in phagocytosis and persistence of symbiotic algae by the scyphistoma stage of the jellyfish *Cassiopeia xamachana*. *Proceedings of the Royal Society of London B Biological Sciences*, **219**, 61–82.

Colley, N.J. and Trench, R.K. (1985) Cellular events in the reestablishment of a symbiosis between a marine dinoflagellate and a coelenterate. *Cell and Tissue Research*, **239**, 93–103.

Colton, J.B. and Temple, R.F. (1961) The enigma of Georges Bank spawning. *Limnology and Oceanography*, **6**, 280–291.

Cones, H.N. (1969) Strobilation of *Chrysaora quinquecirrha* polyps in the laboratory. *Virginia Journal of Science*, **20**, 16–18.

Cones, H.N. and Haven, D.S. (1969) Distribution of *Chrysaora quinquecirrha* in the York River. *Chesapeake Science*, **10**, 75–84.

Conklin, E.G. (1908) The habits and early development of *Linerges mercurius*. *Carnegie Institution of Washington Publication*, **103**, 155–70, pls 1–8. Papers from the Tortugas Laboratory of the Carnegie Institution of Washington.

Conway Morris, S. (1985) Non-skeletalized lower invertebrate fossils: a review, in *The Origins and Relationships of Lower Invertebrates*, (eds S. Conway Morris, J.D. George, R. Gibson, *et al.*), Clarendon Press, Oxford, pp. 343–359.

Conway Morris, S. (1991) Problematic taxa: a problem for biology or biologists?, in *The Early Evolution of Metazoa and the Significance of Problematic Taxa*, (eds A.M. Simonetta and S. Conway Morris), Cambridge University Press, Cambridge, pp. 19–24.

Cormier, M.J. (1978) Comparative biochemistry of animal systems, in *Bioluminescence in Action*, (ed. P.J. Herring), Academic Press, London, pp. 75–108.

Cornelius, P.F.S. (1972) Second occurrence of *Stygiomedusa fabulosa* (Scyphozoa). *Journal of the Marine Biological Association of the United Kingdom*, **52**, 487–488.

Cornelius, P.F.S. (1973) A taxonomic appraisal of the two species of *Stygiomedusa* (Coelenterata, Scyphozoa), with the first southern ocean records. *Journal of the Marine Biological Association of the United Kingdom*, **53**, 609–613.

Cornelius, P.F.S. (1992) Medusa loss in leptolid Hydrozoa (Cnidaria), hydroid rafting, and abbreviated life-cycles among their remote-island faunae: an interim review. *Scientia Marina*, **56**, 245–261.

Cornelius, P.F.S. (in press) Keys to the genera of Cubomedusae and Scyphomedusae, in *Proceedings of the 6th International Conference on Coelenterate Biology*, (ed. J.C. den Hartog), Nationaal Natuurhistorisch Museum, Leiden.

Costello, J. (1992) Foraging mode and energetics of hydrozoan medusae. *Scientia Marina*, **56**, 185–191.

Costello, J.H. and Colin, S.P. (1994) Morphology, fluid motion and predation by the scyphomedusa *Aurelia aurita*. *Marine Biology (Berlin)*, **121**, 327–334.

Costello, J.H. and Kremer, P.M. (1989) Circadian rhythmicity in the location of zooxanthellae of the scyphomedusa *Linuche unguiculata*. *Marine Ecology Progress Series*, **57**, 279–286.

Cowan, J.H. and Houde, E.D. (1992) Size-dependent predation on marine fish larvae by ctenophores, scyphomedusae, and planktivorous fish. *Fisheries Oceanography*, **1**, 113–126.

Cowan, J.H. and Houde, E.D. (1993) Relative predation potentials of scyphomedusae, ctenophores and planktivorous fish on ichthyoplankton in Chesapeake Bay. *Marine Ecology Progress Series*, **95**, 55–65.

Coyne, J.A. (1973) An investigation of the dynamics of population growth and control in scyphistomae of the scyphozoan *Aurelia aurita*. *Chesapeake Science*, **14**, 55–58.

Crafts, C.B. and Tuliszewski, J.R. (1995) Motility rhythms in cultured zooxanthellae isolated from the scyphomedusa *Linuche unguiculata*. *Bulletin of Marine Science*, **56**, 822–825.

Crawford, M.A. and Webb, K.L. (1972) An ultrastructural study of strobilation in *Chrysaora quinquecirrha* with special reference to neurosecretion. *Journal of Experimental Zoology*, **182**, 251–270.

Curl, H. (1962) Standing crops of carbon, nitrogen, and phosphorus and transfer between trophic levels, in continental shelf waters south of New York. *Rapports et Proces-verbaux des Réunions Conseil international pour l'Exploration de la Mer*, **153**, 183–189.

Curtis, S.K. and Cowden, R.R. (1971) Normal and experimentally modified development of buds in *Cassiopea* (Phylum Coelenterata; Class Scyphozoa). *Acta Embryologiae Experimentalis*, **3**, 239–259.

Curtis, S.K. and Cowden, R.R. (1972) Regenerative capacities of the scyphistoma of *Cassiopea* (Phylum, Coelenterata; Class, Scyphozoa). *Acta Embryologiae Experimentalis*, **Suppl. 1972**, 429–454.

Curtis, S.K. and Cowden, R.R. (1974) Some aspects of regeneration in the scyphistoma of *Cassiopea* (Class Scyphozoa) as revealed by the use of antimetabolites and microspectrophotometry. *American Zoologist*, **14**, 851–866.

Czarnetzki, B.M., Thiele, T. and Rosenbach, T. (1990) Evidence for leukotrienes in animal venoms. *Journal of Allergy and Clinical Immunology*, **85**, 505–509.

Czeczuga, B. (1970) Some carotenoids in the jelly-fish *Aurelia aurita* (Scyphozoa: Discomedusae). *Marine Biology (Berlin)*, **5**, 141–144.

Dahl, E. (1959a) The amphipod, *Hyperia galba*, an ectoparasite of the jelly-fish, *Cyanea capillata*. *Nature (London)*, **183**, 1749.

Dahl, E. (1959b) The hyperiid amphipod, *Hyperia galba*, a true ecto-parasite on jelly-fish. *Universitetet i Bergen Arbok Naturvitenskapelig Rekke*, **9**, 1–8.

Dahl, E. (1961) The association between young whiting, *Gadus merlangus*, and the jelly-fish *Cyanea capillata*. *Sarsia*, **3**, 47–55.

Dahlgren, U. (1916) The production of light by animals: Porifera and Coelenterata. *Journal of the Franklin Institute*, **181**, 243–261.

Daniel, T.L. (1983) Mechanics and energetics of medusan jet propulsion. *Canadian Journal of Zoology*, **61**, 1406–1420.

Daumas, R., Ceccaldi, H.J. (1965) Contribution à l'étude biochimique d'organismes marins I Acides amines libres et proteiques chez *Beroe ovata* (Eschscholtz), *Ciona intestinalis* (L.), *Cymbulia peroni* (De Blainville) et *Rhizostoma pulmo* (Agassiz). *Recueil des Travaux de la Station Marine d'Endoume Faculté des Sciences de Marseille*, **38**, 3–14. In French; English abstract.

Davenport, J. (1988) Do diving leatherbacks pursue glowing jelly? *British Herpetological Society Bulletin*, **24**, 20–21.

Davenport, J. and Trueman, E.R. (1985) Oxygen uptake and buoyancy in zooplanktonic organisms from the tropical eastern Atlantic. *Comparative Biochemistry and Physiology*, **81A**, 857–863.

de Beer, G.R. and Huxley, J.S. (1924) Studies in dedifferentation 5 Dedifferentation and reduction in *Aurelia*. *Quarterly Journal of Microscopical Science*, **68**, 471–479.

de Couet, H.G., Mazander, K.-D. and Gröschel-Stewart, U. (1980) A study of invertebrate actins by isoelectric focusing and immunodiffusion. *Experientia (Basel)*, **36**, 404–405.

de Lafontaine, Y. and Leggett, W.C. (1987) Effect of container size on estimates of mortality and predation rates in experiments with macro-zooplankton and larval fish. *Canadian Journal of Fisheries and Aquatic Sciences*, **44**, 1534–1543.

de Lafontaine, Y. and Leggett, W.C. (1988) Predation by jellyfish on larval fish: an experimental evaluation employing in situ enclosures. *Canadian Journal of Fisheries and Aquatic Sciences*, **45**, 1173–1190.

Delap, M.J. (1906) Notes on the rearing, in an aquarium, of *Aurelia aurita*, L. and *Pelagia perla* (Slabber). *Report on the Sea and Inland Fisheries of Ireland II Scientific Investigations*, **1905**, 22–26, Pl. I–II.

Den Hartog, J.C. (1980) Notes on the food of sea turtles: *Eretmochelys imbricata* (Linnaeus) and *Dermochelys coriacea* (Linnaeus). *Netherlands Journal of Zoology*, **30**, 595–610.

Den Hartog, J.C. and Van Nierop, M.M. (1984) A study on the gut contents of six leathery turtles *Dermochelys coriacea* (Linnaeus) (Reptilia:

Testudines: Dermochelyidae) from British waters and from the Netherlands. *Zoologische Verhandelingen uitgegeven door het Rijksmuseum van Natuurlijke Historie te Leiden*, **209**, 1–36.

de Wolf, P. (1989) The price of patchiness. *Helgolaender Meeresuntersuchungen*, **43**, 263–273.

Digby, P.S.B. (1967) Pressure sensitivity and its mechanism in the shallow marine environment. *Symposia of the Zoological Society of London*, **19**, 159–188.

Dillon, T.M. (1977) Effects of acute changes in temperature and salinity on pulsation rates in ephyrae of the scyphozoan *Aurelia aurita*. *Marine Biology (Berlin)*, **42**, 31–35.

Ding, G. and Chen, J. (1981) The life history of *Rhopilema esculenta* Kishinouye. *Journal of Fisheries of China*, **5**, 93–104. In Chinese; English abstract.

Dittrich, B. (1987) Postembryonic development of the parasitic amphipod *Hyperia galba*. *Helgolaender Meeresuntersuchungen*, **41**, 217–232.

Dittrich, B. (1988) Studies on the life cycle and reproduction of the parasitic amphipod *Hyperia galba* in the North Sea. *Helgolaender Meeresuntersuchungen*, **42**, 79–98.

Dollfus, R.P. (1963) Liste des coelenteres marins, palearctiques et indiens, ou ont été trouves des trematodes digenetiques. *Bulletin de l'Institut des Peches Maritimes du Maroc*, **9–10**, 33–57. In French; English abstract.

Dolmer, P. and Svane, I. (1993) Settlement patterns of the Scyphozoan *Cyanea capillata* (L.) planula: Effects of established scyphistomae and water flow. *Ophelia*, **38**, 117–126.

Doores, S. and Cook, T.M. (1976) Occurrence of *Vibrio* and other bacteria on the sea nettle *Chrysaora quinquecirrha*. *Microbial Ecology*, **3**, 31–40.

dos Santos, M.E. and Lacerda, M. (1987) Preliminary observations of the bottlenose dolphin (*Tursiops truncatus*) in the Sado Estuary (Portugal). *Aquatic Mammals*, **13**, 65–80.

Doyle, A.C. (1930) The Adventure of the Lion's Mane, in *The Complete Sherlock Holmes*, Doubleday and Company, New York, pp. 1083–1095.

Drew, E.A. (1972) The biology and physiology of alga–invertebrate symbioses I Carbon fixation in *Cassiopea* sp. at Aldabra Atoll. *Journal of Experimental Marine Biology and Ecology*, **9**, 65–69.

Dubois, J.M., Tanguy, J. and Burnett, J.W. (1983) Ionic channels induced by sea nettle toxin in the nodal membrane. *Biophysical Journal*, **42**, 199–202.

Duffy, D.C. (1988) Predator–prey interactions between common terns and butterfish. *Ornis Scandinavica*, **19**, 160–163.

Dunn, D.F. (1982) Cnidaria, in *Synopsis and Classification of Living Organisms*, (ed. S.P. Parker), McGraw-Hill Book Co., New York, pp. 669–706.

Duron, M., Quero, J.C. and Duron, P. (1983) Présence dans les eaux cotières de France et de Guyane fréquentées par *Dermochelys coriacea* L., de *Remora remora* L., et de *Rhizostoma pulmo* L. *Annales de la Société des Sciences Naturelles de la Charente-Maritime*, **7**, 147–151. In French; English abstract.

Dykens, J.A. (1984) Enzymic defenses against oxygen toxicity in marine cnidarians containing endosymbiotic algae. *Marine Biology Letters*, **5**, 291–301.

Eckelbarger, K.J. and Larson, R.J. (1988) Ovarian morphology and oogenesis in *Aurelia aurita* (Scyphozoa: Semaeostomae): ultrastructural evidence of heterosynthetic yolk formation in a primitive metazoan. *Marine Biology (Berlin)*, **100**, 103–115.

Eckelbarger, K.J. and Larson, R.J. (1992) Ultrastructure of the ovary and oogenesis in the jellyfish *Linuche unguiculata* and *Stomolophus meleagris*, with a review of ovarian structure in the Scyphozoa. *Marine Biology (Berlin)*, **114**, 633–643.

Eckelbarger, K.J. and Larson, R.J. (1993) Ultrastructural study of the ovary of the sessile scyphozoan *Haliclystus octoradiatus* (Cnidaria: Stauromedusae). *Journal of Morphology*, **218**, 225–236.

Eimer, T. (1874) Ueber künstliche Theilbarkeit von *Aurelia aurita* und *Cyanea capillata* in physiologische Individuen. *Verhandlungen der Physikalisch-Medizinischen Gesellschaft zu Wurzburg*, **6**, 137–161, pl. 2. In German.

Eimer, T. (1877) Ueber künstliche Theilbarkeit und über das Nervensystem der Medusen. *Archiv fuer Mikroskopische Anatomie*, **14**, 394–408. In German.

Eisenberg, J.F. and Frazier, J. (1983) A leatherback turtle (*Dermichelys coriacea*) feeding in the wild. *Journal of Herpetology*, **17**, 81–82.

Elder, H.Y. (1966) The fine structure of some invertebrate fibrillar and lamellar elastica. *Proceedings of the Royal Microscopical Society*, **1**, 99–100.

Elder, H.Y. (1973) Distribution and functions of elastic fibers in the invertebrates. *Biological Bulletin (Woods Hole)*, **144**, 43–63.

Elder, H.Y. and Owen, G. (1967) Occurrence of "elastic" fibres in the invertebrates. *Journal of Zoology (London)*, **152**, 1–8.

El-Duweini, A.K. (1945) Cannibalism in *Aurelia*. *Nature (London)*, **155**, 337.

Elofsson, R. and Carlberg, M. (1989) Gland cells in the tentacles of the jellyfish *Cyanea lamarcki* reactive with an antibody against 5-hydroxytryptamine. *Cell and Tissue Research*, **255**, 419–422.

Erokhin, V.Y. (1979) The use of antibiotics when studying the feeding of invertebrates on exometabolites. *Hydrobiological Journal (English Translation of Gidrobiologicheskii Zhurnal)*, **14**, 77–80.

Fancett, M.S. (1988) Diet and prey selectivity of scyphomedusae from Port Phillip Bay, Australia. *Marine Biology (Berlin)*, **98**, 503–509.

Fancett, M.S. and Jenkins, G.P. (1988) Predatory impact of scyphomedusae on ichthyoplankton and other zooplankton in Port Phillip Bay. *Journal of Experimental Marine Biology and Ecology*, **116**, 63–77.

Fautin, D.G. and Fitt, W.K. (1991) A jellyfish-eating sea anemone (Cnidaria, Actinaria) from Palau: *Entacmaea medusivora* sp. nov. *Hydrobiologia*, **216/217**, 453–461.

Fautin, D.G. and Lowenstein, J.M. (1992) Scyphomedusae and their polyps are the same immunologically: Implications for systematics. *Comparative Biochemistry and Physiology*, **102B**, 13–14.

Feigenbaum, D. and Kelly, M. (1984) Changes in the lower Chesapeake Bay food chain in presence of the sea nettle *Chrysaora quinquecirrha* (Scyphomedusa). *Marine Ecology Progress Series*, **19**, 39–47.

Feldmann, R.M. and Babcock, L.E. (1986) Exceptionally preserved conulariids from Ohio – reinterpretation of their anatomy. *National Geographic Research*, **2**, 464–472.

Ferguson, J.C. (1988) Autoradiographic demonstration of the use of free amino acid by Sargasso Sea zooplankton. *Journal of Plankton Research*, **10**, 1225–1238.

Fernandez, G.A., Daniel, A. and Nicket, P. (1977) Oil spills in the marine environment at Madras and its impact on neuston and plankton during 1975–1976, in *Proceedings of the Symposium on the Warm Water Zooplankton*, National Institute of Oceanography, Goa, pp. 432–439.

Fitt, W.K. (1984) The role of chemosensory behavior of *Symbiodinium microadriaticum*, intermediate hosts, and host behavior in the infection of coelenterates and molluscs with zooxanthellae. *Marine Biology (Berlin)*, **81**, 9–17.

Fitt, W.K. (1985) Effect of different strains of the zooxanthella *Symbiodinium microadriaticum* on growth and survival of their crustacean and molluscan hosts, in *Proceedings of the Fifth International Coral Reef Congress Tahiti, 27 May–1 June 1985*, Antenne Museum-Ephe, Moorea, pp. 131–136.

Fitt, W.K. and Hoffmann, D.K. (1985) Chemical induction of settlement and metamorphosis of planulae and buds of the reef-dwelling coelenterate *Cassiopea andromeda*, in *Proceedings of the Fifth International Coral Reef Congress Tahiti, 27 May–1 June 1985*, Antenne Museum-Ephe, Moorea, pp. 239–244.

Fitt, W.K. and Trench, R.K. (1983) Endocytosis of the symbiotic dinoflagellate *Symbiodinium microadriaticum* Freudenthal by endodermal cells of the scyphistomae of *Cassiopeia xamachana* and resistance of the algae to host digestion. *Journal of Cell Science*, **64**, 195–212.

Fitt, W.K., Chang, S.S. and Trench, R.K. (1981) Motility patterns of different strains of the symbiotic dinoflagellate *Symbiodinium (=Gymnodinium) microadriaticum* Freudenthal in culture. *Bulletin of Marine Science*, **31**, 436–443.

Fitt, W.K., Hofmann, D.K., Wolk, M. *et al.* (1987) Requirement of exogenous inducers for metamorphosis of axenic larvae and buds of *Cassiopea andromeda* (Cnidaria: Scyphozoa). *Marine Biology (Berlin)*, **94**, 415–422.

Fleck, J. and Bischoff, A. (1993) Protein kinase C is possibly involved in chemical induction of metamorphosis in *Cassiopea* spp. (Cnidaria: Scyphozoa), in *Proceedings 7th International Coral Reef Symposium*, (ed. R.H. Richmond), University of Guam Press, Mangilao, Guam, pp. 456–462.

Fleck, J. and Hofmann, D.K. (1990) The efficiency of metamorphosis inducing oligopeptides in *Cassiopea* species depends on both primary structure and amino- and carboxiterminal substituents. *Verhandlungen der Deutschen Zoologischen Gesellschaft*, **83**, 452–453.

Fleck, J. and Hofmann, D.K. (1995) In vivo binding of a biologically active oligopeptide in vegetative buds of the scyphozoan *Cassiopea andromeda*: demonstration of receptor-mediated induction of metamorphosis. *Marine Biology (Berlin)*, **122**, 447–451.

Fosså, J.H. (1992) Mass occurrence of *Periphylla periphylla* (Scyphozoa, Coronatae) in a Norwegian fjord. *Sarsia*, 77, 237–251.

Fox, D.L. and Millott, N. (1954) The pigmentation of the jellyfish, *Pelagia noctiluca* (Forskål) var. *panopyra* Péron and Lesueur. *Proceedings of the Royal Society of London B Biological Sciences*, **142**, 392–408.

Fox, D.L. and Pantin, C.F.A. (1944) Pigments in the coelenterata. *Biological Reviews of the Cambridge Philosophical Society*, **19**, 121–134.

Franc, A. (1993) Classe des Scyphozoaires, in *Traité de Zoologie. Anatomie, systématique, biologie III Cnidaires, Cténaires 2*, (ed. D. Doumenc), Masson, Paris, pp. 597–884. In French.

Franc, S. (1988) Le Collagène des cnidaires. *Bulletin de la Societe Zoologique de France*, **113**, 337–346. In French; English abstract.

Frank, K.T. and Leggett, W.C. (1985) Reciprocal oscillations in densities of larval fish and potential predators: a reflection of past or present predation? *Canadian Journal of Fisheries and Aquatic Sciences*, **42**, 1841–1849.

Franqueville, C. (1971) Macroplancton profond (invertébrés) de la Méditerranée Nord-occidentale. *Tethys*, **3**, 11–56. In French; English abstract.

Fraser, J.H. (1969) Experimental feeding of some medusae and Chaetognatha. *Journal of the Fisheries Research Board of Canada*, **26**, 1743–1762.

Freudenthal, H.D. (1962) *Symbiodinium* gen. nov. and *Symbiodinium microadriaticum* sp. nov., a zooxanthella: taxonomy, life cycle, and morphology. *Journal of Protozoology*, **9**, 45–52.

Fry, B. and Sherr, E.B. (1984) $\delta^{13}C$ measurements as indicators of carbon flow in marine and freshwater ecosystems. *Contributions in Marine Science*, **27**, 13–47.

Galil, B.S., Spanier, E. and Ferguson, W.W. (1990) The scyphomedusae of the Mediterranean coast of Israel, including two lessepsian migrants new to the Mediterranean. *Zoologische Mededelingen (Leiden)*, **64**, 95–105.

Galt, C.P. (1989) Bioluminescence of gelatinous zooplankton. *Oceanis*, **15**, 51–59.

Gamble, J.C. and Hay, S.J. (1989) Predation by the scyphomedusan *Aurelia aurita* on herring larvae in large enclosures: effects of predator size and prey starvation. *Rapports et Proces-verbaux des Réunions Conseil international pour l'Exploration de la Mer*, **191**, 366–375.

García, J.R. (1990) Population dynamics and production of *Phyllorhiza punctata* (Cnidaria: Scyphozoa) in Laguna Joyuda, Puerto Rico. *Marine Ecology Progress Series*, **64**, 243–251.

García, J.R. and Durbin, E. (1993) Zooplanktivorous predation by large scyphomedusae *Phyllorhiza punctata* (Cnidaria: Scyphozoa) in Laguna Joyuda. *Journal of Experimental Marine Biology and Ecology*, **173**, 71–93.

Gatz, A.J., Kennedy, V.S. and Mihursky, J.A. (1973) Effects of temperature on activity and mortality of the scyphozoan medusa, *Chrysaora quinquecirrha*. *Chesapeake Science*, **14**, 171–180.

Gemmill, J.F. (1921) Notes on food-capture and ciliation in the ephyrae of *Aurelia*. *Proceedings of the Royal Physical Society for the Promotion of Zoology and other Branches of Natural History, Edinburgh*, **20**, 222–225.

George, R.Y. and Marum, J.P. (1974) The effects of hydrostatic pressure on living aquatic organisms III Behavior and tolerance of euplanktonic organisms to increased hydrostatic pressure. *Internationale Revue der gesamten Hydrobiologie*, **59**, 175–186.

Gerritsen, J. (1980) Adaptive responses to encounter problems. *American Society of Limnology and Oceanography Special Symposium*, **3**, 52–62.

Gerritsen, J. and Srickler, J.R. (1977) Encounter probabilities and community structure in zooplankton: a mathematical model. *Journal of the Fisheries Research Board of Canada*, **34**, 73–82.

Gilchrist, F.G. (1937) Budding and locomotion in the scyphistomas of *Aurelia*. *Biological Bulletin (Woods Hole)*, **72**, 99–124.

Giorgi, R., Avian, M., De Olazabal, S. *et al.* (1991) Feeding of *Pelagia noctiluca* in open sea, in *UNEP: Jellyfish Blooms in the Mediterranean. Proceedings of the II Workshop on Jellyfish in the Mediterranean Sea*, MAP Technical Reports Series No. 47, UNEP, Athens, pp. 102–111.

Gladfelter, W.B. (1972) Structure and function of the locomotory system of the scyphomedusa *Cyanea capillata*. *Marine Biology (Berlin)*, **14**, 150–160.

Gladfelter, W.B. (1973) A comparative analysis of the locomotory systems of medusoid Cnidaria. *Helgolaender Wissenschaftliche Meeresuntersuchungen*, **25**, 228–272.

Glaessner, M.F. (1984) *The Dawn of Animal Life: A Biohistorical Study*, Cambridge University Press, Cambridge.

Gnaiger, E. (1983a) Calculation of energetic and biochemical equivalents of respiratory oxygen consumption, in *Polarographic Oxygen Sensors; Aquatic and Physiological Applications*, (eds E. Gnaiger and H. Forstner), Springer-Verlag, Berlin, pp. 337–345.

Gnaiger, E. (1983b) Symbols and units; toward standardization, in *Polarographic Oxygen Sensors; Aquatic and Physiological Applications*, (eds E. Gnaiger and H. Forstner), Springer-Verlag, Berlin, pp. 352–358.

Gnaiger, E. and Bitterlich, G. (1984) Proximate biochemical composition and caloric content calculated from elemental CHN analysis: a stoichiometric concept. *Oecologia (Berlin)*, **62**, 289–298.

Goette, A. (1893) Vergleichende Entwicklungsgeschichte von *Pelagia noctiluca* Pér. *Zeitschrift fuer wissenschaftliche Zoologie*, **55**, 645–695, pls 28–31. In German.

Gohar, H.A.F. and Eisawy, A.M. (1961a) The biology of *Cassiopea andromeda* (from the Red Sea) (with a note on the species problem). *Publications Marine Biological Station Al Ghardaqa*, **11**, 3–42.

Gohar, H.A.F. and Eisawy, A.M. (1961b) The development of *Cassiopea andromeda* (Scyphomedusae). *Publications Marine Biological Station Al Ghardaqa*, **11**, 147–190.

Goldberg, R.B., Crain, W.R., Ruderman, J.V. *et al.* (1975a) DNA sequence organization in the genomes of five marine invertebrates. *Chromosoma (Berlin)*, **51**, 225–251.

Goldberg, R.B., Crain, W.R., Ruderman, J.V. *et al.* (1975b) DNA sequence organization in the genomes of five marine invertebrates. *Carnegie Institution of Washington Year Book*, **74**, 669–672.

Gomoiu, M.-T. (1980) Ecological observations on the jellyfish *Aurelia aurita* (L.) populations from the Black Sea. *Cercetari marine*, **13**, 91–102.

Gomoiu, M.-T. (1981) Some problems concerning actual ecological changes in the Black Sea. *Cercetari marine*, **14**, 109–127.

Gordon, J.D.M. and Mauchline, J. (1990) Depth-related trends in diet of a deep-sea bottom-living fish assemblage of the Rockall Trough, in *Trophic Relationships in the Marine Environment*, (eds M. Barnes and R.N. Gibson), Aberdeen University Press, Aberdeen, pp. 439–452.

Gorsky, G., Dallot, S., Sardou, J. *et al.* (1988) C and N composition of some northwestern Mediterranean zooplankton and micronekton species. *Journal of Experimental Marine Biology and Ecology*, **124**, 133–144.

Goy, J. (1990) Preliminary observations on the zooplankton of Shark Bay, Western Australia with emphasis on medusae and description of a new rhizostome jellyfish (Cnidaria, Scyphozoa), in *Research in Shark Bay*, (eds P.F. Berry, S.D. Bradshaw and B.R. Wilson), Western Australian Museum, Perth, pp. 107–113.

Goy, J., Morand, P. and Etienne, M. (1989) Long-term fluctuations of *Pelagia noctiluca* (Cnidaria, Scyphomedusa) in the western Mediterranean Sea. Prediction by climatic variables. *Deep-sea Research*, **36**, 269–279.

Grasshoff, M. (1984) Cnidarian phylogeny – a biomechanical approach. Proceedings Fourth International Symposium on Fossil Cnidaria. *Palaeontographica Americana*, **54**, 127–135.

Greve, W. and Parsons, T.R. (1977) Photosynthesis and fish production: hypothetical effects of climatic change and pollution. *Helgolaender Wissenschaftliche Meeresuntersuchungen*, **30**, 666–672.

Grimm, P. (1984) Gonaden der Ohrenqualle (*Aurelia aurita*) als Nahrung des Sanderlings (*Calidris alba*). *Seavogel*, **5**, 24. In German.

Grimmelikhuijzen, C.J.P., Graff, D. and Spencer, A.N. (1988) Structure, location and possible actions of Arg-Phe-amide peptides in coelenterates, in *Neurohormones in Invertebrates*, (eds M.C. Thorndyke and G.J. Goldsworthy), Cambridge University Press, Cambridge, pp. 199–217.

Grimmelikhuijzen, C.J.P. and Westfall, J.A. (1995) The nervous systems of Cnidarians, in *The Nervous Systems of Invertebrates: An evolutionary and comparative approach*, (eds O. Breidbach and W. Kutsch), Birkhaeuser Verlag, Basel, Switzerland, pp. 7–24.

Groat, C.S., Thomas, C.R. and Schurr, K. (1980) Improved culture of *Aurelia aurita* scyphistomae for bioassay and research. *Ohio Journal of Science*, **80**, 83–87.

Gröndahl, F. (1988a) A comparative ecological study on the scyphozoans *Aurelia aurita*, *Cyanea capillata* and *C. lamarckii* in the Gullmar Fjord, western Sweden, 1982 to 1986. *Marine Biology (Berlin)*, **97**, 541–550.

Gröndahl, F. (1988b) Interactions between polyps of *Aurelia aurita* and planktonic larvae of scyphozoans: an experimental study. *Marine Ecology Progress Series*, **45**, 87–93.

Gröndahl, F. (1989) Evidence of gregarious settlement of planula larvae of the scyphozoan *Aurelia aurita*: an experimental study. *Marine Ecology Progress Series*, **56**, 119–25.

Gröndahl, F. and Hernroth, L. (1987) Release and growth of *Cyanea capillata* (L.) ephyrae in the Gullmar Fjord, western Sweden. *Journal of Experimental Marine Biology and Ecology*, **106**, 91–101.

Gubareva, A.E., Pustovalova, L.M., Ryzhkov, Y.D. *et al.* (1983) The features of the chemical composition of medusae in the Sea of Azov. *Izvestiya Severo-Kavkazskogo Nauchnogo Tsentra Vysshei Shkoly Estestvennye Nauki*, **0**, 22–25. In Russian.

Guo, P. (1990) Effect of nutritional condition on the formation and germination of the podocyst of scyphistomae of *Rhopilema esculenta* Kishinouye. *Journal of Fisheries of China*, **14**, 206–211. In Chinese; English summary.

Gushchin, A.V. and Podrazhanskaya, S.G. (1984) Feeding of Roundnose Grenadier *Coryphaenoides rupestris* and its trophic relationships in the North Atlantic. *NAFO Science Council Studies*, **7**, 53–59.

Gutmann, W.F. (1965) Rückstoss und Ruderschlag der Quallen. *Natur und Museum*, **95**, 455–462. In German.

Gutmann, W.F. (1966) Zu Bau und Leistung von Tierkonstruktionen 2 Die Glocke von *Rhizostoma octopus*. *Natur und Museum*, **96**, 103–108.

Gwilliam, G.F. (1960) Neuromuscular physiology of a sessile scyphozoan. *Biological Bulletin (Woods Hole)*, **119**, 454–473.

Haahtela, I. and Lassig, J. (1967) Records of *Cyanea capillata* (Scyphozoa) and *Hyperia galba* (Amphipoda) from the Gulf of Finland and the northern Baltic. *Annales Zoologici Fennici*, **4**, 469–471.

Hadzi, J. (1963) Bemerkungen zur Entwicklungsgeschichte der Bathypelagischen, Viviparen Scyphomedusenart *Stygiomedusa fabulosa* Russell 1959. *Bulletin Scientifique Conseil des Academies de la RSF de Yougoslavie*, **8**, 68–69. In German.

Hallock, P. (1981) Algal symbiosis: A mathematical analysis. *Marine Biology (Berlin)*, **62**, 249–255.

Hamner, W.M. (1995) Sensory ecology of scyphomedusae. *Marine and Freshwater Behavior and Physiology*, **26**, 101–118.

Hamner, W.M. and Hauri, I.R. (1981) Long-distance horizontal migrations of zooplankton (Scyphomedusae: *Mastigias*). *Limnology and Oceanography*, **26**, 414–423.

Hamner, W.M. and Jenssen, R.M. (1974) Growth, degrowth, and irreversible cell differentiation in *Aurelia aurita*. *American Zoologist*, **14**, 833–849.

Hamner, W.M. and Schneider, D. (1986) Regularly spaced rows of medusae in the Bering Sea: role of Langmuir circulation. *Limnology and Oceanography*, **31**, 171–177.

Hamner, W.M., Gilmer, R.W. and Hamner, P.P. (1982) The physical, chemical, and biological characteristics of a stratified, saline, sulfide lake in Palau. *Limnology and Oceanography*, **27**, 896–909.

Hamner, W.M., Hamner, P.P. and Strand, S.W. (1994) Sun-compass migration by *Aurelia aurita* (Scyphozoa): population retention and reproduction in Saanich Inlet, British Columbia. *Marine Biology (Berlin)*, **119**, 347–356.

Hanaoka, K.-I. (1934) Notes on the early development of a stalked medusa. *Proceedings of the Imperial Academy of Japan*, **10**, 117–120.

Hand, C. (1955) A study of the structure, affinities, and distribution of *Tetraplatia volitans* Busch (Coelenterata: Hydrozoa: Pteromedusae). *Pacific Science*, **9**, 332–348.

Hansson, L.J. and Kultima, K. (1995) Behavioural response of the scyphozoan jellyfish *Aurelia aurita* (L.) upon contact with the predatory jellyfish *Cyanea capillata* (L.). *Marine and Freshwater Behavior and Physiology*, **26**, 131–137.

Harbison, G.R., Smith, K.L. and Backus, R.H. (1973) *Stygiomedusa fabulosa* from the North Atlantic: its taxonomy, with a note on its natural history. *Journal of the Marine Biological Association of the United Kingdom*, **53**, 615–617.

Hargitt, C.W. (1904) Regeneration in scyphomedusae. *American Naturalist*, **38**, 517–519.

Hargitt, C.W. and Hargitt, G.T. (1910) Studies in the development of scyphomedusae. *Journal of Morphology*, **21**, 217–262.

Harrison, N.M. (1984) Predation on jellyfish and their associates by seabirds. *Limnology and Oceanography*, **29**, 1335–1337.

Harrison, N.M. (1990) Gelatinous zooplankton in the diet of the parakeet auklet: Comparisons with other auklets. *Studies in Avian Biology*, **14**, 114–124.

Harvey, E.N. (1926) Oxygen and luminescence, with a description of methods for removing oxygen from cells and fluids. *Biological Bulletin (Woods Hole)*, **51**, 89–97.

Harvey, E.N. (1952) *Bioluminescence*, Academic Press, New York.

Harvey, E.N. (1957) *A History of Luminescence from the Earliest Times until 1900*, The American Philosophical Society, Philadelphia.

Hatai, S. (1917) On the composition of the medusa, *Cassiopea xamachana* and the changes in it after starvation. *Proceedings of the National Academy of Sciences of the United States of America*, **3**, 22–24.

Haury, L.R., McGowan, J.A. and Wiebe, P.H. (1978) Patterns and processes in the time–space scales of plankton distributions, in *Spatial Pattern in Plankton Communities*, (ed. J.H. Steele), Plenum Press, New York, pp. 277–327.

Hay, S.J., Hislop, J.R.G. and Shanks, A.M. (1990) North Sea scyphomedusae; summer distribution, estimated biomass and significance particularly for 0-group gadoid fish. *Netherlands Journal of Sea Research*, **25**, 113–130.

Hedwig, M. and Schäfer, W. (1986) Vergleichende Untersuchungen zur Ultrastruktur und zur phylogenetischen Bedeutung der Spermien der Scyphozoa. *Zeitschrift fuer Zoologische Systematik und Evolutionsforschung*, **24**, 109–122. In German; English abstract.

Heeger, T. and Möller, H. (1987) Ultrastructural observations on prey capture and digestion in the scyphomedusa *Aurelia aurita*. *Marine Biology (Berlin)*, **96**, 391–400.

Heeger, T., Möller, H. and Mrowietz, U. (1992) Protection of human skin against jellyfish (*Cyanea capillata*) stings. *Marine Biology (Berlin)*, **113**, 669–678.

Heeger, T., Piatkowski, U. and Möller, H. (1992) Predation on jellyfish by the cephalopod *Argonauta argo*. *Marine Ecology Progress Series*, **88**, 293–296.

Hein, W. (1900) Untersuchungen über die Entwicklung von *Aurelia aurita*. *Zeitschrift fuer wissenschaftliche Zoologie*, **67**, 401–438, pls 14–15. In German.

Hein, W. (1902) Untersuchungen über die Entwicklung von *Cotylorhiza tuberculata*. *Zeitschrift fuer wissenschaftliche Zoologie*, **73**, 302–320, pls 20–21. In German.

Hela, I. (1952) On the occurrence of the jellyfish, *Aurelia aurita* L., on the south coast of Finland. *Archivum Societatis Zoologicae-Botanicae Fennicae 'Vanamo'*, **6**, 71–78.

Henschel, J. (1935) Untersuchungen über den chemischen Sinn der Scyphomedusen *Aurelia aurita* und *Cyanea capillata* und der Hydromeduse *Sarsia tubulosa*. *Wissenschaftliche Meeresuntersuchungen der Kommission zur Wissenschaften Untersuchung der Deutschen Meere. Kiel.*, **22**, 25–42. In German.

Hentschel, J. and Hündgen, M. (1980) The morphology and ultrastructure of the scyphistoma of *Aurelia aurita* (Scyphozoa, Semaeostomae). *Zoologische Jahrbuecher Abteilung fuer Anatomie und Ontogenie der Tiere*, **104**, 295–316. In German; English abstract.

Hernroth, L. (1986) On the reproduction and occurrence of *Aurelia aurita* in Gullmar Fjord, Western Sweden. *Nova Thalassia*, **8**, **Suppl. 2**, 35–44.

Hernroth, L. and Ackefors, H. (1979) The zooplankton of the Baltic proper. *Report Institute of Marine Research Fishery Board of Sweden*, **2**, 1–60.

Hernroth, L. and Gröndahl, F. (1983) On the biology of *Aurelia aurita* (L.) 1. Release and growth of *Aurelia aurita* (L.) ephyrae in the Gullmar Fjord, Western Sweden, 1982–1983. *Ophelia*, **22**, 189–199.

Hernroth, L. and Gröndahl, F. (1985a) On the biology of *Aurelia aurita* (L.) 2. Major factors regulating the occurrence of ephyrae and young medusae in the Gullmar Fjord, Western Sweden. *Bulletin of Marine Science*, **37**, 567–576.

Hernroth, L. and Gröndahl, F. (1985b) On the biology of *Aurelia aurita* (L.) 3. Predation by *Coryphella verrucosa* (Gastropoda, Opisthobranchia), a major factor regulating the development of *Aurelia* populations in the Gullmar Fjord, Western Sweden. *Ophelia*, **24**, 37–45.

Herring, P.J. (1972) Porphyrin pigmentation in deep-sea medusae. *Nature (London)*, **238**, 276–277.

Herring, P.J. (1983) The spectral characteristics of luminous marine organisms. *Proceedings of the Royal Society of London B Biological Sciences*, **220**, 183–217.

Herring, P.J. (1990) Bioluminescent responses of the deep-sea scyphozoan *Atolla wyvillei*. *Marine Biology (Berlin)*, **106**, 413–417.

Herrnkind, W., Halusky, J. and Kanciruk, P. (1976) A further note on phyllosoma larvae associated with medusae. *Bulletin of Marine Science*, **26**, 110–112.

Hertwig, O. and Hertwig, R. (1878) *Das Nervensystem und die Sinnesorgane der Medusen*, F.C.W. Vogel, Leipzig. In German.

Hesse, R. (1895) Über das Nervensystem und die Sinnesorgane von *Rhizostoma Cuvieri*. *Zeitschrift fuer wissenschaftliche Zoologie*, **60**, 411–457, pls 20–22. In German.

Hessinger, D.A. (1988) Nematocyst venoms and toxins, in *The Biology of Nematocysts*, (eds D.A. Hessinger and H.M. Lenhoff), Academic Press, San Diego, pp. 333–368.

Heymans, C. and Moore, A.R. (1924) Luminescence in *Pelagia noctiluca*. *Journal of General Physiology*, **6**, 273–280.

Hidaka, M. (1993) Mechanism of nematocyst discharge and its cellular control. *Advances in Comparative and Environmental Physiology*, **15**, 45–76.

Hinsch, G.W. (1974) Comparative ultrastructure of Cnidarian sperm. *American Zoologist*, **14**, 457–465.

Hinsch, G.W. and Clark, W.H. (1973) Comparative fine structure of Cnidaria spermatozoa. *Biology of Reproduction*, **8**, 62–73.

Hirai, E. (1958) On the developmental cycles of *Aurelia aurita* and *Dactylometra pacifica*. *Bulletin of the Marine Biological Station of Asamushi*, **9**, 81.

Hirano, Y.M. (1986a) Species of Stauromedusae from Hokkaido, with notes on their metamorphosis. *Journal of the Faculty of Science Hokkaido University Series VI Zoology*, **24**, 182–201.

Hirano, Y.M. (1986b) Studies on Stauromedusae (Coelenterata, Scyphozoa) in Hokkaido – Distribution and life history (Abstract). *Benthos Research*, **30**, 10–13. In Japanese.

Hiromi, J., Yamamoto, T., Koyama, Y. *et al.* (1995) Experimental study of predation on scyphopolyp *Aurelia aurita*. *Bulletin of the College of Agriculture and Veterinary Medicine Nihon University*, **52**, 126–130. In Japanese; English abstract.

Hochachka, P.W. (1991) Design of energy metabolism, in *Environmental and Metabolic Animal Physiology*, Comparative Animal Physiology, 4, (ed. C.L. Prosser), Wiley-Liss, New York, pp. 325–351.

Hoffmann, R.J., Bishop, S.H. and Sassaman, C. (1978) Glutamate dehydrogenase from coelenterates NADP specific. *Journal of Experimental Zoology*, **203**, 165–170.

Hofmann, D.K. and Brand, U. (1987) Induction of metamorphosis in the symbiotic scyphozoan *Cassiopea andromeda*: Role of marine bacteria and of biochemicals. *Symbiosis*, **4**, 99–116.

Hofmann, D.K. and Gottlieb, M. (1991) Bud formation in the scyphozoan *Cassiopea andromeda*: epithelial dynamics and fate map. *Hydrobiologia*, **216/217**, 53–59.

Hofmann, D.K. and Hellman, M. (1995) Studies in the reef-dwelling cnidarian *Cassiopea* spp.: RF-amide positive elements of the nervous system at different stages of development. *Beitraege zur Palaeontologie von Oesterreich*, **20**, 21–29.

Hofmann, D.K. and Honegger, T.G. (1990) Bud formation and metamorphosis in *Cassiopea andromeda* (Cnidaria: Scyphozoa): a developmental and ultrastructural study. *Marine Biology (Berlin)*, **105**, 509–518.

Hofmann, D.K. and Kremer, B.P. (1981) Carbon metabolism and strobilation in *Cassiopea andromeda* (Cnidaria: Scyphozoa): significance of endosymbiotic dinoflagellates. *Marine Biology (Berlin)*, **65**, 25–33.

Hofmann, D.K., Manitz, B. and Reckenfelderbäumer, B. (1993) Metamorphosis in the scyphozoan *Cassiopea andromeda*: cellular and morphogenetic responses to exogenous inducers, in *Proceedings of the Seventh International Coral Reef Symposium*, (ed. R.H. Richmond), University of Guam Press, Mangilao, Guam, pp. 463–470.

Hofmann, D.K., Neumann, R. and Henne, K. (1978) Strobilation, budding and initiation of scyphistoma morphogenesis in the rhizostome *Cassiopea andromeda* (Cnidaria: Scyphozoa). *Marine Biology (Berlin)*, **47**, 161–176.

Holland, D.L., Davenport, J. and East, J. (1990) The fatty acid composition of the leatherback turtle *Dermochelys coriacea* and its jellyfish prey. *Journal of the Marine Biological Association of the United Kingdom*, **70**, 761–770.

Holstein, T. and Hausmann, K. (1988) The cnidocil apparatus of hydrozoans: a progenitor of higher metazoan mechanoreceptors?, in *The Biology of Nematocysts*, (eds D.A. Hessinger and H.M. Lenhoff), Academic Press, San Diego, pp. 53–73.

Hooper, S.N. and Ackman, R.G. (1972) Presence of *trans*-6-hexadecenoic acid in the white jellyfish *Aurelia aurita* Lamarck and in a Caribbean gorgonian. *Lipids*, 7, 624–626.

Hori, H. and Osawa, S. (1987) Origin and evolution of organisms as deduced from 5S ribosomal RNA sequences. *Molecular Biology and Evolution*, **4**, 445–472.

Hori, H. and Satow, Y. (1991) Dead-end evolution of the Cnidaria as deduced from 5S ribosomal RNA sequences. *Hydrobiologia*, **216/217**, 505–508.

Hori, H., Ohama, T., Kumazaki, T. *et al.* (1982) Nucleotide sequences of 5S rRNAs from four jellyfishes. *Nucleic Acids Research*, **10**, 7405–7408.

Horridge, A. (1953) An action potential from the motor nerves of the jellyfish *Aurelia aurita* Lamarck. *Nature (London)*, **171**, 400.

Horridge, G.A. (1954) The nerves and muscles of medusae. I. Conduction in the nervous system of *Aurelia aurita* Lamarck. *Journal of Experimental Biology*, **31**, 594–600.

Horridge, G.A. (1955) The nerves and muscles of medusae. III. A decrease in the refractory period following repeated stimulation of the muscle of *Rhizostoma pulmo*. *Journal of Experimental Biology*, **32**, 636–641.

Horridge, G.A. (1956a) The nerves and muscles of medusae. V. Double innervation in scyphozoa. *Journal of Experimental Biology*, **33**, 366–383.

Horridge, G.A. (1956b) The nervous system of the ephyra larva of *Aurelia aurita*. *Quarterly Journal of Microscopical Science*, **97**, 59–74.

Horridge, G.A. (1959) The nerves and muscles of medusae. VI. The rhythm. *Journal of Experimental Biology*, **36**, 72–91.

Horridge, G.A. (1969) Statocysts of medusae and evolution of stereocilia. *Tissue and Cell*, **1**, 341–353.

Horridge, G.A. and Mackay, B. (1962) Naked axons and symmetrical synapses in coelenterates. *Quarterly Journal of Microscopical Science*, **103**, 531–541.

Horridge, G.A., Chapman, D.M. and MacKay, B. (1962) Naked axons and symmetrical synapses in an elementary nervous system. *Nature (London)*, **193**, 899–900.

Huang, M., Hu, J., Wang, Y. *et al.* (1985) Preliminary study on the breeding habits of edible jellyfish in Hangzhou Wan Bay. *Journal of Fisheries of China*, **9**, 239–246. In Chinese; English abstract.

Huang, Y.-W. (1988) Cannonball jellyfish (*Stomolophus meleagris*) as a food resource. *Journal of Food Science*, **53**, 341–343.

Hündgen, M. and Biela, C. (1982) Fine structure of touch-plates in the scyphomedusen *Aurelia aurita*. *Journal of Ultrastructure Research*, **80**, 178–184.

Hunter, J.R. (1984) Inferences regarding predation on the early life stages of cod and other fishes. *Flodevigen Rapportser*, **1**, 533–562.

Hüsing, J.O. (1956) Zur Frage der Ernährung der Ohrenqualle *Aurelia aurita* L. (Scyphozoa, Semaeostomae). *Wissenschaftliche Zeitschrift der Martin-Luther-Universität Halle-Wittenberg Mathematisch-Naturwissenschaftliche Reihe*, **5**, 479–482. In German.

Husson, B. and Fay, S. (1984) Rèpartition, frèquence et régime alimentaire de la méduse *Rhizostoma pulmo*. *Annales de la Sociètè des Sciences Naturelles de la Charente-Maritime*, **7**, 231–252. In French.

Hutton, C.H., DeLisle, P.F., Roberts, M.H. *et al.* (1986) *Chrysaora quinquecirrha*; a predator on mysids (*Mysidopsis bahia*) in culture. *Progressive Fish-Culturist*, **48**, 154–155.

Hyde, I.H. (1895) Entwicklungsgeschichte einiger Scyphomedusen. *Zeitschrift fuer wissenschaftliche Zoologie*, **58**, 531–565, pls 32–37. In German.

Hyman, L.H. (1938) The water content of medusae. *Science (Washington D. C.)*, **87**, 166–167.

Hyman, L.H. (1940) Observations and experiments on the physiology of medusae. *Biological Bulletin (Woods Hole)*, **79**, 282–296.

Iglesias-Prieto, R. and Trench, R.K. (1994) Acclimation and adaptation to irradiance in symbiotic dinoflagellates I Responses of the photosynthetic unit to changes in photon flux density. *Marine Ecology Progress Series*, **113**, 163–175.

Ikeda, T. (1974) Nutritional ecology of marine zooplankton. *Memoirs of the Faculty of Fisheries Hokkaido University*, **22**, 1–97.

Ikeda, T. (1977) The effect of laboratory conditions on the extrapolation of experimental measurements to the ecology of marine zooplankton IV Changes in respiration and excretion rates of boreal zooplankton species maintained under fed and starved conditions. *Marine Biology (Berlin)*, **41**, 241–252.

Inagaki, T. and Toyokawa, M. (1991) Studies on detecting methods of jellyfish *Aurelia aurita* by acoustic means. *Bulletin of the Japanese Society of Fisheries Oceanography*, **55**, 25–34. In Japanese; English abstract.

Irisawa, H., Irisawa, T. and Nishida, Y. (1956) Responses of the receptor of a medusa *Aurelia aurita* to light stimuli. *Kagaku Nan'yo*, **26**, 312–313. In Japanese.

Ivleva, I.V. and Litovchenko, T.I. (1978) Respiration rate of coelenterata and ctenophora depending on the habitat temperature. *Biologiya Morya (Kiev)*, **46**, 3–25. In Russian; English abstract.

James, D.B., Vivekanandan, E. and Srinivasarengan, S. (1985) Menace from medusae off Madras with notes on their utility and toxicity. *Journal of the Marine Biological Association of India*, **27**, 170–174.

Jarms, G. (1990) Neubeschreibung dreier Arten der Gattung *Nausithoe* (Coronata, Scyphozoa) sowie Wiederbeschreibung der Art *Nausithoe marginata* Kölliker, 1853. *Mitteilungen aus dem Hamburgischen Zoologischen Museum und Institut*, **87**, 7–39. In German; English abstract.

Jarms, G. (1991) Taxonomic characters from the polyp tubes of coronate medusae (Scyphozoa, Coronatae). *Hydrobiologia*, **216/217**, 463–470.

Jenkins, R.J.F. (1992) Functional and ecological aspects of Ediacaran assemblages, in *Origin and Early Evolution of the Metazoa*, (eds J.H. Lipps and P.W. Signor), Plenum Press, New York, pp. 131–176.

Johnson, M.C. and Wuensch, K.L. (1994) An investigation of habituation in the jellyfish *Aurelia aurita*. *Behavioral and Neural Biology*, **61**, 54–59.

Jokiel, P.L. and York, R.H. (1982) Solar ultraviolet photobiology of the reef coral *Pocillopora damicornis* and symbiotic zooxanthellae. *Bulletin of Marine Science*, **32**, 301–315.

Jones, S. (1960) Notes on animal associations 2 The scyphomedusa, *Acromitus flagellatus* Stiasny and young *Seleroides leptolepis* (Cuvier & Valenciennes) with the latter forming a vanguard. *Journal of the Marine Biological Association of India*, **2**, 51–52, pl. 1.

Joseph, J.D. (1979) Lipid composition of marine and estuarine invertebrates; Porifera and Cnidaria. *Progress in Lipid Research*, **18**, 1–30.

Kakinuma, Y. (1967) Development of a scyphozoan, *Dactylometra pacifica* Goette. *Bulletin of the Marine Biological Station of Asamushi*, **13**, 29–32.

Kakinuma, Y. (1975) An experimental study of the life cycle and organ differentiation of *Aurelia aurita* Lamarck. *Bulletin of the Marine Biological Station of Asamushi*, **15**, 101–113, pls 3–6.

Kakinuma, Y. and Sugiura, Y. (1980) Organ differentiation in *Aurelia aurita*, in *Developmental and Cellular Biology of Coelenterates*, (eds P. Tardent and R. Tardent), Elsevier/North-Holland Biomedical Press, Amsterdam, pp. 257–262.

Kanazawa, N., Sato, O. and Takano-Ohmuro, H. *et al.* (1993) Sea anemone (*Actinia equina*) myosin. *Comparative Biochemistry and Physiology*, **104B**, 509–514.

Kassianow, N. (1901) Studien über das Nervensystem der Lucernariden, nebst sonstigen histologischen Beobachtungen über diese Gruppe. *Zeitschrift fuer wissenschaftliche Zoologie*, **69**, 287–377, pls 22–25. In German.

Kato, K.-I., Aochi, M. and Ozato, K. (1973) Developmental aspects of strobilation in *Aurelia aurita*. *Publications of the Seto Marine Biological Laboratory*, **20**, 179–194. Proc. Second Int. Symp. Cnidaria.

Kato, K.-I., Aochi, M. and Sakaguchi, K. (1973) Morphogenesis of the small pieces taken from polyps and strobilae in *Aurelia aurita*. *Memoirs of Osaka Kyoiku University III Natural Science and Applied Science*, **22**, 117–131.

Kato, K.-I., Tomioka, T. and Sakagami, K. (1980) Morphogenetic patterns of scyphozoan strobilation, in *Developmental and Cellular Biology of Coelenterates*, (eds P. Tardent and R. Tardent), Elsevier/North-Holland Biomedical Press, Amsterdam, pp. 245–250.

Kawaguti, S. and Yoshimoto, F. (1973) Electron microscopy on a scyphozoan, *Stephanoscyphus racemosus*. *Biological Journal of Okayama University*, **16**, 47–66.

Keen, S. (1987) Recruitment of *Aurelia aurita* (Cnidaria: Scyphozoa) larvae is position-dependent, and independent of conspecific density, within a settling surface. *Marine Ecology Progress Series*, **38**, 151–160.

Keen, S.L. and Gong, A.J. (1989) Genotype and feeding frequency affect clone formation in a marine cnidarian (*Aurelia aurita* Lamarck 1816). *Functional Ecology*, **3**, 735–745.

Kem, W.R. and Östman, C. (1991) Methods for isolating the tentacular nematocysts of the Lion's Mane Jellyfish, *Cyanea capillata*, in *UNEP: Jellyfish Blooms in the Mediterranean. Proceedings of the II Workshop on Jellyfish in the Mediterranean Sea*, MAP Technical Reports Series No. 47, UNEP, Athens, pp. 241–252.

Kennedy, G.Y. (1979) Pigments of marine invertebrates. *Advances in Marine Biology*, **16**, 309–381.

Kerstan, M. (1977): Untersuchungen zur Nahrungsökologie von *Aurelia aurita* Lam. Diploma Thesis, Christian-Albrechts-Universität, Kiel. 95 pp. In German.

Kevin, M.J., Hall, W.T., McLaughlin, J.J.A. *et al.* (1969) *Symbiodinium microadriaticum* Freudenthal, a revised taxonomic description, ultrastructure. *Journal of Phycology*, **5**, 341–350.

Khlebovich, V.V. (1973) Activity of *Hydrobia ulvae* (Pennant) and planulae of *Aurelia aurita* (L.) following separate change of ionic strength and total osmotic concentration of the medium. *Doklady Akademii Nauk SSSR Seriya Biologiya*, **280**, 1221–1223. In Russian.

Khlebovich, V.V. and Lukanin, V.V. (1980) Dynamics of RNA and mitochondrial synthesis in the process of adaptation of *Aurelia aurita* planulas to reduced salinity as revealed by inhibitory analysis, in *The Theoretical and Practical Importance of the Coelenterates*, (eds D.V. Naumov and S.D. Stepanjants), Zoological Institute Academy Science USSR, Leningrad, pp. 127–130. In Russian.

Kihara, H., Anraku, M., Ohno, M. *et al.* (1988) Tetrodotoxin-unaffected depolarization of frog muscles induced by the venom of jellyfish (Genus *Aurelia*). *Japanese Journal of Physiology*, **38**, 839–849.

Kikinger, R. (1986) Development cycle of *Cotylorhiza tubercolata* (Scyphozoa: Rhizostomae). *Nova Thalassia*, **8 Suppl. 2**, 73–82.

Kikinger, R. (1992) *Cotylorhiza tuberculata* (Cnidaria: Scyphozoa) – Life history of a stationary population. *Marine Ecology (Pubblicazioni della Stazione Zoologica di Napoli I)*, **13**, 333–362.

Kikinger, R. and Salvini-Plawen, L.von (1995) Development from polyp to stauromedusa in *Stylocoronella* (Cnidaria: Scyphozoa). *Journal of the Marine Biological Association of the United Kingdom*, **75**, 899–912.

Kimura, S., Miura, S. and Park, Y.–H. (1983) Collagen as the major edible component of jellyfish (*Stomolophus nomurai*). *Journal of Food Science*, **48**, 1758–1760.

Kingsford, M.J., Wolanski, E. and Choat, J.H. (1991) Influence of tidally induced fronts and Langmuir circulations on distribution and movements of presettlement fishes around a coral reef. *Marine Biology (Berlin)*, **109**, 167–180.

Kirschner, L.B. (1991) Water and Ions, in *Environmental and Metabolic Animal Physiology*, Comparative Animal Physiology, 4, (ed. C.L. Prosser), Wiley-Liss, New York, pp. 13–107.

Knight-Jones, E.W. and Morgan, E. (1966) Responses of marine animals to changes in hydrostatic pressure. *Oceanography and Marine Biology; an Annual Review*, **4**, 267–299.

Koizumi, T. and Hosoi, K. (1936) On the inorganic composition of the medusae: *Aequorea coerulescens* (Brandt), *Dactylometra pacifica* Goette and *Cyanea capillata* Eschscholtz. *Science Reports of the Tohoku University Fourth Series (Biology)*, **10**, 709–719.

Kokina, N.N. (1971) On regulatory interaction of nervous elements of the jellyfish *Cyanea arctica* in relation to rhythmic pulsation of the umbrella. *Zhurnal Evolyutsionnoi Biokhimii i Fiziologii*, **7**, 262–269. In Russian; English abstract.

Komai, T. (1935) On *Stephanoscyphus* and *Nausithoe*. *Memoirs of the College of Science, Kyoto Imperial University, Series B*, **10**, 289–339, pls XXI, XXII.

Komai, T. and Tokuoka, Y. (1939) Further observations on the strobilation of *Stephanoscyphus*. *Memoirs of the College of Science, Kyoto Imperial University, Series B*, **15**, 127–133.

Kon, T. and Honma, Y. (1972) Studies on gonad maturity in some marine invertebrates V. Seasonal changes in the gonads of the moon jelly. *Bulletin of the Japanese Society of Scientific Fisheries*, **38**, 545–553.

Konstantinova, M.I. (1966) Characteristics of motion of pelagic larvae of marine invertebrates. *Doklady Akademii Nauk SSSR*, **170**, 726–729. In Russian; Translation in *Doklady Biological Sciences*, **170**, 753–755, 1966.

Korn, H. (1966) Zur Ontogenetischen Differenzierung der Coelenteratengewebe (Polyp-Stadium) unter Besonderer Berucksichtigung des Nervensystems. *Zeitschrift fuer Morphologie und Oekologie der Tiere*, **57**, 1–118. In German.

Kraeuter, J.N. and Setzler, E.M. (1975) The seasonal cycle of scyphozoa and cubozoa in Georgia estuaries. *Bulletin of Marine Science*, **25**, 66–74.

Kramp, P.L. (1947) Medusae. Part III. Trachylina and Scyphozoa with zoogeographical remarks on all the medusae of the northern Atlantic. *Danish Ingolf-Expedition*, **5**, 1–66, pls 1–6.

Kramp, P.L. (1961) Synopsis of the Medusae of the World. *Journal of the Marine Biological Association of the United Kingdom*, **40**, 1–469.

Kramp, P.L. (1965) Some medusae (mainly scyphomedusae) from Australian coastal waters. *Transactions Royal Society of South Australia*, **89**, 257–278, pls 1–3.

Kramp, P.L. (1970) Zoogeographical studies on Rhizostomeae (Scyphozoa). *Videnskabelige Meddelelser fra Dansk Naturhistorisk Forening*, **133**, 7–30.

Krasińska, S. (1914) Beiträge zur Histologie der Medusen. *Zeitschrift fuer wissenschaftliche Zoologie*, **109**, 256–348, pls VII–VIII. In German.

Kremer, P., Costello, J., Kremer, J. *et al.* (1990) Significance of photosynthetic endosymbionts to the carbon budget of the scyphomedusa *Linuche unguiculata*. *Limnology and Oceanography*, **35**, 609–624.

Krüger, F. (1968) Stoffwechsel und Wachstum bei Scyphomedusen. *Helgolaender Wissenschaftliche Meeresuntersuchungen*, **18**, 367–383. In German; English abstract; Fisheries Research Board of Canada Translation Series No. 2197.

Kühl, H. (1972) Hydrography and biology of the Elbe estuary. *Oceanography and Marine Biology; an Annual Review*, **10**, 225–309.

Kuwabara, R., Sato, S. and Noguchi, N. (1969) Ecological studies on the medusa, *Aurelia aurita* Lamarck I Distribution of *Aurelia* patches in the north-east region of Tokyo Bay in summer 1966 and 1967. *Bulletin of the Japanese Society of Scientific Fisheries*, **35**, 156–162. In Japanese; English abstract.

Kuzmicheva, V.I. (1980) Respiration level of Black Sea jellyfish, in *Ekosistemy Pelagiaii Chernogo Morya*, (ed. M.E. Vinogradov), NAUKA, Moscow, pp. 219–223. In Russian.

Lal, D.M., Calton, G.J., Neeman, I. *et al.* (1981) Characterization of *Chrysaora quinquecirrha* (sea nettle) nematocyst venom collagenase. *Comparative Biochemistry and Physiology B Comparative Biochemistry*, **69**, 529–533.

Lambert, F.J. (1935) Observations on the scyphomedusae of the Thames estuary and their metamorphoses. *Travaux de la Station Zoologique de Wimereux.*, **12**, 281–307.

Larson, R.J. (1976) Cubomedusae: Feeding – functional morphology, behavior and phylogenetic position, in *Coelenterate Ecology and Behavior*, (ed. G.O. Mackie), Plenum Press, New York, pp. 237–245.

Larson, R.J. (1979) Feeding in coronate medusae (Class Scyphozoa, Order Coronatae). *Marine Behaviour and Physiology*, **6**, 123–129.

Larson, R.J. (1980) A new stauromedusa, *Kishinouyea corbini* (Scyphozoa, Stauromedusae) from the tropical Western Atlantic. *Bulletin of Marine Science*, **30**, 102–107.

Larson, R.J. (1986a) The feeding and growth of the sea nettle, *Chrysaora quinquecirrha* (Desor), in the laboratory. *Estuaries*, **9**, 376–379.

Larson, R.J. (1986b) Pelagic Scyphomedusae (Scyphozoa: Coronatae and Semaeostomae) of the Southern Ocean. *Antarctic Research Series*, **41**, 59–165.

Larson, R.J. (1986c) Seasonal changes in the standing stocks, growth rates, and production rates of gelatinous predators in Saanich Inlet, British Columbia. *Marine Ecology Progress Series*, **33**, 89–98.

Larson, R.J. (1986d) Water content, organic content, and carbon and nitrogen composition of medusae from the northeast Pacific. *Journal of Experimental Marine Biology and Ecology*, **99**, 107–120.

Larson, R.J. (1987a) Costs of transport for the scyphomedusa *Stomolophus meleagris* L. Agassiz. *Canadian Journal of Zoology*, **65**, 2690–2695.

Larson, R.J. (1987b) Daily ration and predation by medusae and ctenophores in Saanich Inlet, BC, Canada. *Netherlands Journal of Sea Research*, **21**, 35–44.

Larson, R.J. (1987c) First report of the little-known scyphomedusa *Drymonema dalmatinum* in the Caribbean Sea, with notes on its biology. *Bulletin of Marine Science*, **40**, 437–441.

Larson, R.J. (1987d) A note on the feeding, growth, and reproduction of the epipelagic scyphomedusa *Pelagia noctiluca* (Forskål). *Biological Oceanography*, **4**, 447–454.

Larson, R.J. (1987e) Respiration and carbon turnover rates of medusae from the NE Pacific. *Comparative Biochemistry and Physiology*, **87A**, 93–100.

Larson, R.J. (1988) *Kyopoda lamberti* gen. nov., sp. nov., an atypical stauromedusa (Scyphozoa, Cnidaria) from the eastern Pacific, representing a new family. *Canadian Journal of Zoology*, **66**, 2301–2303.

Larson, R.J. (1990) Scyphomedusae and cubomedusae from the eastern Pacific. *Bulletin of Marine Science*, **47**, 546–556.

Larson, R.J. (1991) Diet, prey selection and daily ration of *Stomolophus meleagris*, a filter-feeding scyphomedusa from the NE Gulf of Mexico. *Estuarine Coastal and Shelf Science*, **32**, 511–525.

Larson, R.J. (1992) Riding Langmuir circulations and swimming in circles: a novel form of clustering behavior by the scyphomedusa *Linuche unguiculata*. *Marine Biology (Berlin)*, **112**, 229–235.

Larson, R.J. and Arneson, A.C. (1990) Two medusae new to the coast of California: *Carybdea marsupialis* (Linnaeus, 1758), a cubomedusa and *Phyllorhiza punctata* von Lendenfeld, 1884, a rhizostome scyphomedusa. *Bulletin Southern California Academy of Sciences*, **89**, 130–136.

Larson, R.J. and Fautin, D.G. (1989) Stauromedusae of the genus *Manania* (=*Thaumatoscyphus*) (Cnidaria, Scyphozoa) in the northeast Pacific, including descriptions of new species *Manania gwilliami* and *Manania handi*. *Canadian Journal of Zoology*, **67**, 1543–1549.

Larson, R.J. and Harbison, G.R. (1989) Source and fate of lipids in polar gelatinous zooplankton. *Arctic*, **42**, 339–346.

Larson, R.J. and Harbison, G.R. (1990) Medusae from McMurdo Sound, Ross Sea including the descriptions of two new species, *Leuckartiara brownei* and *Benthocodon hyalinus*. *Polar Biology*, **11**, 19–25.

Larson, R.J., Madin, L.P. and Harbison, G.R. (1988) *In situ* observations of deepwater medusae in the genus *Deepstaria*, with a description of *D. reticulum*, sp. nov. *Journal of the Marine Biological Association of the United Kingdom*, **68**, 689–699.

Larson, R.J., Mills, C.E. and Harbison, G.R. (1991) Western Atlantic midwater hydrozoan and scyphozoan medusae: *in situ* studies using manned submersibles. *Hydrobiologia*, **216/217**, 311–317.

Lauckner, G. (1980) Diseases of Cnidaria, in *Diseases of Marine Animals*, (ed. O. Kinne), John Wiley and Sons, Chichester, pp. 167–237.

Laval, P. (1980) Hyperiid amphipods as crustacean parasitoids associated with gelatinous zooplankton. *Oceanography and Marine Biology; an Annual Review*, **18**, 11–56.

Lebedeva, L.P. and Shushkina, E.A. (1991) The estimation of population characteristics of the medusa *Aurelia aurita* in the Black Sea. *Oceanology*, **31**(3), 314–319.

Lebedeva, L.P. and Shushkina, E.A. (1994) Modelling the effect of *Mnemiopsis* on the Black Sea plankton community. *Oceanology*, **34**, 72–80.

Lebour, M.V. (1922) The food of plankton organisms. *Journal of the Marine Biological Association of the United Kingdom*, **12**, 644–677.

Lebour, M.V. (1923) The food of plankton organisms II. *Journal of the Marine Biological Association of the United Kingdom*, **13**, 70–92.

Lederhendler, I., Bell, L. and Tobach, E. (1975) Preliminary observations of the behavior of *Aplysia dactylomela* (Rang, 1828) in Bimini waters. *Veliger*, **17**, 347–353.

Legovic, T. (1987) A recent increase in jellyfish populations: a predator–prey model and its implications. *Ecological Modelling*, **38**, 243–256.

Legovic, T. (1991) Causes, consequences and possible control of massive occurrences of jellyfish *Pelagia noctiluca* in the Adriatic Sea, in *UNEP: Jellyfish Blooms in the Mediterranean. Proceedings of the II Workshop on Jellyfish in the Mediterranean Sea*, MAP Technical Reports Series No. 47, UNEP, Athens, pp. 128–132.

Lenhoff, H.M. (1971) Research on feeding, digestion, and metabolism in coelenterates: some reflections, in *Experimental Coelenterate Biology*, (eds H.M. Lenhoff, L. Muscatine, and L.V. Davis), University of Hawaii Press, Honolulu, pp. 75–83.

Lerner, J., Mellen, S.A., Waldron, I. *et al.* (1971) Neural redundancy and regularity of swimming beats in scyphozoan medusae. *Journal of Experimental Biology*, **55**, 177–184.

Lesh-Laurie, G.E. and Corriel, R. (1973) Scyphistoma regeneration from isolated tentacles in *Aurelia aurita*. *Journal of the Marine Biological Association of the United Kingdom*, **53**, 885–894.

Lesh-Laurie, G.E. and Suchy, P.E. (1986) Regulation in metagenic cnidarian planulae. *Progress in Clinical and Biological Research*, **217A**, 53–56.

Lesh-Laurie, G.E. and Suchy, P.E. (1991) Cnidaria; Scyphozoa and Cubozoa, in *Microscopic Anatomy of Invertebrates, Volume 2: Placozoa, Porifera, Cnidaria, and Ctenophora*, (eds F.W. Harrison and J.A. Westfall), Wiley-Liss, Inc., New York, pp. 185–266.

Lesh-Laurie, G.E., Hujer, A. and Suchy, P. (1991) Polyp regeneration from isolated tentacles of *Aurelia* scyphistomae: a role for gating mechanisms and cell division. *Hydrobiologia*, **216/217**, 91–97.

Li, M. *et al.* (1992) A study on temperature tolerance of marine animals for ten species. *Marine Environmental Science/ Haiyang Huanjing Kexue. Dalian*, **11**, 29–32. In Chinese.

Li, P., Tan, K. and Ye, C. (1988) Growth pattern and characteristics of edible jellyfish (*Rhopilema esculenta*) in Liaodong Bay. *Journal of Fisheries of China*, **12**, 243–250. In Chinese; English abstract.

Lin, A.L. and Zubkoff, P.L. (1973) Malate dehydrogenase and tetrazolium oxidase of scyphistomae of *Aurelia aurita, Chrysaora quinquecirrha,* and *Cyanea capillata* (Scyphozoa: Semaeostomeae). *Helgolaender Meeresuntersuchungen*, **25**, 206–213.

Lin, A.L. and Zubkoff, P.L. (1976a) Malate dehydrogenase isozymes of different stages of Chesapeake Bay jellyfish. *Biological Bulletin (Woods Hole)*, **150**, 268–278.

Lin, A.L. and Zubkoff, P.L. (1976b) Pentose phosphate pathway in the scyphistomae of *Chrysaora quinquecirrha. Comparative Biochemistry and Physiology*, **54B**, 33–36.

Lin, A.L. and Zubkoff, P.L. (1977) Enzymes associated with carbohydrate metabolism of scyphistomae of *Aurelia aurita* and *Chrysaora quinquecirrha* (Scyphozoa: Semaeostomae). *Comparative Biochemistry and Physiology*, **57B**, 303–308.

Lin, W.W., Lee, C.Y. and Burnett, J.W. (1988) Effect of sea nettle (*Chrysaora quinquecirrha*) venom on isolated rat aorta. *Toxicon*, **26**, 1209–1212.

Littleford, R.A. (1939) The life cycle of *Dactylometra quinquecirrha* L. Agassiz in the Chesapeake Bay. *Biological Bulletin (Woods Hole)*, 77, 368–381.

Loeb, M.J. (1972) Strobilation in the Chesapeake Bay sea nettle *Chrysaora quinquecirrha* I The effects of environmental temperature changes on strobilation and growth. *Journal of Experimental Zoology*, **180**, 279–292.

Loeb, M.J. (1973) The effect of light on strobilation in the Chesapeake Bay sea nettle *Chrysaora quinquecirrha. Marine Biology (Berlin)*, **20**, 144–147.

Loeb, M.J. (1974a) Strobilation in the Chesapeake Bay sea nettle *Chrysaora quinquecirrha* II Partial characterization of the neck-inducing factor from strobilating polyps. *Comparative Biochemistry and Physiology*, **47A**, 291–301.

Loeb, M.J. (1974b) Strobilation in the Chesapeake bay sea nettle *Chrysaora quinquecirrha* III Dissociation of the neck-inducing factor from strobilating polyps. *Comparative Biochemistry and Physiology*, **49A**, 423–432.

Loeb, M.J. and Blanquet, R.S. (1973) Feeding behavior in polyps of the Chesapeake Bay sea nettle, *Chrysaora quinquecirrha* (Desor, 1848). *Biological Bulletin (Woods Hole)*, **145**, 150–158.

Loeb, M.J. and Gordon, C.M. (1975) Strobilation in the Chesapeake Bay sea nettle, *Chrysaora quinquecirrha* IV Tissue levels of iodinated high molecular weight component and NIF in relation to temperature change-induced behavior. *Comparative Biochemistry and Physiology*, **51A**, 37–42.

Loeb, M.J. and Hayes, D.K. (1981) Strobilation in the Chesapeake Bay Sea Nettle, *Chrysaora quinquecirrha* V. Neurons and neurosecretion. *Transactions of the American Microscopical Society*, **100**, 264–270.

Loeblich, A.R. and Sherley, J.L. (1979) Observations on the theca of the motile phase of free-living and symbiotic isolates of *Zooxanthella microadriatica* Freudenthal comb. nov. *Journal of the Marine Biological Association of the United Kingdom*, **59**, 195–205.

Loginova, N.P. and Perzova, N.M. (1967) Some data on ecology of feeding of pelagic Coelenterata in the White Sea. *Issledovanie Fauny Morei*, 7, 21–28. In Russian.

Long, K.O. and Burnett, J.W. (1989) Isolation, characterization, and comparison of hemolytic peptides in nematocyst venoms of two species of jellyfish (*Chrysaora quinquecirrha*) and (*Cyanea capillata*). *Comparative Biochemistry and Physiology B Comparative Biochemistry*, **94**, 641–646.

Longhurst, A.R. (1985) The structure and evolution of plankton communities. *Progress in Oceanography*, **15**, 1–35.

Long-Rowe, K.O. and Burnett, J.W. (1994) Characteristics of hyaluronidase and hemolytic activity in fishing tentacle nematocyst venom of *Chrysaora quinquecirrha*. *Toxicon*, **32**, 165–174.

Lotan, A., Ben-Hillel, R. and Loya, Y. (1992) Life cycle of *Rhopilema nomadica*: a new immigrant scyphomedusan in the Mediterranean. *Marine Biology (Berlin)*, **112**, 237–242.

Lotan, A., Fine, M. and Ben-Hillel, R. (1994) Synchronization of the life cycle and dispersal pattern of the tropical invader scyphomedusan *Rhopilema nomadica* is temperature dependent. *Marine Ecology Progress Series*, **109**, 59–65.

Lotan, A., Fishman, L., Loya, Y. *et al.* (1995) Delivery of a nematocyst toxin. *Nature (London)*, **375**, 456.

Low, J.W. (1921) Variation in ephyrae of *Aurelia aurita*. *Proceedings Royal Physical Society of Edinburgh*, **20**, 226–235.

Lowndes, A.G. (1942) Percentage of water in jelly-fish. *Nature (London)*, **150**, 234–235.

Lowndes, A.G. (1943) Some applications of the displacement method of weighing living aquatic organisms. *Proceedings of the Zoological Society of London*, **113**, 28–43.

Lu, N., Liu, C. and Guo, P. (1989) Effect of salinity on larva of edible medusae (*Rhopilema esculenta* Kishinouye) at different developmental phases and a review on the cause of jellyfish resources falling greatly in Liaodong Bay. *Acta Ecologica Sinica*, **9**, 304–309. In Chinese; English abstract.

Lucas, C.H. (1994) Biochemical composition of *Aurelia aurita* in relation to age and sexual maturity. *Journal of Experimental Marine Biology and Ecology*, **183**, 179–192.

Lucas, C.H. and Williams, J.A. (1994) Population dynamics of the scyphomedusa *Aurelia aurita* in Southampton Water. *Journal of Plankton Research*, **16**, 879–895.

Ludwig, F.-D. (1969) Die Zooxanthellen bei *Cassiopea andromeda* Eschscholtz 1829 (Polyp-Stadium) und ihre Bedeutung für die Strobilation. *Zoologische Jahrbuecher Abteilung fuer Anatomie*, **86**, 238–277. In German; English abstract.

Lukanin, V.V. (1976) Adaptive response to changing water salinity in the larvae of the scyphoid jellyfish *Aurelia aurita*. *Zhurnal Obshchei Biologii*, **37**, 776–787. In Russian; English abstract.

Lutcavage, M. and Lutz, P.L. (1986) Metabolic rate and food energy requirements of the leatherback sea turtle, *Dermochelys coriacea*. *Copeia*, **1986**, 796–798.

Ma, N.C. (1960) *Mrs. Ma's Chinese Cookbook*, Charles E. Tuttle, Tokyo.

Maaden, H.V.D. (1939) Über das Sinnesgrübchen von *Aurelia aurita* Linné. *Zoologischer Anzeiger*, **125**, 29–35. In German.

Maas, O. (1903) Die Scyphomedusen der Siboga-Expedition. *Siboga-Expeditie*, **11**, 1–91. In German.

Mackie, G.O. (1974) Locomotion, flotation, and dispersal, in *Coelenterate Biology: Reviews and New Perspectives*, (eds L. Muscatine and H.M. Lenhoff), Academic Press, New York, pp. 313–357.

Mackie, G.O. (1990) The elementary nervous system revisited. *American Zoologist*, **30**, 907–920.

Mackie, G.O. and Mills, C.E. (1983) Use of the *Pices IV* submersible for zooplankton studies in coastal waters of British Columbia. *Canadian Journal of Fisheries and Aquatic Sciences*, **40**, 763–776.

Mackie, G.O., Anderson, P.A.V. and Singla, C.L. (1984) Apparent absence of gap junctions in two classes of cnidaria. *Biological Bulletin (Woods Hole)*, **167**, 120–123.

Mackie, G.O., Larson, R.J., Larson, K.S. *et al.* (1981) Swimming and vertical migration of *Aurelia aurita* (L) in a deep tank. *Marine Behaviour and Physiology*, 7, 321–329.

Macon McDermott, A. and Blanquet, R.S. (1991) Glucose and glycerol uptake by isolated zooxanthellae from *Cassiopea xamachana*: transport mechanisms and regulation by host homogenate fractions. *Marine Biology (Berlin)*, **108**, 129–136.

Madin, L.P. (1988) Feeding behavior of tentaculate predators: in situ observations and a conceptual model. *Bulletin of Marine Science*, **43**, 413–429.

Magnusen, J.E. (1980) Epidermal cell movement during podocyst formation in *Chrysaora quinquecirrha*, in *Developmental and Cellular Biology of Coelenterates*, (eds P. Tardent and R. Tardent), Elsevier/North-Holland Biomedical Press, Amsterdam, pp. 435–440.

Malej, A. (1989a) Behaviour and trophic ecology of the jellyfish *Pelagia noctiluca* (Forsskål, 1775). *Journal of Experimental Marine Biology and Ecology*, **126**, 259–270.

Malej, A. (1989b) Respiration and excretion rates of *Pelagia noctiluca* (Semaeostomae, Scyphozoa), in *Proceedings of the Twenty First European Marine Biology Symposium*, (eds R.Z. Klekowski, E. Styczynska-Jurewicz, and L. Falkowski), Institute of Oceanology, Polish Academy of Sciences, Gdansk, pp. 107–113.

Malej, A. (1991) Rates of metabolism of jellyfish as related to body weight, chemical composition and temperature, in *UNEP: Jellyfish Blooms in the Mediterranean. Proceedings of the II Workshop on Jellyfish in the Mediterranean Sea*, MAP Technical Reports Series No. 47, UNEP, Athens, pp. 253–259.

Malej, A. and Malej, M. (1992) Population dynamics of the jellyfish *Pelagia noctiluca* (Forsskål, 1775), in *Marine Eutrophication and Population Dynamics*, (eds G. Colombo, I. Ferrari, V.U. Ceccherelli, *et al.*), Olsen and Olsen, ISSS, Fredensborg, pp. 215–219.

Malej, A. and Vukovic, A. (1986) Some data on the metabolism of *Pelagia noctiluca* in the Gulf of Trieste. *Nova Thalassia*, **8 Suppl. 2**, 107–111.

Malej, A., Faganeli, J. and Pezdic, J. (1993) Stable isotope and biochemical fractionation in the marine pelagic food chain: the jellyfish *Pelagia noctiluca* and net zooplankton. *Marine Biology (Berlin)*, **116**, 565–570.

Manchenko, G.P. and Zaslavskaya, N.I. (1980) Genetic variability of hexokinase and leucine aminopeptidase in the scyphozoan medusa *Cyanea capillata*. *Biologiya Morya (Vladivostok)*, **6**, 65–68. In Russian; English abstract.

Mangum, C.P., Oakes, M.J. and Shick, J.M. (1972) Rate-temperature responses in scyphozoan medusae and polyps. *Marine Biology (Berlin)*, **15**, 298–303.

Mansueti, R. (1963) Symbiotic behavior between small fishes and jellyfishes, with new data on that between the stromateid, *Peprilus alepidotus*, and the scyphomedusa, *Chrysaora quinquecirrha*. *Copeia*, **1963**, 40–80.

Mariscal, R.N. (1971) Effect of a disulfide reducing agent on the nematocyst capsules from some coelenterates, with an illustrated key to nematocyst classification, in *Experimental Coelenterate Biology*, (eds H.M. Lenhoff, L. Muscatine, and L.V. Davis), University of Hawaii Press, Honolulu, pp. 157–168.

Mariscal, R.N. (1984) Cnidaria: Cnidae, in *The Biology of the Integument*, (eds J. Bereiter-Hahn, A.O. Matolsky, and K.S. Richards), Springer-Verlag, Berlin, pp. 57–68.

Mariscal, R.N. (1988) X-ray microanalysis and perspectives on the role of calcium and other elements in cnidae, in *The Biology of Nematocysts*, (eds D.A. Hessinger and H.M. Lenhoff), Academic Press, San Diego, pp. 95–113.

Mariscal, R.N. and Bigger, C.H. (1976) A comparison of putative sensory receptors associated with nematocysts in an anthozoan and a scyphozoan, in *Coelenterate Ecology and Behavior*, (ed. G.O. Mackie), Plenum Press, New York, pp. 559–568.

Mariscal, R.N. and Lenhoff, H.M. (1969) Effect of a disulphide reducing agent on coelenterate nematocyst capsules. *Experientia (Basel)*, **25**, 330–331.

Markell, D.A. and Trench, R.K. (1993) Macromolecules exuded by symbiotic dinoflagellates in culture: amino acid and sugar composition. *Journal of Phycology*, **29**, 64–68.

Markell, D.A., Trench, R.K. and Iglesias-Prieto, R. (1992) Macromolecules associated with the cell walls of symbiotic dinoflagellates. *Symbiosis*, **12**, 19–31.

Martin, J.W. and Kuck, H.G. (1991) Faunal associates of an undescribed species of *Chrysaora* (Cnidaria, Scyphozoa) in the Southern California Bight, with notes on unusual occurrences of other warm water species in the area. *Bulletin Southern California Academy of Sciences*, **90**, 89–101.

Martin, S.M. and Spencer, A.N. (1983) Neurotransmitters in coelenterates. *Comparative Biochemistry and Physiology C Comparative Pharmacology*, **74**, 1–14.

Martin, V.J. and Chia, F.-S. (1982) Fine structure of a scyphozoan planula, *Cassiopeia xamachana*. *Biological Bulletin (Woods Hole)*, **163**, 320–328.

Maso, M. and Castellon, A. (1985) The vertical distribution of fish eggs and larvae and *Pelagia noctiluca* obtained in a cycle of 24 hours. *Rapports et Proces-Verbaux des Reunions Commission Internationale pour l'Exploration Scientifique de la Mer Mediterranee Monaco*, **29**, 185–188.

Matsakis, S. and Conover, R.J. (1991) Abundance and feeding of medusae and their potential impact as predators on other zooplankton in Bedford Basin (Nova Scotia, Canada) during spring. *Canadian Journal of Fisheries and Aquatic Sciences*, **48**, 1419–1430.

Matsuno, A. (1981a) Electron microscopic investigations on the muscular systems of the polyp of *Atorella* species. *Annotationes Zoologicae Japonenses*, **54**, 171–181.

Matsuno, A. (1981b) Electron microscopic investigations on the striated muscle of an ephyra of *Atorella* sp. *Zoological Magazine (Tokyo)*, **90**, 152–156.

Matsuno, A. (1983) An electron microscopic study on the development of cross-striated muscles in ephyrae of *Aurelia aurita. Zoological Magazine (Tokyo)*, **92**, 416–422.

Matsuno, A. and Hisamatsu, T. (1982) Electron microscopy on striated muscles in the ephyra of *Aurelia aurita. Memoirs of the Faculty of Science Shimane University*, **16**, 61–67.

Matsuno, A. and Kawaguti, S. (1991) An ultrastructural study of the polyp and strobila of *Atorella japonica* (Cnidaria, Coronatae) with respect to muscles and nerves. *Hydrobiologia*, **216/217**, 39–43.

Mauchline, J. and Gordon, J.D.M. (1983) Diets of clupeoid, stomiatoid and salmonoid fish of the Rockall Trough, northeastern Atlantic Ocean. *Marine Biology (Berlin)*, **77**, 67–78.

Mauchline, J. and Harvey, P.F. (1983) The scyphomedusae of the Rockall Trough, northeastern Atlantic Ocean. *Journal of Plankton Research*, **5**, 881–890.

Mayer, A.G. (1906) Rhythmical pulsation in scyphomedusae. *Carnegie Institution of Washington Publication*, **47**, 1–62.

Mayer, A.G. (1910) *Medusae of the World III The Scyphomedusae*, Carnegie Institution of Washington, Washington.

Mayer, A.G. (1914a) The effects of temperature upon tropical marine animals. *Carnegie Institution of Washington Publication*, **183**, 3–24. Papers from the Tortugas Laboratory of the Carnegie Institution of Washington 6.

Mayer, A.G. (1914b) The law governing the loss of weight in starving *Cassiopea*. *Carnegie Institution of Washington Publication*, **183**, 55–83. Papers from the Tortugas Laboratory of the Carnegie Institution of Washington 6.

Mayer, A.G. (1917) Report upon the scyphomedusae collected by the United States Bureau of Fisheries Steamer 'Albatross' in the Philippine Islands and Malay Archipelago. *U S National Museum Bulletin 100*, **1** (3), 174–233.

McCloskey, L.R., Muscatine, L. and Wilkerson, F.P. (1994) Daily photosynthesis, respiration, and carbon budgets in a tropical marine jellyfish (*Mastigias* sp.). *Marine Biology (Berlin)*, **119**, 13–22.

McKay, M.C. and Anderson, P.A.V. (1988) On the preparation and properties of isolated cnidocytes and cnidae, in *The Biology of Nematocysts*, (eds

D.A. Hessinger and H.M. Lenhoff), Academic Press, San Diego, pp. 273–293.

McLaughlin, J.A. and Zahl, P. (1966) Endozoic algae, in *Symbiosis*, (ed. S.M. Henry), Academic Press, New York, pp. 257–297.

Mergner, H. (1971) Cnidaria, in *Experimental embryology of marine and fresh-water invertebrates*, (ed. G. Reverberi), North-Holland Publishing Company, pp. 1–83.

Mergner, H. and Svoboda, A. (1977) Productivity and seasonal changes in selected reef areas in the Gulf of Aqaba (Red Sea). *Helgolaender Wissenschaftliche Meeresuntersuchungen*, **30**, 383–399.

Metz, P. (1967) On the relations between *Hyperia galba* Montagu (Amphipoda, Hyperiidae) and its host *Aurelia aurita* in the Isefjord Area (Sjaelland, Denmark). *Videnskabelige Meddelelser fra Dansk Naturhistorisk Forening*, **130**, 85–108.

Mihursky, J.A. and Kennedy, V.S. (1967) Water temperature criteria to protect aquatic life. *American Fisheries Society Special Publication*, **4**, 20–32.

Mikhailov, B.N. (1962) On the feeding of the Black Sea medusa *Aurelia aurita* L. *Zoologicheskii Zhurnal*, **41**, 286–287. In Russian; English abstract.

Millott, N. and Fox, D.L. (1954) Pigmentation of the jellyfish, *Pelagia noctiluca*. *Nature (London)*, **173**, 169.

Mills, C.E. (1995) Medusae, siphonophores, and ctenophores as planktivorous predators in changing global ecosystems. *ICES (International Council for the Exploration of the Sea) Journal of Marine Science*, **52**, 575–581.

Mills, E.L., Pittman, K. and Tan, F.C. (1984) Food web structure on the Scotian shelf, eastern Canada: a study using ^{13}C as a food-chain tracer. *Rapports et Proces-verbaux des Réunions Conseil international pour l'Exploration de la Mer*, **183**, 111–118.

Minchin, E.A. (1889) Note on the mode of attachment of the embryos to the oral arms of *Aurelia aurita*. *Proceedings of the Zoological Society of London*, **39**, 583–85, pls LVII–LVIII.

Mironov, G.N. (1967) Feeding of planktonic predators III Food utilization and daily rations of *Aurelia aurita* (L.), in *Biology and Distribution of Plankton in Southern Seas*, Academy Nauka, Moscow, pp. 124–137. In Russian.

Mironov, G.N. (1971) The biomass and distribution of the medusa *Aurelia aurita* (L.) after data from trawler catches between 1949–1962 in the Black Sea. *Biologiya Morya (Kiev)*, **24**, 49–69. In Russian.

Miura, S. and Kimura, S. (1985) Jellyfish mesogloea collagen. Characterization of molecules as α1α2α3 heterotrimers. *Journal of Biological Chemistry*, **260**, 15352–15356.

Moestafa, S.H. and McConnaughey, B.H. (1966) *Catostylus ouwensi* (Rhizostomeae, Catostylidae), a new jellyfish from Irian (New Guinea) and *Ouwensia catostyli* n.gen., n.sp., parasitic in *C. ouwensi*. *Treubia*, **27**, 1–9.

Möller, H. (1980a) Population dynamics of *Aurelia aurita* medusae in Kiel Bight, Germany (FRG). *Marine Biology (Berlin)*, **60**, 123–128.

Möller, H. (1980b) Scyphomedusae as predators and food competitors of larval fish. *Meeresforschung*, **28**, 90–100.

Möller, H. (1984a) *Daten zur Biologie der Quallen und Jungfische in der Kieler Bucht. Mit einer Bibliographie zur Biologie der Kieler Bucht*; H. Möller, Kiel. In German.

Möller, H. (1984b) Effects of jellyfish predation on fishes, in *UNEP: Workshop on jellyfish blooms in the Mediterranean, Athens, 31 October–4 November 1983*, UNEP, Athens, pp. 45–59.

Möller, H. (1984c) Reduction of a larval herring population by jellyfish predator. *Science (Washington DC)*, **224**, 621–622.

Montgomery, M.K. and Kremer, P.M. (1995) Transmission of symbiotic dinoflagellates through the sexual cycle of the host scyphozoan *Linuche unguiculata*. *Marine Biology (Berlin)*, **124**, 147–155.

Moore, A.R. (1926) Galvanic stimulation of luminescence in *Pelagia noctiluca*. *Journal of General Physiology*, **9**, 375–379.

Moore, P.G. and Rainbow, P.S. (1989) Feeding biology of the mesopelagic Gammaridean Amphipod *Parandania boecki* (Stebbing, 1888) (Crustacea: Amphipoda: Stegocephalidae) from the Atlantic Ocean. *Ophelia*, **30**, 1–19.

Moore, P.G., Rainbow, P.S. and Larson, R.J. (1993) The mesopelagic shrimp *Notostomus robustus* Smith (Decapoda: Oplophoridae) observed in situ feeding on the medusan *Atolla wyvillei* Haeckel in the northwest Atlantic, with notes on gut contents and mouthpart morphology. *Journal of Crustacean Biology*, **13**, 690–696.

Morand, P., Carré, C. and Biggs, D.C. (1987) Feeding and metabolism of the jellyfish *Pelagia noctiluca* (Scyphomedusae, Semaeostomae). *Journal of Plankton Research*, **9**, 651–665.

Morand, P., Goy, J. and Dallot, S. (1992) Recrutement et fluctuations à long-terme de *Pelagia noctiluca* (Cnidaria, Scyphozoa). *Annales de l'Institut Oceanographique*, **68**, 151–158. In French; English abstract.

Morin, J.G. (1974) Coelenterate bioluminescence, in *Coelenterate Biology: Reviews and New Perspectives*, (eds L. Muscatine and H.M. Lenhoff), Academic Press, New York, pp. 397–438.

Morin, J.G. and Hastings, J.W. (1971a) Biochemistry of the bioluminescence of colonial hydroids and other coelenterates. *Journal of Cellular Physiology*, 77, 305–311.

Morin, J.G. and Hastings, J.W. (1971b) Energy transfer in a bioluminescent system. *Journal of Cellular Physiology*, **77**, 313–318.

Morin, J.G. and Reynolds, G.T. (1972) Spectral and kinetic characteristics of bioluminescence in *Pelagia noctiluca* and other coelenterates. *Biological Bulletin (Woods Hole)*, **143**, 470–471. Abstract.

Moshina, N.I. (1974) On mass reproduction of *Rhizostoma pulmo* and *Aurelia aurita* (Discomedusae) in the Azov Sea. *Zoologicheskii Zhurnal*, **53**, 1248–1251. In Russian; English abstract.

Murray, M.J. (1977) Systems of mutually-triggering event generators: basic properties and functions in information transmission and rhythm generation. *Journal of Comparative Physiology A Sensory Neural and Behavioral Physiology*, **117**, 63–98.

Muscatine, L. and Marian, R.E. (1982) Dissolved inorganic nitrogen flux in symbiotic and nonsymbiotic medusae. *Limnology and Oceanography*, **27**, 910–917.

Muscatine, L., Wilkerson, F.P. and McCloskey, L.R. (1986) Regulation of population density of symbiotic algae in a tropical marine jellyfish (*Mastigias* sp.). *Marine Ecology Progress Series*, **32**, 279–290.

Mutlu, E., Bingel, F., Gücü, A.C. *et al.* (1994) Distribution of the new invader *Mnemiopsis* sp. and the resident *Aurelia aurita* and *Pleurobrachia pileus* populations in the Black Sea in the years 1991–1993. *ICES (International Council for the Exploration of the Sea) Journal of Marine Science*, **51**, 407–421.

Nagabhushanam, A.K. (1964) On the biology of the whiting, *Gadus merlangus*, in Manx waters. *Journal of the Marine Biological Association of the United Kingdom*, **44**, 177–202.

Nagabhushanam, A.K. (1965) On the biology of the commoner gadoids in Manx waters. *Journal of the Marine Biological Association of the United Kingdom*, **45**, 615–657.

Napara, T.O. and Chaga, O.Y. (1992a) Proliferative activity and growth of a mesogleal cell population in the scyphomedusa *Aurelia aurita* (L.) I Mesogleal cells of medusae. *Tsitologiya*, **34**, 59–66. In Russian; English abstract.

Napara, T.O. and Chaga, O.Y. (1992b) Proliferative activity and growth of the mesogleal cell population in scyphomedusa *Aurelia aurita* (L.) II Mesogleal cells of strobilae and ephyrae. *Tsitologiya*, **34**, 33–37. In Russian; English abstract.

Naumov, D.V. (1961) Stsifoidnye meduzy morei SSSR (Scyphomedusae of the seas of the USSR). *Opredeliteli po Faune SSSR*, **75**, 1–98, pls 1–3. In Russian.

Neeman, I., Calton, G.J. and Burnett, J.W. (1981) Purification of an endonuclease present in *Chrysaora quinquecirrha* venom. *Proceedings of the Society for Experimental Biology and Medicine*, **166**, 374–382.

Nemazie, D.A., Purcell, J.E. and Glibert, P.M. (1993) Ammonium excretion by gelatinous zooplankton and their contribution to the ammonium requirements of microplankton in Chesapeake Bay. *Marine Biology (Berlin)*, **116**, 451–458.

Neumann, R. (1979) Bacterial induction of settlement and metamorphosis in the planula larvae of *Cassiopea andromeda* (Cnidaria: Scyphozoa, Rhizostomeae). *Marine Ecology Progress Series*, **1**, 21–28.

Neumann, R., Schmahl, G. and Hofmann, D.K. (1980) Bud formation and control of polyp morphogenesis in *Cassiopea andromeda* (Scyphozoa), in *Developmental and Cellular Biology of Coelenterates*, (eds P. Tardent and R. Tardent), Elsevier/North-Holland Biomedical Press, Amsterdam, pp. 217–223.

Newton, C. and Potts, W.T.W. (1993) Ionic regulation and buoyancy in some planktonic organisms. *Journal of the Marine Biological Association of the United Kingdom*, **73**, 15–23.

Nicol, J.A.C. (1958) Observations on luminescence in pelagic animals. *Journal of the Marine Biological Association of the United Kingdom*, **37**, 705–752.

Nizhegorodova, L.E. and Nidzvetskaya, L.M. (1983) Microbial flora of the medusa *Aurelia aurita*. *Biologiya Morya (Vladivostok)*, **0**(4), 33–37. In Russian; English abstract.

Norris, R.D. (1989) Cnidarian taphonomy and affinities of the Ediacara biota. *Lethaia*, **22**, 381–393.

Oakes, M.J. and Haven, D.S. (1971) Some predators of polyps of *Chrysaora quinquecirrha* (Scyphozoa, Semaeostomae) in the Chesapeake Bay. *Virginia Journal of Science*, **22**, 45–46.

Ohtsuki, T. (1930) Digestive enzymes of some marine animals. *Zoological Magazine (Tokyo)*, **42**, 411–421. In Japanese.

Oiestad, V. (1985) Predation on fish larvae as a regulatory force, illustrated in mesocosm studies with large groups of larvae. *NAFO (Northwest Atlantic Fisheries Organization) Scientific Council Studies*, **8**, 25–32.

Olesen, N.J., Frandsen, K. and Riisgard, H.U. (1994) Population dynamics, growth and energetics of jellyfish *Aurelia aurita* in a shallow fjord. *Marine Ecology Progress Series*, **105**, 9–18.

Olmon, J.E. and Webb, K.L. (1974) Metabolism of ^{131}I in relation to strobilation of *Aurelia aurita* L. (Scyphozoa). *Journal of Experimental Marine Biology and Ecology*, **16**, 113–122.

Olsson, P., Granéli, E., Carlsson, P. *et al.* (1992) Structuring of a postspring phytoplankton community by manipulation of trophic interactions. *Journal of Experimental Marine Biology and Ecology*, **158**, 249–266.

Omori, M. (1981) Edible jellyfish (Scyphomedusae: Rhizostomeae) in the Far East waters: A brief review of the biology and fishery. *Bulletin of Plankton Society of Japan*, **28**, 1–11. In Japanese; English abstract.

Omori, M. and Hamner, W.M. (1982) Patchy distribution of zooplankton: behavior, population assessment and sampling problems. *Marine Biology (Berlin)*, **72**, 193–200.

Ortiz-Corp's, E., Cutress, C.E. and Cutress, B.M. (1987) Life history of the coronate scyphozoan *Linuche unguiculata* (Swartz, 1788). *Caribbean Journal of Science*, **23**, 432–443.

Orton, J.H. (1922) The mode of feeding of the jelly-fish, *Aurelia aurita*, on the smaller organisms in the plankton. *Nature (London)*, **110**, 178–179.

Östman, C. (1991) Scanning electron microscopic observations on Scandinavian scyphozoans with special reference to their nematocysts, in *UNEP: Jellyfish blooms in the Mediterranean. Proceedings of the II Workshop on Jellyfish in the Mediterranean Sea*, MAP Techniucal Reports Series no. 47,UNEP, Athens, pp. 283–292.

Ostopenya, A.P. (1965) Completeness of oxidation of organic matter in aquatic invertebrates by the bichromate oxidation method. *Doklady Akademii Nauk Belorusskii SSR*, **9**, 273–276. In Russian.

Otto, J.J. (1976) Early development and planula movement in *Haliclystus* (Scyphozoa: Stauromedusae), in *Coelenterate Ecology and Behavior*, (ed. G.O. Mackie), Plenum Press, New York, pp. 319–329.

Otto, J.J. (1978) The settlement of *Haliclystus* planulae, in *Settlement and Metamorphosis of Marine Invertebrate Larvae*, (eds F.S. Chia and M. Rice), Elsevier, New York, pp. 13–22.

Pagès, F. (1992) Mesoscale coupling between planktonic cnidarian distribution and water masses during a temporal transition between active upwelling and abatement in the Northern Benguela System. *South African Journal of Marine Science*, **12**, 41–52.

Pagès, F., Gili, J.-M. and Bouillon, J. (1992) Medusae (Hydrozoa, Scyphozoa, Cubozoa) of the Benguela Current (southeastern Atlantic). *Scientia Marina*, **56 Suppl. 1**, 1–64.

Palmén, E. (1953) Seasonal occurrence of ephyrae and subsequent instars of *Aurelia aurita* (L.) in the shallow waters of Tvärminne, S. Finland. *Archivum Societatis Zoologicae-Botanicae Fennicae 'Vanamo'*, **8**, 122–131.

Panayotidis, P., Anagnostaki, G. and Siokou-Frangou, I. (1986) Variations saisonnières du diamètre et de la biomasse de la scyphoméduse *Aurelia aurita* Lam., dans la Baie d'Elefsis (Saronikos, Mer Egée). *Nova Thalassia*, **8 Suppl. 2**, 89–92. In French; English abstract.

Panceri, M. (1872) Études sur la phosphorescence des animaux marins. *Annales des Sciences Naturelles Zoologie et Paleontologie Series 5*, **16**, 1–68. In French.

Pantin, C.F.A. and Vianna Dias, M. (1952) Rhythm and afterdischarge in medusae. *Anais da Academia Brasileira de Ciencias*, **24**, 351–364.

Papathanassiou, E., Panayotidis, P. and Anagnostaki, K. (1986) Reproduction and growth of *Aurelia aurita* in Elefsis Bay. *Nova Thalassia*, **8 Suppl. 2**, 83–88.

Papathanassiou, E., Panayotidis, P. and Anagnostaki, K. (1987) Notes on the biology and ecology of the jellyfish *Aurelia aurita* Lam. in Elefsis Bay (Saronikos Gulf, Greece). *Marine Ecology (Pubblicazioni della Stazione Zoologica di Napoli I)*, **8**, 49–58.

Papenfuss, E.J. (1936) The utility of the nematocysts in the classification of certain Scyphomedusae I *Cyanea capillata, Cyanea palmstruchii, Dactylometra quinquecirrha* var *chesapeakei*, and *Chrysaora hysocella*. *Lunds Universitets Arsskrift N.F., Avd. 2,*, **31**, 1–26.

Paspaleff, G.W. (1938) Über die Entwicklung von *Rhizostoma pulmo* Agass. *Arbeiten aus der Biologischen Meeresstation am Schwarzen Meer in Varna*, **7**, 1–25. In German; Bulgarian abstract.

Passano, K.N. and Passano, L.M. (1971) The endodermal nerve net of scyphozoa. *Journal of Morphology*, **133**, 105–124.

Passano, L.M. (1965) Pacemakers and activity patterns in medusae: Homage to Romanes. *American Zoologist*, **5**, 465–481.

Passano, L.M. (1973) Behavioral control systems in medusae: A comparison between hydro- and scyphomedusae. *Publications of the Seto Marine Biological Laboratory*, **20**, 615–645. Proc. Second Int. Symp. Cnidaria.

Passano, L.M. (1982) Scyphozoa and Cubozoa, in *Electrical Conduction and Behaviour in 'Simple' Invertebrates*, (ed. G.A.B. Shelton), Clarendon Press, Oxford, pp. 149–202.

Passano, L.M. (1988) Variability in the initiation of diffuse nerve-net impulses in the mangrove jellyfish *Cassiopea xamachana* (Coelenterate: Scyphozoa).

Comparative Biochemistry and Physiology C Comparative Pharmacology, **91**, 273–279.

Pavlova, Y.V. (1968) Respiration of Black Sea plankton animals, in *Biologicheskiye issledovaniya Chernogo morya, i yego promyslovykh resursov (Biological Investigation of the Black Sea and its commercial fishing resources)*, NAUKA, Moscow, pp. 71–75. In Russian.

Pearre, S. (1982) Estimating prey preference by predators: uses of various indices, and a proposal of another based on x^2. *Canadian Journal of Fisheries and Aquatic Sciences*, **39**, 914–923.

Pedersen, K.J. (1991) Invited review: structure and composition of basement membranes and other basal matrix systems in selected invertebrates. *Acta Zoologica (Stockholm)*, **72**, 181–201.

Pepin, P., Shears, T.H. and de Lafontaine, Y. (1992) Significance of body size to the interaction between a larval fish (*Mallotus villosus*) and a vertebrate predator (*Gasterosteus aculeatus*). *Marine Ecology Progress Series*, **81**, 1–12.

Percival, E. (1923) On the strobilization of *Aurelia*. *Quarterly Journal of Microscopical Science*, **67**, 85–100, pl. 6.

Perkins, F.O., Ramsey, R.W. and Street, S.F. (1971) The ultrastructure of fishing tentacle muscle in the jellyfish *Chrysaora quinquecirrha*: a comparison of contracted and relaxed states. *Journal of Ultrastructure Research*, **35**, 431–450.

Petersen, K.W. (1979) Development of coloniality in hydrozoa, in *Biology and Systematics of Colonial Organisms*, (eds G. Larwood and B.R. Rosen), Academic Press, London, pp. 105–139.

Phillips, P.J., Burke, W.D. and Keener, E.J. (1969) Observations on the trophic significance of jellyfishes in Mississipi Sound with quantitative data on the associative behavior of small fishes with medusae. *Transactions of the American Fisheries Society*, **98**, 703–712.

Phillips, P.J. and Levin, N.L. (1973) Cestode larvae from scyphomedusae of the Gulf of Mexico. *Bulletin of Marine Science*, **23**, 574–584.

Plotnikova, E.D. (1961) On the diet of medusae in the littoral of eastern Murman, in *Hydrological and Biological Features of the Shore Waters of Murman*, (ed. M.M. Kamshilov), Kolski Filial, Akademia Nauk SSSR, Murmansk, pp. 153–156. In Russian.

Pollack, G.H. (1990) *Muscles and Molecules: Uncovering the Principles of Biological Motion*, Ebner and Sons, Seattle.

Pollmans, D. and Hündgen, M. (1981) A light and electron microscope study of the rhopalia in *Aurelia aurita* (Scyphozoa, Semaeostomae). *Zoologische Jahrbuecher Abteilung fuer Anatomie*, **105**, 508–525. In German; English abstract.

Polteva, D.G., Znidaric, D. and Lui, A. (1985) Observations of asexual reproduction and regeneration in *Cassiopea* (Scyphozoa, Coelenterata). *Zoologicheskii Zhurnal*, **64**, 172–180. In Russian; English abstract.

Powers, D.A., Lenhoff, H.M. and Leone, C.A. (1968) Glucose-6-phosphate dehydrogenase and 6-phosphogluconate dehydrogenase activities in coelenterates. *Comparative Biochemistry and Physiology*, **27**, 139–144.

Purcell, J.E. (1985) Predation on fish eggs and larvae by pelagic cnidarians and ctenophores. *Bulletin of Marine Science*, **37**, 739–755.

Purcell, J.E. (1990) Soft-bodied zooplankton predators and competitors of larval herring (*Clupea harengus pallasi*) at herring spawning grounds in British Columbia. *Canadian Journal of Fisheries and Aquatic Sciences*, **47**, 505–515.

Purcell, J.E. (1991) A review of cnidarians and ctenophores feeding on competitors in the plankton. *Hydrobiologia*, **216/217**, 335–342.

Purcell, J.E. (1992) Effects of predation by the scyphomedusan *Chrysaora quinquecirrha* on zooplankton populations in Chesapeake Bay, USA. *Marine Ecology Progress Series*, **87**, 65–76.

Purcell, J.E. and Cowan, J.H. (1995) Predation by the scyphomedusan *Chrysaora quinquecirrha* on *Mnemiopsis leidyi* ctenophores. *Marine Ecology Progress Series*, **129**, 63–70.

Purcell, J.E., White, J.R. and Roman, M.R. (1994) Predation by gelatinous zooplankton and resource limitation as potential controls of *Acartia tonsa* copepod populations in Chesapeake Bay. *Limnology and Oceanography*, **39**, 263–278.

Purcell, J.E., Cresswell, F.P., Cargo, D.G. *et al.* (1991) Differential ingestion and digestion of bivalve larvae by the scyphozoan *Chrysaora quinquecirrha* and the ctenophore *Mnemiopsis leidyi*. *Biological Bulletin (Woods Hole)*, **180**, 103–111.

Purcell, J.E., Nemazie, D.A., Dorsey, S.E. *et al.* (1994) Predation mortality of bay anchovy *Anchoa mitchilli* eggs and larvae due to scyphomedusae and ctenophores in Chesapeake Bay. *Marine Ecology Progress Series*, **114**, 47–58.

Quensen, J.M., Black, R.E. and Webb, K.L. (1981) Composition of gelatin from developmental stages of the sea nettle, *Chrysaora quinquecirrha*. *Comparative Biochemistry and Physiology B Comparative Biochemistry*, **70**, 649–651.

Rahat, M. and Adar, O. (1980) Effect of symbiotic zooxanthellae and temperature on budding and strobilation in *Cassiopeia andromeda* (Eschscholz). *Biological Bulletin (Woods Hole)*, **159**, 394–401.

Rahat, M. and Hofmann, D.K. (1987) Bacterial and algal effects on metamorphosis in the life cycle of *Cassiopea andromeda*. *Annals of the New York Academy of Sciences*, **503**, 449–458.

Ralph, P.M. (1959) Notes on the species of the pteromedusan genus *Tetraplatia* Busch, 1851. *Journal of the Marine Biological Association of the United Kingdom*, **38**, 369–379.

Ralph, P.M. (1960) *Tetraplatia*, a coronate scyphomedusan. *Proceedings of the Royal Society of London B Biological Sciences*, **152**, 263–281.

Rasmussen, E. (1973) Systematics and ecology of the Isefjord marine fauna (Denmark) with a survey of the eelgrass (*Zostera*) vegetation and its communities. *Ophelia*, **11**, 1–507.

Raymont, J.E.G., Krishnaswamy, S. and Tundisi, J. (1967) Biochemical studies on marine zooplankton IV Investigations on succinic dehydrogenase activity in zooplankton with special reference to *Neomysis integer*.

Journal du Conseil Conseil International pour l'Exloration de la Mer, **31**, 164–169.

Read, L.K. (1986) Effect of varying solar radiation intensities and ultraviolet radiation on growth rates of *Symbiodinium microadriaticum* isolated from different hosts. *Hawaii Institute of Marine Biology Technical Report*, **37**, 430–439.

Reinhardt, S.B. and Van Vleet, E.S. (1986) Lipid composition of twenty-two species of Antarctic midwater zooplankton and fish. *Marine Biology (Berlin)*, **91**, 149–159.

Repelin, R. (1965) La meduse *Paraphyllina ransoni* dans la Vallée du Trou sans Fond (Côte d'Ivoire). *Cahiers ORSTOM (office de la recherche scientifique et technique outre-mer) Serie Oceanographie*, **3**, 81–86. In French; English abstract.

Repelin, R. (1966) Scyphomeduses Atollidae du Bassin de Guinée. *Cahiers ORSTOM (office de la recherche scientifique et technique outre-mer) Serie Oceanographie*, **4**, 21–33. In French; English abstract.

Repelin, R. (1967) *Stygiomedusa stauchi* n. sp. scyphoméduse gèante des profondeurs. *Cahiers ORSTOM (office de la recherche scientifique et technique outre-mer) Serie Oceanographie*, **5**, 23–28. In French.

Rice, A.L. (1964) Observations on the effects of changes of hydrostatic pressure on the behaviour of some marine animals. *Journal of the Marine Biological Association of the United Kingdom*, **44**, 163–175.

Rigby, B.J. and Hafey, M. (1972) Thermal properties of the collagen of jellyfish (*Aurelia coerulea*) and their relation to its thermal behaviour. *Australian Journal of Biological Sciences*, **25**, 1361–1363.

Robertson, J.D. (1949) Ionic regulation in some marine invertebrates. *Journal of Experimental Biology*, **26**, 182–200.

Robson, E.A. (1985) Speculations on Coelenterates. *Systematics Association Special Volume*, **28**, 60–77.

Roden, C.L., Lohoefener, R.R., Rogers, C.M. *et al.* (1990) Aspects of the ecology of the moon jellyfish, *Aurelia aurita*, in the Northern Gulf of Mexico. *Northeast Gulf Science*, **11**, 63–67.

Roe, H.S.J., James, P.T. and Thurston, M.H. (1984) The diel migrations and distributions within a mesopelagic community in the North East Atlantic 6 Medusae, ctenophores, amphipods and euphausiids. *Progress in Oceanography*, **13**, 425–460.

Romanes, G.J. (1876) Preliminary observations on the locomotor system of medusae. *Philosophical Transactions of the Royal Society of London*, **166**, 269–313.

Romanes, G.J. (1877) Further observations on the locomotor system of medusae. *Philosophical Transactions of the Royal Society of London*, **167**, 659–752.

Romeo, M., Gnassia-Barelli, M. and Carré, C. (1987) Trace metals: Cd, Cu, Pb and Zn in gelatinous macroplankton from the Northwestern Mediterranean. *Water Research*, **21**, 1287–1292.

Rottini-Sandrini, L. (1982) Effect of water temperature on the motility of *Pelagia noctiluca* (Forskal). *Experientia (Basel)*, **38**, 453–454.

Rottini-Sandrini, L. and Avian, M. (1983) Biological cycle of *Pelagia noctiluca*: morphological aspects of the development from planula to ephyra. *Marine Biology (Berlin)*, **74**, 169–174.

Rottini-Sandrini, L. and Avian, M. (1989) Feeding mechanism of *Pelagia noctiluca* (Scyphozoa: Semaeostomeae); laboratory and open sea observations. *Marine Biology (Berlin)*, **102**, 49–55.

Rottini-Sandrini, L. and Avian, M. (1991) Reproduction of *Pelagia noctiluca* in the central and northern Adriatic. *Hydrobiologia*, **216/217**, 197–202.

Rottini-Sandrini, L., Avian, M. and Zanelli, R. (1985) Influence de la température sur la biologie de *Pelagia noctiluca* (Forskål). *Rapports et Proces-Verbaux des Reunions Commission Internationale pour l'Exploration Scientifique de la Mer Mediterranée Monaco*, **29**, 199–201. In French; English abstract.

Rottini-Sandrini, L., Bratina, F. and Avian, M. (1983) Remarques histologiques sur les gonades mûres de *Pelagia noctiluca*. *Rapports et Proces-Verbaux des Reunions Commission Internationale pour l'Exploration Scientifique de la Mer Mediterranée Monaco*, **28**, 125–126. In French; English abstract.

Rottini-Sandrini, L., Bratina, F. and Avian, M. (1986) Aspetti ultrastrutturali della gametogenesi in *Discomedusa lobata* (Claus). *Nova Thalassia*, **8 Suppl. 2**, 59–66. In Italian; English abstract.

Rottini-Sandrini, L., Stravisi, F., Avian, M. *et al.* (1991) MED POL II Programme 'Jellyfish in the Mediterranean Sea' CIMAM final report, in *UNEP: Jellyfish Blooms in the Mediterranean. Proceedings of the II Workshop on Jellyfish in the Mediterranean Sea*, UNEP, Athens, pp. 147–162.

Rowan, R., Powers, D.A. (1991a) A molecular genetic classification of zooxanthellae and the evolution of animal–algal symbiosis. *Science (Washington DC)*, **251**, 1348–1351.

Rowan, R. and Powers, D.A. (1991b) Molecular genetic identification of symbiotic dinoflagellates (zooxanthellae). *Marine Ecology Progress Series*, **71**, 65–73.

Russell, F.S. (1956) On a new scyphomedusa *Paraphyllina ransoni* n.sp. *Journal of the Marine Biological Association of the United Kingdom*, **35**, 105–111, pls 1–2.

Russell, F.S. (1959a) Some observations on the scyphomedusa *Atolla*. *Journal of the Marine Biological Association of the United Kingdom*, **38**, 33–40.

Russell, F.S. (1959b) A viviparous deep-sea jellyfish. *Nature (London)*, **184**, 1527–1529.

Russell, F.S. (1970) *The Medusae of the British Isles II Pelagic Scyphozoa with a supplement to the first volume on hydromedusae*, Cambridge University Press, London.

Russell, F.S. and Rees, W.J. (1960) The viviparous scyphomedusa *Stygiomedusa fabulosa* Russell. *Journal of the Marine Biological Association of the United Kingdom*, **39**, 303–322.

Ryndina, D.D. and Polikarpov, G.G. (1988) Chemical composition of mass Black Sea medusa *Aurelia aurita* (L.) during intensive growth. *Ekologiya Morya (Kiev)*, **29**, 26–31. In Russian; English abstract.

Sadak, A., Hekim, N. and Güzel, O. (1980) Determination of testosterone levels in *Aurelia aurita* Per. Les. *Doga Bilim Dergisi Seri A Temel Bilimler*, **1**, 28–30. In Turkish; English abstract.

Salleo, A. (1984) Discharge mechanism of the nematocysts of *Pelagia noctiluca*, in *Toxins, Drugs, and Pollutants in Marine Animals*, (eds L. Bolis, J. Zadunaisky, and R. Gilles), Springer-Verlag, Berlin, pp. 63–76.

Salleo, A., La Spada, G. and Alfa, M. (1983) Blockage of trypsin-induced discharge of nematocysts of *Pelagia noctiluca* by Ca^{2+}. *Molecular Physiology*, **3**, 89–97.

Salleo, A., La Spada, G. and Barbera, R. (1994) Gadolinium is a powerful blocker of the activation of nematocysts of *Pelagia noctiluca*. *Journal of Experimental Biology*, **187**, 201–206.

Salleo, A., La Spada, G. and Denaro, M.G. (1991) Events associated with the discharge in nematocysts of *Pelagia noctiluca*, in *UNEP: Jellyfish Blooms in the Mediterranean Sea. Proceedings of the II Workshop on Jellyfish in the Mediterranean Sea*, MAP Technical Reports Series No. 47, UNEP, Athens, pp. 293–297.

Salleo, A., La Spada, G., Denaro, M.G. *et al.* (1986) Effects produced by SCN^- and thioglycolate on isolated nematocysts of *Pelagia noctiluca*. *Cellular and Molecular Biology*, **32**, 661–666.

Salleo, A., La Spada, G., Denaro, M.G. *et al.* (1988) Dynamics of release of free calcium during the discharge of holotrichous isorhiza of nematocysts of *Pelagia noctiluca*, in *The Biology of Nematocysts*, (eds D.A. Hessinger and H.M. Lenhoff), Academic Press, San Diego, pp. 551–565.

Salleo, A., La Spada, G., Falzea, G. *et al.* (1984a) Discharging effect of anions and inhibitory effect of divalent cations on isolated nematocysts of *Pelagia noctiluca*. *Molecular Physiology*, **5**, 25–34.

Salleo, A., La Spada, G., Falzea, G. *et al.* (1984b) Discharging effectiveness of lyotropic anions on nematocysts of *Pelagia noctiluca*. *Molecular Physiology*, **6**, 19–26.

Salvini-Plawen, L. (1966) Zur Kenntnis der Cnidaria des nordadriatischen Mesopsammon. *Veroeffentlichungen des Instituts fuer Meeresforschung in Bremerhaven*, **Suppl. 2**, 165–186. In German.

Salvini-Plawen, L. (1978) On the origin and evolution of the lower metazoa. *Zeitschrift fuer Zoologische Systematik und Evolutionsforschung*, **16**, 40–88.

Salvini-Plawen, L. (1987) Mesopsammic cnidaria from Plymouth (with systematic notes). *Journal of the Marine Biological Association of the United Kingdom*, **67**, 623–637.

Sato, M., Takeuchi, M., Kanno, N. *et al.* (1993) Distribution of opine dehydrogenases and lactate dehydrogenase activities in marine animals. *Comparative Biochemistry and Physiology*, **106B**, 955–960.

Satterlie, R.A. and Spencer, A.N. (1980) Satterlie and Spencer Reply. *Nature (London)*, **284**, 377.

Scemes, E. (1989) Rethinking the role of cholinergic neurotransmission in the Cnidaria, in *Evolution of the First Nervous Systems*, (ed. P.A.V. Anderson), Plenum Press, New York, pp. 157–166.

Schäfer, E.A. (1878) Observations on the nervous system of *Aurelia aurita*. *Philosophical Transactions of the Royal Society of London*, **169**, 563–575.

Schewiakoff, W. (1889) Beiträge zur Kenntnis des Acalephenauges. *Morphologisches Jahrbuch*, **15**, 21–60, pls 1–3. In German.

Schlichter, D. (1980) Adaptations of cnidarians for integumentary absorption of dissolved organic material. *Revue Canadienne de Biologie*, **39**, 259–282.

Schmahl, G. (1980) Morphogenesis in isolated apical and basal parts of strobilating scyphistomae in *Aurelia aurita* and *Chrysaora* sp. (Scyphozoa), in *Developmental and Cellular Biology of Coelenterates*, (eds P. Tardent and R. Tardent), Elsevier/North-Holland Biomedical Press, Amsterdam, pp. 251–256.

Schmahl, G. (1985a) Bacterially induced stolon settlement in the scyphopolyp of *Aurelia aurita* (Cnidaria, Scyphozoa). *Helgolaender Meeresuntersuchungen*, **39**, 33–42.

Schmahl, G. (1985b) Induction of stolon settlement in the scyphopolyps of *Aurelia aurita* (Cnidaria, Scyphozoa, Semaeostomeae) by glycolipids of marine bacteria. *Helgolaender Meeresuntersuchungen*, **39**, 117–128.

Schmahl, G. (1986) Investigations concerning the interactions between pedal stolons and apical as basal structures in the scyphopolyp of *Aurelia aurita* Lamarck (Scyphozoa, Semaeostomeae). *Zoologische Jahrbuecher Abteilung fuer Anatomie*, **114**, 353–369. In German; English abstract.

Schmid, V., Stidwill, R., Bally, A. *et al.* (1981) Heat dissociation and maceration of marine cnidaria. *Wilhelm Roux's Archives of Developmental Biology*, **190**, 143–149.

Schneider, G. (1988a) Chemical composition and biomass parameters of the common jellyfish *Aurelia aurita*. *Helgolaender Meeresuntersuchungen*, **42**, 319–327. In German; English abstract.

Schneider, G. (1988b) Larvae production of the common jellyfish *Aurelia aurita* in the Western Baltic 1982–1984. *Kieler Meeresforschungen*, **6**, 295–300.

Schneider, G. (1989a) The common jellyfish *Aurelia aurita*: standing stock, excretion, and nutrient regeneration in the Kiel Bight, Western Baltic. *Marine Biology (Berlin)*, **100**, 507–514.

Schneider, G. (1989b) Estimation of food demands of *Aurelia aurita* medusae populations in the Kiel Bight/Western Baltic. *Ophelia*, **31**, 17–27.

Schneider, G. (1990) A comparison of carbon based ammonia excretion rates between gelatinous and non-gelatinous zooplankton: implications and consequences. *Marine Biology (Berlin)*, **106**, 219–225.

Schneider, G. (1992) A comparison of carbon-specific respiration rates in gelatinous and non-gelatinous zooplankton: a search for general rules in zooplankton metabolism. *Helgolaender Meeresuntersuchungen*, **46**, 377–388.

Schneider, G. and Behrends, G. (1994) Population dynamics and the trophic role of *Aurelia aurita* medusae in the Kiel Bight and western Baltic. *ICES*

(International Council for the Exploration of the Sea) Journal of Marine Science, **51**, 359–367.

Schneider, G. and Weisse, T. (1985) Metabolism measurements of *Aurelia aurita* planulae larvae, and calculation of maximal survival period of the free swimming stage. *Helgolaender Meeresuntersuchungen*, **39**, 43–47.

Schoenberg, D.A. and Trench, R.K. (1976) Specificity of symbioses between marine cnidarians and zooxanthellae, in *Coelenterate Ecology and Behavior*, (ed. G.O. Mackie), Plenum Press, New York, pp. 423–432.

Schoenberg, D.A. and Trench, R.K. (1980a) Genetic variation in *Symbiodinium (=Gymnodinium) microadriaticum* Freudenthal, and specificity in its symbiosis with marine invertebrates III Specificity and infectivity of *Symbiodinium microadriaticum. Proceedings of the Royal Society of London B Biological Sciences*, **207**, 445–460.

Schoenberg, D.A. and Trench, R.K. (1980b) Genetic variation in *Symbiodinium (=Gymnodinium) microadriaticum* Freudenthal, and specificity in its symbiosis with marine invertebrates II Morphological variation in *Symbiodinium microadriaticum. Proceedings of the Royal Society of London B Biological Sciences*, **207**, 429–444.

Schoenberg, D.A. and Trench, R.K. (1980c) Genetic variation in *Symbiodinium (=Gymnodinium) microadriaticum* Freudenthal, and specificity in its symbiosis with marine invertebrates I Isoenzyme and soluble protein patterns of axenic cultures of *Symbiodinium microadriaticum. Proceedings of the Royal Society of London B Biological Sciences*, **207**, 405–427.

Schriever, G. (1975) In-situ observations of *Hyperia galba* Montagu (Amphipoda, Hyperiidea) in the western Baltic. *Kieler Meeresforschungen*, **31**, 107–10. In German; English abstract.

Schuchert, P. (1993) Phylogenetic analysis of the Cnidaria. *Zeitschrift fuer Zoologische Systematik und Evolutionsforschung*, **31**, 161–173.

Schulz, S. (1989) Changes in the Baltic pelagic ecosystem, in *Proceedings of the Twentyfirst European Marine Biology Symposium*, (eds R.Z. Klekowski, E. Styczyñska-Jurewicz, and L. Falkowski), Institute of Oceanology, Polish Academy of Sciences, Warsaw, pp. 463–471.

Schwab, W.E. (1977a) The ontogeny of swimming behavior in the scyphozoan, *Aurelia aurita*. I Electrophysiological analysis. *Biological Bulletin (Woods Hole)*, **152**, 233–250.

Schwab, W.E. (1977b) The ontogeny of swimming behavior in the scyphozoan, *Aurelia aurita* II The effects of ions and drugs. *Biological Bulletin (Woods Hole)*, **152**, 251–262.

Scrutton, C.T. (1979) Early fossil cnidarians, in *The Origin of Major Invertebrate Groups*, (ed. M.R. House), Academic Press, London, pp. 161–207.

Sebens, K.P. (1987) Coelenterata, in *Animal Energetics*, (eds F.J. Vernberg and T.J. Pandian), Academic Press, San Diego, pp. 55–120.

Segerstråle, S.G. (1951) The recent increase in salinity off the coasts of Finland and its influence upon the fauna. *Journal du Conseil International pour l'Exploration de la Mer*, **17**, 103–110.

Segerstråle, S.G. (1953) Further notes on the increase in salinity of the inner Baltic and its influence on the fauna. *Suomen Tiedeseura Commentationes Biologicae*, **13**, 1–7.

Seravin, L.N. (1987a) Orientation of invertebrates in three-dimensional space 3 Tumbling of hydroid (Hydrozoa) and scyphoid (Scyphozoa) medusae. *Zoologicheskii Zhurnal*, **66**, 1685–1692. In Russian; English abstract.

Seravin, L.N. (1987b) Upheaval movement of marine jellyfishes as the adaptation to planktonic life. *Trudy Zoologicheskogo Instituta Akademii Nauk SSSR*, **160**, 127–141. In Russian.

Seravin, L.N. (1991) Role of oral arms in feeding and digestion in the Lion's Mane jellyfish *Cyanea capillata* (L.). *Doklady Akademii Nauk SSSR*, **321**, 1301–1303. In Russian: journal issue translated to English 1992.

Seravin, L.N. (1995) The peculiarities of nutrition and feeding behaviour of scyphoid medusae *Aurelia aurita* (L.) and *Cyanea capillata* (L.), in *Cnidaria. Modern and Perspective Investigations II*, (ed. S.D. Stepanjants), Zoological Institute Russian Academy of Sciences, St. Petersburg, pp. 136–154. In Russian; English abstract.

Seravin, L.N., Railkin, A.I., Letunov, V.N. *et al.* (1979) Chemical inductors of feeding behaviour in coelenterates. *Vestnik Leningradskogo Universiteta Biologiya*, **1979**, 17–22. In Russian: English abstract.

Severin, S.E., Boldyrev, A.A. and Lebedev, A.V. (1972) Nitrogenous extractive compounds of muscle tissue of invertebrates. *Comparative Biochemistry and Physiology*, **43B**, 369–381.

Shanks, A.L. and Graham, W.M. (1987) Orientated swimming in the jellyfish *Stomolophus meleagris* L. Agassiz (Scyphozoan: Rhizostomida). *Journal of Experimental Marine Biology and Ecology*, **108**, 159–169.

Shanks, A.L. and Graham, W.M. (1988) Chemical defense in a scyphomedusa. *Marine Ecology Progress Series*, **45**, 81–86.

Shenker, J.M. (1984) Scyphomedusae in surface waters near the Oregon Coast, May–August, 1981. *Estuarine Coastal and Shelf Science*, **19**, 619–632.

Shenker, J.M. (1985) Carbon content of the neritic scyphomedusa *Chrysaora fuscescens*. *Journal of Plankton Research*, **7**, 169–173.

Shick, J.M. (1973) Effects of salinity and starvation on the uptake and utilization of dissolved glycine by *Aurelia aurita* polyps. *Biological Bulletin (Woods Hole)*, **144**, 172–179.

Shick, J.M. (1975) Uptake and utilization of dissolved glycine by *Aurelia aurita* scyphistomae: temperature effects on the uptake process; nutritional role of dissolved amino acids. *Biological Bulletin (Woods Hole)*, **148**, 117–140.

Shick, J.M. (1976) Free amino acids in *Aurelia aurita* scyphistomae from Corpus Christi, Texas. *Comparative Biochemistry and Physiology*, **53B**, 1–2.

Shick, J.M. (1991) *A Functional Biology of Sea Anemones*, Chapman & Hall, London.

Shushkina, E.A. and Musayeva, E.I. (1982) Heat content of marine zooplankton. *Oceanology*, **22**, 73–77. In Russian; English abstract and journal translation.

Shushkina, E.A. and Musayeva, E.I. (1983) The role of jellyfish in the energy system of Black Sea plankton communities. *Oceanology*, **23**, 92–96. In Russian; English abstract and journal translation.

Shushkina, E.A. and Musayeva, E.I. (1990) Structure of planktonic community of the Black Sea epipelagic zone and its variation caused by invasion of a new ctenophore species. *Oceanology*, **30**, 225–228.

Shushkina, E.A. and Vinogradov, M.Y. (1991) Long-term changes in the biomass of plankton in open areas of the Black Sea. *Oceanology*, **31**, 716–721.

Siddiqui, P.J.A., Akbar, Z. and Qasim, R. (1988) Distribution of macro and micro-elements in marine invertebrates from Karachi. *Pakistan Journal of Scientific and Industrial Research*, **31**, 326–329.

Silverstone, M., Galton, V.A. and Ingbar, S.H. (1978) Observations concerning the metabolism of iodine by polyps of *Aurelia aurita*. *General and Comparative Endocrinology*, **34**, 132–140.

Silverstone, M., Tosteson, T.R. and Cutress, C.E. (1977) The effect of iodide and various iodocompounds on initiation of strobilation in *Aurelia*. *General and Comparative Endocrinology*, **32**, 108–113.

Singla, C.L. (1976) Ultrastructure and attachment of the basal disc of *Haliclystus*, in *Coelenterate Ecology and Behavior*, (ed. G.O. Mackie), Plenum Press, New York, pp. 533–540.

Sipos, J.C. and Ackman, R.G. (1968) Jellyfish (*Cyanea capillata*) lipids: fatty acid composition. *Journal of the Fisheries Research Board of Canada*, **25**, 1561–1569.

Sloan, N.A. (1986) World jellyfish and tunicate fisheries, and the Northeast Pacific echinoderm fishery. *Canadian Special Publication of Fisheries and Aquatic Sciences*, **92**, 23–33.

Sloan, N.A. and Gunn, C.R. (1985) Fishing, processing, and marketing of the jellyfish, *Aurelia aurita* (L.), from southern British Columbia. *Canadian Industry Report of Fisheries and Aquatic Sciences*, **157**, 1–29.

Smith, F. (1891) Gastrulation of *Aurelia flavidula*, Pér. & Les. *Bulletin of the Museum of Comparative Zoology at Harvard University*, **22**, 115–125, 2 pls.

Smith, H.G. (1936) Contribution to the anatomy and physiology of *Cassiopea frondosa*. *Carnegie Institution of Washington Publication*, **475**, 19–52.

Smith, K.L. (1982) Zooplankton of a bathyal benthic boundary layer: In situ rates of oxygen consumption and ammonium excretion. *Limnology and Oceanography*, **27**, 461–471.

Southcott, R.V. and Glover, C.J.M. (1987) The occurence of *Desmonema gaudichaudi* (Lesson) (Scyphozoa, Semaeostomeae) in South Australian waters with records of fish–jellyfish symbioses. *Transactions Royal Society of South Australia*, **111**, 131–132.

Southward, A.J. (1955) Observations on the ciliary currents of the jellyfish *Aurelia aurita*. *Journal of the Marine Biological Association of the United Kingdom*, **34**, 210–216.

Spangenberg, D.B. (1964) New observations on *Aurelia*. *Transactions of the American Microscopical Society*, **83**, 448–455.

Spangenberg, D.B. (1965a) Cultivation of the life stages of *Aurelia aurita* under controlled conditions. *Journal of Experimental Zoology*, **159**, 303–318.

Spangenberg, D.B. (1965b) A study of strobilation in *Aurelia aurita* under controlled conditions. *Journal of Experimental Zoology*, **160**, 1–10.

Spangenberg, D.B. (1967) Iodine induction of metamorphosis in *Aurelia*. *Journal of Experimental Zoology*, **165**, 441–450.

Spangenberg, D.B. (1968a) Recent studies of strobilation in jellyfish. *Oceanography and Marine Biology; an Annual Review*, **6**, 231–247.

Spangenberg, D.B. (1968b) Statolith differentiation in *Aurelia aurita*. *Journal of Experimental Zoology*, **169**, 487–500.

Spangenberg, D.B. (1971) Thyroxine induced metamorphosis in *Aurelia*. *Journal of Experimental Zoology*, **178**, 183–194.

Spangenberg, D.B. (1976) Intracellular statolith synthesis in *Aurelia aurita*. *Belle W. Baruch Library in Marine Science*, **5**, 231–247.

Spangenberg, D.B. (1977) *Aurelia* metamorphosis: Model system for study of thyroxine action, in *Thyroid Hormones and Brain Development*, (ed. G.D. Grave), Raven Press, New York, pp. 151–167.

Spangenberg, D.B. (1979) Statolith synthesis and ephyra development in *Aurelia* metamorphosing in strontium and low calcium containing sea water. *Scanning Electron Microscopy*, **2**, 433–438.

Spangenberg, D.B. (1981) The effect of monobasic sodium phosphate on statolith synthsis in *Aurelia*. *Scanning Electron Microscopy*, **3**, 355–362.

Spangenberg, D.B. (1984) Use of the *Aurelia* metamorphosis test system to detect subtle effects of selected hydrocarbons and petroleum oil. *Marine Environmental Research*, **14**, 281–303.

Spangenberg, D.B. (1986) Statolith formation in Cnidaria: Effects of cadmium on *Aurelia* statoliths. *Scanning Electron Microscopy*, **4**, 1609–1618.

Spangenberg, D.B. (1987) Effects of energy-related contamination on ephyra development in *Aurelia*, in *Physiological Responses of Marine Organisms to Environmental Stressors*, (eds J.V. Dorigan and F.L. Harrison), US Department of Energy, Washington, pp. 423–427.

Spangenberg, D.B. (1991) Rhopalium development in *Aurelia aurita* ephyrae. *Hydrobiologia*, **216/217**, 45–49.

Spangenberg, D.B. and Beck, C.W. (1968) Calcium sulfate dihydrate statoliths in *Aurelia*. *Transactions of the American Microscopical Society*, **87**, 329–335.

Spangenberg, D.B. and Kuenning, W. (1976) SEM studies of strobilating *Aurelia*, in *Coelenterate Ecology and Behavior*, (ed. G.O. Mackie), Plenum Press, New York, pp. 377–386.

Spangenberg, D.B., Ives, K. and Patten, M. (1980) Strobilation aberrations in *Aurelia* induced by petroleum-related aniline and phenol, in *Developmental and Cellular Biology of Coelenterates*, (eds P. Tardent and R. Tardent), Elsevier/North-Holland Biomedical Press, Amsterdam, pp. 263–269.

Spanier, E. and Galil, B.S. (1991) Lessepsian migration: a continuous biogeographical process. *Endeavour (Oxford)*, **15**, 102–106.

Spencer, A.N. (1982) The physiology of a coelenterate neuromuscular synapse. *Journal of Comparative Physiology A Sensory Neural and Behavioral Physiology*, **148**, 353–363.

Spencer, A.N. (1989) Chemical and electrical synaptic transmission in the Cnidaria, in *Evolution of the First Nervous Systems*, (ed. P.A.V. Anderson), Plenum Press, New York, pp. 33–53.

Spencer Davies, P. (1992) Endosymbiosis in marine cnidarians, in *Plant–Animal Interactions in the Marine Benthos*, (eds D.M. John, S.J. Hawkins, and J.H. Price), Clarendon Press, Oxford, pp. 511–540.

Stepanjants, S.D. (1988) Cnidaria origin and the possible way of the Hydrozoa evolution, in *Porifera and Cnidaria: Modern and Perspective Investigations*, (eds V.M. Koltun and S.D. Stepanjants), USSR Academy of Sciences, Moscow, pp. 130–144. In Russian; English abstract.

Stepanjants, S.D. and Sheiko, O.V. (1989) The Identification Key of the Classes, Orders and Families, in *Marine Plankton. Taxonomy, Ecology, and Distribution*, (eds M.G. Petrushevska and S.D. Stepanjants), Zoological Institute USSR Academy of Sciences, Leningrad, pp. 100–131.

Stewart, B.C. and Lakshmanan, S. (1975) Some properties of the acid phosphatase and the alkaline phosphatase of the summer jellyfish of the Chesapeake Bay, *Chrysaora quinquecirrha* Desor. *Comparative Biochemistry and Physiology*, **50A**, 319–326.

Stiasny, G. (1922) Papers from Dr. Th. Mortensen's Pacific Expedition 1914–16 XII Zur Kenntnis der Entwicklung von *Stomolophus meleagris* L. Agassiz. *Videnskabelige Meddelelser fra Dansk Naturhistorisk Forening*, **73**, 499–511. In German.

Stockard, C.R. (1908) Studies on tissue growth I On experimental study of the rate of regeneration in *Cassiopea xamachana. Carnegie Institution of Washington Publication*, **103**, 61–102. Papers from the Tortugas Laboratory of the Carnegie Institution of Washington 2.

Stoecker, D.K., Michaels, A.E. and Davis, L.H. (1987) Grazing by the jellyfish, *Aurelia aurita*, on microzooplankton. *Journal of Plankton Research*, **9**, 901–915.

Stone, J.H., Burnett, J.W. and Goldner, R. (1970) The amino acid content of sea nettle (*Chrysaora quinquecirrha*) nematocysts. *Comparative Biochemistry and Physiology*, **33**, 707–710.

Strand, S.W. and Hamner, W.M. (1988) Predatory behavior of *Phacellophora camtschatica* and size-selective predation upon *Aurelia aurita* (Scyphozoa: Cnidaria) in Saanich Inlet, British Columbia. *Marine Biology (Berlin)*, **99**, 409–414.

Stunkard, H.W. (1969) The morphology and life-history of *Neopechona pyriforme* (Linton, 1900) n.gen., n. comb. (Trematoda: Lepocreadiidae). *Biological Bulletin (Woods Hole)*, **136**, 96–113.

Stunkard, H.W. (1972) Observations on the morphology and life-history of the digenetic trematode, *Lepocreadium setiferoides* (Miller and Northrup, 1926) Martin, 1938. *Biological Bulletin (Woods Hole)*, **142**, 326–334.

Suchanek, T.H. (1993) Oil impacts on marine invertebrate populations and communities. *American Zoologist*, **33**, 510–523.

Suckow, R. (1971) Die Larvenentwicklung bei der Ohrenqualle *Aurelia aurita*. *Mikrokosmos*, **60**, 328–332. In German.

Sugiura, Y. (1963) On the life-history of rhizostome medusae I *Mastigias papua* L. Agassiz. *Annotationes Zoologicae Japonenses*, **36**, 194–202.

Sugiura, Y. (1964) On the life-history of rhizostome medusae II Indispensability of *Zooxanthellae* for strobilation in *Mastigias papua*. *Embryologia*, **8**, 223–233.

Sugiura, Y. (1965) On the life-history of rhizostome medusae III On the effects of temperature on the strobilation of *Mastigias papua*. *Biological Bulletin (Woods Hole)*, **128**, 493–496.

Sugiura, Y. (1966) On the life history of rhizostome medusae IV *Cephea cephea*. *Embryologia*, **9**, 105–122.

Sugiura, Y. (1969) On the life-history of rhizostome medusae V. On the relation between zooxanthellae and the strobilation of *Cephea cephea*. *Bulletin of the Marine Biological Station of Asamushi*, **13**, 227–233.

Sullivan, B.K., Garcia, J.R. and Klein-MacPhee, G. (1994) Prey selection by the scyphomedusan predator *Aurelia aurita*. *Marine Biology (Berlin)*, **121**, 335–341.

Sun, W.-G. (1986) Late Precambrian scyphozoan medusa *Mawsonites randellensis* sp. nov. and its significance in the Ediacara metazoan assemblage, South Australia. *Alcheringa*, **10**, 169–181.

Sutton, J.S. and Burnett, J.W. (1969) A light and electron microscopic study of nematocysts of *Chrysaora quinquecirrha*. *Journal of Ultrastructure Research*, **28**, 214–234.

Svane, I. and Dolmer, P. (1995) Perception of light at settlement: a comparative study of two invertebrate larvae, a scyphozoan planula and a simple ascidian tadpole. *Journal of Experimental Marine Biology and Ecology*, **187**, 51–61.

Sveshnikov, V.A. (1963) The feeding habits of jellyfish as possible rivals of the White Sea herring. *Trudy Belomorskoi Biologicheskoi Stantsii Moskovskogo Gosudarstvennogo Universiteta*, **2**, 246–249. In Russian; English abstract.

Sviderskaya, G.E., Polyakova, L.A. and Voskresensky, V.O. (1990) Endogenous rhythms of excitation in the marginal tentacle of the scyphomedusae. *Biologiya Morya (Vladivostok)*, **0**, 51–54. In Russian; English abstract.

Svoboda, A. (1978) In situ monitoring of oxygen production and respiration in Cnidaria with and without zooxanthellae, in *Physiology and Behaviour of Marine Organisms*, (eds D.S. McLusky and A.J. Berry), Pergamon Press, Oxford, pp. 75–82.

Tabachnik, K.R. (1986) Parasitism of *Alepas pacifica* (Cirripedia, Thoracica) on scyphozoan jellyfish. *Zoologicheskii Zhurnal*, **65**, 1731–1733. In Russian; English abstract.

Takasu, N. and Yoshida, M. (1984) Freeze-fracture and histofluorescence studies on photoreceptive membranes of medusan ocelli. *Zoological Science (Tokyo)*, **1**, 367–374.

Tardent, P. (1984) The significance of the metagenic life cycles of Cnidaria for genetic diversification and adaptability, in *Advances in Invertebrate Reproduction 3*, (ed. W. Engels *et al.*), Elsevier Science Publ., pp. 269–278.

Tardent, P. (1988) History and current state of knowledge concerning discharge of cnidae, in *The Biology of Nematocysts*, (eds D.A. Hessinger and H.M. Lenhoff), Academic Press, San Diego, pp. 309–332.

Tardent, P., Zierold, K., Klug, M. *et al.* (1990) X-ray microanalysis of elements present in the matrix of cnidarian nematocysts. *Tissue and Cell*, **22**, 629–643.

Taylor, D.L. (1974) Symbiotic marine algae: taxonomy and biological fitness, in *Symbiosis in the Sea*, (ed. W.B. Vernberg), University of South Carolina Press, Columbia, pp. 245–262.

Teissier, G. (1926) Sur la teneur en eau et en substances organiques de *Chrysaora hysocella* (L.) aux différents stades de son ontogénése. *Bulletin de la Societe Zoologique de France*, **51**, 266–273. In French.

Teissier, G. (1929) La croissance embryonnaire de *Chrysaora hysocella*. *Archives de Zoologie Experimentale et Generale*, **69**, 137–178. In French.

Teissier, G. (1932) Sur la composition chimique des planulas de *Chrysaora hysocella*. *Bulletin de la Societe Zoologique de France*, **57**, 160–162. In French.

Théodoridès, J. (1989) Parasitology of marine zooplankton. *Advances in Marine Biology*, **25**, 117–177.

Thiel, H. (1963a) Teil- und Spiralephyren von *Aurelia aurita* und ihre Regulation. *Zoologischer Anzeiger*, **171**, 303–327. In German.

Thiel, H. (1963b) Untersuchungen über die Entstehung abnormer Scyphistomae, Strobilae und Ephyrae von *Aurelia aurita* Lam. und ihre theoretische Bedeutung. *Zoologische Jahrbuecher Abteilung fuer Anatomie*, **81**, 311–358. In German.

Thiel, H. (1966) The evolution of scyphozoa. A review, in *The Cnidaria and their Evolution*, (ed. W.J. Rees), Academic Press, London, pp. 77–117.

Thiel, M.E. (1964) Untersuchungen über die Ernährungsweise und den Nahrungskreislauf bei *Rhizostoma octopus* L. Ag. *Mitteilungen aus dem Hamburgischen Zoologischen Museum und Institut*, **61**, 247–269. In German.

Thiel, M.E. (1966) Untersuchungen über die Herkunft, das Auftreten, das Wachstum und die Fortpflanzung von *Rhizostoma octopus* L. Ag. im Elbmündungsgebiet. *Abhandlungen und Verhandlungen des Naturwissenschaftlichen Vereins in Hamburg*, **10**, 59–88. In German.

Thiel, M.E. (1970a) Das Zusammenleben von Jung- und Kleinfischen mit Rhizostomeen (Scyphomedusae). *Berichte der Deutschen Wissenschaftlichen Kommission fuer Meeresforschung*, **21**, 444–473. In German; English abstract.

Thiel, M.E. (1970b) Über das stoffweise Schwimmen eines Schwarmes von *Caranx leptolepis* (Cuv. et Val.) (Pisces) synchron mit dem Schlagrhythmus der Meduse *Acromitus flagellatus* Stiasny (Scyphomedusae). *Abhandlungen und Verhandlungen des Naturwissenschaftlichen Vereins in Hamburg*, **14**, 71–79. In German; English abstract.

Thiel, M.E. (1971) Tintenfische 'reiten' auf Quallen (Cephalopoda, Coelenterata). *Abhandlungen und Verhandlungen des Naturwissenschaftlichen Vereins in Hamburg*, **15**, 37–41. In German; English abstract.

Thiel, M.E. (1976a) Die Cassiopeida (Arcadomyariae MAAS, 1903, Kampylomyariae STIASNY, 1921) als dritte Ordnung der Rhizostomae (Scyphomedusae) und die Erhebung der letzteren zu der Überordnung Rhizostomoidea (Wurzelmundartige) Supraord. nov. *Mitteilungen aus dem Hamburgischen Zoologischen Museum und Institut*, **73**, 1–16. In German; English abstract.

Thiel, M.E. (1976b) Wirbellose Meerestiere als Parasiten, Kommensalen oder Symbionten in oder an Scyphomedusen. *Helgolaender Wissenschaftliche Meeresuntersuchungen*, **28**, 417–446. In German; English abstract.

Thiel, M.E. (1978) Das Zusammenleben von Jung- und Kleinfischen mit Semaeostomen (Scyphomedusae). *Mitteilungen aus dem Hamburgischen Zoologischen Museum und Institut*, **75**, 19–47. In German.

Thill, H. (1937) Beiträge zur Kenntnis der *Aurelia aurita* (L.). *Zeitschrift fuer wissenschaftliche Zoologie*, **150**, 51–96. In German; English translation Fisheries Research Board of Canada Translation Series no. 2296.

Thompson, W.E., Gold, B., Spangenberg, D.B. *et al.* (1991) Monospecifc antibody against chicken sarcomeric actin does not recognize actin in jellyfish, *Aurelia aurita*. *Journal of Cell Biology*, **115**, 162a. Abstract.

Thuesen, E.V. and Childress, J.J. (1994) Oxygen consumption rates and metabolic enzyme activities of oceanic California medusae in relation to body size and habitat depth. *Biological Bulletin (Woods Hole)*, **187**, 84–98.

Thurston, M.H. (1977) Depth distributions of *Hyperia spinigera* Bovallius, 1889 (Crustacea: Amphipoda) and medusae in the North Atlantic Ocean, with notes on the associations between *Hyperia* and coelenterates, in *A Voyage of Discovery*, (ed. M. Angel), Pergamon Press, Oxford, pp. 499–536.

Titova, L.K., Vinnikov, Y.A. and Kharkejevich, T.A. (1979) The structural and functional organization of the gravity receptor – rhopalia in medusa *Cyanea arctica*. *Tsitologiya*, **21**, 995–1001. In Russian; English abstract.

Tolley, S.G. (1987) Association of young *Chloroscombrus chrysurus* (Pisces: Carangidae) with the jellyfish *Aurelia aurita*. *Copeia*, **1987**, 216–219.

Toonen, R.J. and Chia, F.-S. (1993) Limitations of laboratory assessments of coelenterate predation: container effects on the prey selection of the Limnomedusa, *Proboscidactyla flavicirrata* (Brandt). *Journal of Experimental Marine Biology and Ecology*, **167**, 215–235.

Torres, J.J. and Mangum, C.P. (1974) Effects of hyperoxia on survival of benthic marine invertebrates. *Comparative Biochemistry and Physiology*, **47A**, 17–22.

Trench, R.K. (1971a) The physiology and biochemistry of zooxanthellae symbiotic with marine coelenterates II Liberation of fixed ^{14}C by zooxanthellae *in vitro*. *Proceedings of the Royal Society of London B Biological Sciences*, **177**, 237–250.

Trench, R.K. (1971b) The physiology and biochemistry of zooxanthellae symbiotic with marine coelenterates III. The effect of homogenates of host tissues on the excretion of photosynthetic products *in vitro* by zooxanthellae from two marine coelenterates. *Proceedings of the Royal Society of London B Biological Sciences*, **177**, 251–264.

Trench, R.K. (1987) Dinoflagellates in non-parasitic symbioses, in *The Biology of Dinoflagellates*, (ed. F.J.R. Tayor), Blackwell Scientific Publications, Oxford, pp. 530–570.

Trench, R.K. and Blank, R.J. (1987) *Symbiodinium microadriaticum* Freudenthal, *S.goreauii* sp. nov., *S. kawagutii* sp. nov. and *S. pilosum* sp. nov.: gymnodinioid dinoflagellate symbionts of marine invertebrates. *Journal of Phycology*, **23**, 469–481.

Trench, R.K. and Thinh, L.-V. (1995) *Gymnodinium linucheae* sp. nov.: The dinoflagellate symbiont of the jellyfish *Linuche unguiculata*. *European Journal of Phycology*, **30**, 149–154.

Trench, R.K., Colley, N.J. and Fitt, W.K. (1981) Recognition phenomena in symbioses between marine invertebrates and 'zooxanthellae'; uptake, sequestration and persistence. *Berichte der Deutschen Botanischen Gesellschaft*, **94**, 529–545.

Tsukaguchi, R. (1914) Über die feinere Struktur des Ovarialeies von *Aurelia aurita* L. *Archiv fuer Mikroskopische Anatomie*, **85**, 114–123, pl. 9. In German.

Uchida, T. (1926) The anatomy and development of a rhizostome medusa, *Mastigias papua* L. Agassiz, with observations on the phylogeny of Rhizostomae. *Journal of the Faculty of Science Imperial University of Tokyo Section IV Zoology*, **1**, 45–95, pl. 6.

Uchida, T. (1929) Studies on the Stauromedusae and Cubomedusae, with special reference to their metamorphosis. *Japanese Journal of Zoology*, **2**, 103–193, pls 3–5.

Uchida, T. (1969) The interrelationships of scyphozoan class. *Bulletin of the Marine Biological Station of Asamushi*, **13**, 247–250.

Uchida, T. (1973) The systematic position of the Stauromedusae. *Publications of the Seto Marine Biological Laboratory*, **20**, 133–139. Proc. Second Int. Symp. Cnidaria.

Uchida, T. and Hanaoka, K.I. (1933) On the morphology of a stalked medusa, *Thaumatoscyphus distinctus*. *Journal of the Faculty of Science Hokkaido University Series VI Zoology*, **2**, 135–153.

Uchida, T. and Nagao, Z. (1963) The metamorphosis of the scyphomedusa, *Aurelia limbata* (Brandt). *Annotationes Zoologicae Japonenses*, **36**, 83–91.

Uchida, T. and Sugiura, Y. (1978) On the polyp of the scyphomedusa, *Sanderia malayensis* and its reproduction. *Journal of the Faculty of Science Hokkaido University Series VI Zoology*, **21**, 279–287.

Ulrich, W. (1972) Ein drittes Exemplar der grossen, viviparen Tiefseemeduse *Stygiomedusa* Russ. (Scyphozoa, Semaeostomeae) erbeutet (1962) nordwestlich Island von Dr. Hans Thiele. *Sitzungsberichte der Gesellschaft Naturforschender Freunde zu Berlin*, **12**, 48–60. In German.

Uriz, M.-J., Rosell, D. and Maldonado, M. (1992) Parasitism, commensalism or mutualism? The case of Scyphozoa (Coronatae) and horny sponges. *Marine Ecology Progress Series*, **81**, 247–255.

van der Linden, A. and Decleir, W. (1982) Catecholamine levels during strobilation of *Aurelia aurita* (Lam). *Archives Internationales de Physiologie et de Biochimie*, **90**, B220. Abstract.

van der Spoel, S. (1987) Medusae of the genera *Paraphyllina, Periphylla* and *Atolla* from the Amsterdam mid North Atlantic Plankton Expeditions (1980–1983). *Bijdragen tot de Dierkunde*, **57**, 42–52.

van der Spoel, S. and Bleeker, J. (1988) Medusae from the Banda Sea and Aru Sea plankton, collected during the Snellius II Expeditions, 1984–1985. *Indo-Malayan Zoology*, **5**, 161–202.

van der Spoel, S. and Schalk, P.H. (1988) Unique deviations in depth distribution of the deep-sea fauna. *Deep-sea Research*, **35**, 1185–1193.

van der Veer, H.W. (1985) Impact of coelenterate predation on larval plaice *Pleuronectes platessa* and flounder *Platichthys flesus* stock in the western Wadden Sea. *Marine Ecology Progress Series*, **25**, 229–238.

van der Veer, H.W. and Oorthuysen, W. (1985) Abundance, growth and food demand of the scyphomedusa *Aurelia aurita* in the western Wadden Sea. *Netherlands Journal of Sea Research*, **19**, 38–44.

Vanhöffen, E. (1900) Über Tiefseemedusen und ihre Sinnesorgane. *Zoologischer Anzeiger*, **23**, 277–279. In German.

Vanhöffen, E. (1902) Die acraspedoten Medusen der deutschen Tiefsee-Expedition 1898–1899. *Wissenschaftliche Ergebnisse der Deutschen Tiefsee-Expedition auf dem Dampfer "Valdivia" 1898–1899*, **3**, 3–86, pls 1–12. In German.

van Hyning, J.M. and Cooney, R.T. (1974) Association of walleye pollock, *Theragra chalcogramma*, with the jellyfish, *Cyanea*. *Copeia*, **1974**, 791.

van Iten, H. (1991a) Anatomy, patterns of occurrence, and nature of the conulariid schott. *Palaeontology (London)*, **34**, 939–954.

van Iten, H. (1991b) Evolutionary affinities of conulariids, in *The Early Evolution of Metazoa and the Significance of Problematic Taxa*, (eds A.M. Simonetta and S. Conway Morris), Cambridge University Press, Cambridge, pp. 145–155.

van Iten, H. (1992a) Microstructure and growth of the conulariid test, implications for conulariid affinities. *Palaeontology (London)*, **35**, 359–372, pl. 1.

van Iten, H. (1992b) Morphology and phylogenetic significance of the corners and midlines of the conulariid test. *Palaeontology (London)*, **35**, 335–358.

van Lieshout, J.S. and Martin, V.J. (1992) Development of planuloid buds of *Cassiopea xamachana* (Cnidaria: Scyphozoa). *Transactions of the American Microscopical Society*, **111**, 89–110.

van Nierop, M.M. and den Hartog, J.C. (1984) A study on the gut contents of five juvenile loggerhead turtles, *Caretta caretta* (Linnaeus) (Reptilia, Cheloniidae), from the south-eastern part of the North Atlantic Ocean, with emphasis on coelenterate identification. *Zoologische Mededelingen (Leiden)*, **59**, 35–54.

Vannucci, M. (1957) Double monster ephyrae of *Aurelia aurita*. *Nature (London)*, **179**, 326–327.

Vernon, H.M. (1895) The respiratory exchange of the lower marine invertebrates. *Journal of Physiology (London)*, **19**, 18–70.

Verwey, J. (1966) The role of some external factors in the vertical migration of marine animals. *Netherlands Journal of Sea Research*, **3**, 245–266.

Vinnikov, Y.A., Aronova, M.Z., Kharkeevich, T.A. *et al.* (1981) Structural and chemical features of the invertebrate otoliths. *Zeitschrift fuer Mikroskopisch-Anatomische Forschung (Leipzig)*, **95**, 127–140.

Vinogradov, A.P. (1953) The elementary chemical composition of marine organisms. *Memoir Sears Foundation for Marine Research*, **2**, 1–647.

Vinogradov, M.E. (1968) *Vertikal'noe raspredelenie okeanicheskogo zooplanktona (Vertical distribution of the oceanic zooplankton)*, Akademiya nauk SSSR, Institut Okeanologii, Moscow. (Transl. from Russian by Israel Program for Scientific Transl. Jerusalem, 1970, 339 pp.)

Vinogradov, M.E. and Semenova, T.N. (1985) A new species of the genus *Hyperia* (Crustacea, Amphipoda) from the coastal waters of Peru. *Zoologicheskii Zhurnal*, **64**, 139–143. In Russian; English abstract.

Vinogradov, M.E., Flint, M.V. and Shushkina, E.A. (1985) Vertical distribution of mesoplankton in the open area of the Black Sea. *Marine Biology (Berlin)*, **89**, 95–107.

Vinogradov, M.Y. and Grinberg, V.M. (1979) 64th cruise of the R/V *Vitaz* – A study of pelagic ecosystems of the Black Sea. *Oceanology*, **19**, 225–228.

Vinogradov, M.Y. and Shushkina, E.A. (1982) Estimate of the concentration of Black Sea jellyfish, ctenophores and *Calanus*, based on observations from the Argus submersible. *Oceanology*, **22**, 351–355.

Voino-Yasenetskii, A.V., Bursian, A.V., Polyakova, L.A. *et al.* (1979) Primary rhythms of excitation in scyphomedusan locomotor activity. *Zhurnal Evolyutsionnoi Biokhimii i Fiziologii*, **15**, 347–356. In Russian; English translation *J. Evol. Biochem. Physiol.*

von Lendenfeld, R. (1882) Über Coelenteraten der Südsee I Mittheilung. *Cyanea annaskala* nov. sp. *Zeitschrift fuer wissenschaftliche Zoologie*, **37**, 465–552. In German.

Vucetic, T. and Alegria Hernandez, V. (1988) Trends of annual catches or stock densities of some pelagic fishes in recent '*Pelagia* years' in the Adriatic. *FAO (Food and Agricultural Organization of the United Nations) Fisheries Reports*, **394**, 133–136.

Wade, M. (1993) Scyphozoa fossiles: Vendimedusae, Cubomedusae, Scyphomedusae, Conulata, in *Traité de Zoologie. Anatomie, systématique, biologie III Cnidaires, Cténaires 2*, (ed. D. Doumenc), Masson, Paris, pp. 923–942. In French.

Walker, M.J.A. (1988) Coelenterate and Echinoderm toxins: mechanisms and actions, in *Handbook of Natural Toxins Volume 3 Marine Toxins and Venoms*, (ed. A.T. Tu), Marcel Dekker, New York, pp. 279–325.

Walker, W.F. and Doolittle, W.F. (1983) 5S rRNA sequences from four marine invertebrates and implications for base pairing models of metazoan sequences. *Nucleic Acids Research*, **11**, 5159–5164.

Wang, W. and Xu, Z. (1990) Nematocysts of some species of siphonophores and scyphomedusae in Xiamen Harbour. *Journal of Xiamen University (Natural Science)*, **29**, 85–88. In Chinese; English abstract.

Watson, G.M. (1988) Ultrastructure and cytochemistry of developing nematocysts, in *The Biology of Nematocysts*, (eds D.A. Hessinger and H.M. Lenhoff), Academic Press, San Diego, pp. 143–164.

Watson, G.M. and Mire-Thibodeaux, P. (1994) The cell biology of nematocysts. *International Review of Cytology*, **156**, 275–300.

Watson, G.M. and Wood, R.L. (1988) Colloquium on terminology, in *The Biology of Nematocysts*, (eds D.A. Hessinger and H.M. Lenhoff), Academic Press, San Diego, pp. 21–23.

Webb, K.L. and Johannes, R.E. (1967) Studies of the release of dissolved free amino acids by marine zooplankton. *Limnology and Oceanography*, **12**, 376–382.

Webb, K.L., Schimpf, A.L. and Olmon, J. (1972) Free amino acid composition of scyphozoan polyps of *Aurelia aurita, Chrysaora quinquecirrha* and *Cyanea capillata* at various salinities. *Comparative Biochemistry and Physiology*, **43B**, 653–663.

Weber, J. (1991) A novel kind of polyanions as principal components of cnidarian nematocysts. *Comparative Biochemistry and Physiology A Comparative Physiology*, **98**, 285–291.

Weiler, K.S. and Black, R.E. (1991) Immediate decrease in apparent translation rate caused by salinity change in scyphistomae of *Aurelia aurita* (Linnaeus). *Comparative Biochemistry and Physiology A Comparative Physiology*, **99**, 199–202.

Weill, R. (1925) Foyers de formation et voies de migration des nèmatocystes de *Halyclistus octoradiatus* Clark. Existence, sur leur parcours, de réservoirs sélectifs. *Comptes Rendus des Seances de la Societe de Biologie et de ses Filiales*, **89**, 327–330. In French.

Weill, R. (1930) Essai d'une classification des nématocystes des cnidaires. *Bulletin Biologique de la France et de la Belgique*, **64**, 141–153, pl. 1. In French.

Weill, R. (1934) Contribution à l'étude des Cnidaires et de leurs nématocysts. *Travaux de la Station Zoologique de Wimereux.*, **10/11**, 1–701. In French.

Weill, R. (1935) Les 'réservoirs sélectifs' et la maturation des nématocystes de *Lucernaria campanulata* Lam. *Comptes Rendus des Seances de la Societe de Biologie et de ses Filiales*, **120**, 774–777. In French.

Weis, V.M., Smith, G.J. and Muscatine, L. (1989) A 'CO_2 supply' mechanism in zooxanthellate cnidarians: role of carbonic anhydrase. *Marine Biology (Berlin)*, **100**, 195–202.

Werner, B. (1966) *Stephanoscyphus* (Scyphozoa, Coronatae) und seine direkte Abstammung von den fossilen Conulata. *Helgolaender Wissenschaftliche Meeresuntersuchungen*, **13**, 317–347. In German; English abstract.

Werner, B. (1967) Morphologie, Systematik und Lebensgeschichte von *Stephanoscyphus* (Scyphozoa Coronatae) sowie seine Bedeutung für die Evolution der Scyphozoa. *Zoologischer Anzeiger*, **30**, 297–319. In German.

Werner, B. (1970) Contribution to the evolution in the genus *Stephanoscyphus* (Scyphozoa, Coronatae) and ecology and regeneration qualities of *Stephanoscyphus racemosus* Komai. *Publications of the Seto Marine Biological Laboratory*, **18**, 1–20.

Werner, B. (1971a) Ein neuer, bisher unbekannter Entwicklungsmodus bei einem Scyphopolypen. *Experientia (Basel)*, **27**, 351–353. In German; English abstract.

Werner, B. (1971b) *Stephanoscyphus planulophorus* n.spec., ein neuer Scyphopolyp mit einem neuen Entwicklungsmodus. *Helgolaender Wissenschaftliche Meeresuntersuchungen*, **22**, 120–140. In German; English abstract.

Werner, B. (1973) New investigations on systematics and evolution of the class Scyphozoa and the phylum Cnidaria. *Publications of the Seto Marine Biological Laboratory*, **20**, 35–61. Proc. Second Int. Symp. Cnidaria.

Werner, B. (1974) *Stephanoscyphus eumedusoides* n. spec. (Scyphozoa, Coronatae) a cave-living polyp with a new mode of development. *Helgolaender Wissenschaftliche Meeresuntersuchungen*, **26**, 434–463. In German; English abstract.

Werner, B. (1975) Bau und Lebensgeschichte des Polypen von *Tripedalia cystophora* (Cubozoa, Class nov., Carybdeidae) und seine Bedeutung für die Evolution der Cnidaria. *Helgolaender Wissenschaftliche Meeresuntersuchungen*, **27**, 461–504. In German; English abstract.

Werner, B. (1979) Coloniality in the Scyphozoa: Cnidaria, in *Biology and Systematics of Colonial Organisms*, (eds G. Larwood and B.R. Rosen), Academic Press, London, pp. 81–103.

Werner, B. (1983) Weitere Untersuchungen zur Morphologie, Verbreitung und Ökologie von *Stephanoscyphus planulophorus* (Scyphozoa, Coronata). *Helgolaender Meeresuntersuchungen*, **36**, 119–135. In German; English abstract.

Werner, B. and Hentschel, J. (1983) Apogamous life cycle of *Stephanoscyphus planulophorus*. *Marine Biology (Berlin)*, **74**, 301–304.

Westfall, J.A. (1966) Fine structure and evolution of nematocysts, in *Proceedings Sixth International Congress for Electron Microscopy, Kyoto*, Maruzen Company, Tokyo, pp. 235–236.

Westfall, J.A. (1973) Ultrastructural evidence for neuromuscular systems in coelenterates. *American Zoologist*, **13**, 237–246.

Westfall, J.A. (1987) Ultrastructure of invertebrate synapses, in *Nervous Systems in Invertebrates*, (ed. M.A. Ali), Plenum Press, New York, pp. 3–28.

Wetochin, I.A. (1930) The capture of food and feeding of the medusa *Aurelia aurita* in connection with the work of ciliated epithelium. *Fisiologicheskii Zhurnal SSSR imeni I. M. Sechenova*, **13**, 585–601. In Russian; German abstract.

White, M.G. and Bone, D.G. (1972) The interrelationship of *Hyperia galba* (Crustacea, Amphipoda) and *Desmonema gaudichaudi* (Scyphomedusae, Semaeostomae) from the Antarctic. *British Antarctic Survey Bulletin*, **27**, 39–49.

Widder, E.A., Latz, M.I. and Case, J.F. (1983) Marine bioluminescence spectra measured with an optical multichannel detection system. *Biological Bulletin (Woods Hole)*, **165**, 791–810.

Widder, E.A., Bernstein, S.A., Bracher, D.F. *et al.* (1989) Bioluminescence in the Monterey Submarine Canyon: image analysis of video recordings from a midwater submersible. *Marine Biology (Berlin)*, **100**, 541–551.

Widersten, B. (1965) Genital organs and fertilization in some scyphozoa. *Zoologiska Bidrag fran Uppsala*, **37**, 45–58, pls 1–3.

Widersten, B. (1966) On the development of septal muscles, supporting fibrils and periderm in the scyphistoma of semaeostome Scyphozoa. *Arkiv for Zoologi*, **18**, 567–574, pl. 1.

Widersten, B. (1968) On the morphology and development in some cnidarian larvae. *Zoologiska Bidrag fran Uppsala*, **37**, 139–182.

Widersten, B. (1969) Development of the periderm and podocysts in *Cyanea palmstruchi* Swartz 1809. *Zoologiska Bidrag fran Uppsala*, **38**, 51–60.

Widersten, B. (1973) Den palmstruchska manaten, *Cyanea palmstruchi* Swartz 1809. *Zoologisk Revy*, **35**, 147–152. In German; English abstract.

Widmark, E.M.P. (1911) Über die Gastrovascularströmungen bei *Aurelia aurita* L. und *Cyanea capillata* Eschz. *Zoologischer Anzeiger*, **38**, 378–382. In German.

Widmark, E.M.P. (1913) Über die Wasserströmungen in dem Gastrovaskularapparat von *Aurelia aurita* L. *Zeitschrift fuer allgemeine Physiologie*, **15**, 33–48. In German.

Wietrzykowski, W. (1912) Recherches sur le développement des Lucernaires. *Archives de Zoologie Experimentale et Generale*, **10**, 1–95 pls 1–3. In French.

Wilkerson, F.P. and Dugdale, R.C. (1984) Possible connections between sewage effluent, nutrient levels and jellyfish blooms, in *Workshop on Jellyfish Blooms in the Mediterranean*, UNEP, Athens, pp. 195–201.

Wilkerson, F.P. and Kremer, P. (1992) DIN, DON and PO_4 flux by a medusa with algal symbionts. *Marine Ecology Progress Series*, **90**, 237–250.

Wilkerson, F.P., Muller-Parker, G. and Muscatine, L. (1983) Temporal patterns of cell division in natural populations of endosymbiotic algae. *Limnology and Oceanography*, **28**, 1009–1014.

Willmer, P. (1990) *Invertebrate Relationships: Patterns in Animal evolution*, Cambridge University Press, Cambridge.

Witham, R. and Futch, C.R. (1977) Early growth and oceanic survival of pen-reared sea turtles. *Herpetologica*, **33**, 404–409.

Wolk, M., Rahat, M., Fitt, W.K. *et al.* (1985) Cholera toxin and thyrotropin can replace natural inducers required for the metamorphosis of larvae and buds of the scyphozoan *Cassiopea andromeda*. *Roux's Archives of Developmental Biology*, **194**, 487–490.

Wootton, M., Buckle, K.A. and Martin, D. (1982) Studies on the preservation of Australian jellyfish (*Catostylus* sp.). *Food Technology in Australia*, **34**, 398–400.

Wu, H.W. (1927) Preliminary observations on the sense organs and the adjacent structures of two scyphomedusae at young stage. *Contributions from the Biological Laboratory of the Science Society of China*, **3**, 1–5.

Yakovleva, K.K. (1964) Intensity of respiration of some Black Sea medusae. *Trudy Sevastopol'skoi Biologicheskoi Stantsii*, **17**, 364–367. In Russian; Fisheries Research Board of Canada Translation Series no. 2218.

Yamashita, T. (1957) Das Aktionspotential der Sinneskörper (Randkörper) der Meduse *Aurelia aurita*. *Zeitschrift fuer Biologie*, **109**, 116–122. In German; English abstract.

Yamasu, T. and Yoshida, M. (1973) Electron microscopy on the photoreceptors of an anthomedusa and a scyphomedusa. *Publications of the Seto Marine Biological Laboratory*, **20**, 757–778. Proc. Second Int. Symp. Cnidaria.

Yasuda, T. (1969) Ecological studies on the jelly-fish, *Aurelia aurita* in Urazoko Bay, Fukui Prefecture III Growth. *Aquiculture (Kagoshima)*, **17**, 145–154. In Japanese.

Yasuda, T. (1971) Ecological studies on the jelly-fish, *Aurelia aurita* in Urazoko Bay, Fukui Prefecture IV Monthly change in the bell-length composition and breeding season. *Bulletin of the Japanese Society of Scientific Fisheries*, **37**, 364–370. In Japanese; English abstract.

Yasuda, T. (1972) Ecological studies on the jelly-fish, *Aurelia aurita* in Urazoko Bay, Fukui Prefecture VII Diel change in the vertical distribution of the medusa observed by underwater-television and net sampling. *Bulletin of the Japanese Society of Scientific Fisheries*, **38**, 1229–1235. In Japanese; English abstract.

Yasuda, T. (1973a) Ecological studies on the jelly-fish, *Aurelia aurita* in Urazoko Bay, Fukui Prefecture IX Diel vertical migration of the medusa in late spring, 1970. *Bulletin of the Japanese Society of Scientific Fisheries*, **39**, 1145–1150. In Japanese; English abstract.

Yasuda, T. (1973b) Ecological studies on the jellyfish, *Aurelia aurita* (Linné), in Urazoko Bay, Fukui Prefecture VIII Diel vertical migration of the medusa in early fall, 1969. *Publications of the Seto Marine Biological Laboratory*, **20**, 491–500. Proc. Second Int. Symp. Cnidaria.

Yasuda, T. (1974) Ecological studies on the jelly-fish, *Aurelia aurita* in Urazoko Bay, Fukui Prefecture X Diel vertical migration of the medusa in summer, 1970. *Bulletin of the Japanese Society of Scientific Fisheries*, **40**, 971–975. In Japanese; English abstract.

Yasuda, T. (1975a) Ecological studies on the jelly-fish *Aurelia aurita* in Urazoko Bay, Fukui Prefecture XII Diel vertical migration of the medusa in summer, 1972 and a note on the relation between bell-diameter and occurrence depth. *Aquiculture (Kagoshima)*, **22**, 128–134. In Japanese; English abstract.

Yasuda, T. (1975b) Ecological studies on the jelly-fish, *Aurelia aurita* (Linné), in Urazoko Bay, Fukui Prefecture XI An observation on ephyra formation. *Publications of the Seto Marine Biological Laboratory*, **22**, 75–80.

Yasuda, T. (1979) Studies on reproductive biology of harmful marine animals – The common jelly-fish, *Aurelia aurita*, along coast of Wakasa Bay, Japan Sea, in *Proceedings 7th Japan–Soviet Joint Symposium Aquaculture, September 1978*, (ed. G. Yamamoto), Tokai Univerisity Press, Tokyo, pp. 185–195.

Yasuda, T. (1982) Biology of the common jelly-fish, *Aurelia aurita* (II) – Horizontal and vertical distribution pattern of the medusa. *Aquabiology (Tokyo)*, **4**, 462–470. In Japanese; English abstract.

Yasuda, T. (1983) Biology of the common jellyfish, *Aurelia aurita* III Notes of the metamorphosis from ephyra to young medusa and survival rate. *Aquabiology (Tokyo)*, **5**, 302–306. In Japanese; English abstract.

Yasuda, T. and Suzuki, Y. (1992) Notes on an edible medusa, *Rhopilema asamushi* Uchida, caught in Wakasa Bay, Japan. *Bulletin of Plankton Society of Japan*, **38**, 147–148.

Yoshida, M. and Yoshino, Y. (1980) Differentiation of ocelli in ephyrae of *Aurelia aurita*, in *Developmental and Cellular Biology of Coelenterates*, (eds P. Tardent and R. Tardent), Elsevier/North-Holland Biomedical Press, Amsterdam, pp. 343–346.

Zahl, P.A. and McLaughlin, J.J.A. (1959) Studies on marine biology IV On the role of algal cells in the tissues of marine invertebrates. *Journal of Protozoology*, **6**, 344–352.

Zahn, M. (1981) Wie alt können Scyphomedusen werden? *Zoologische Beitraege*, **27**, 491–495. In German; English abstract.

Zaika, V.E. (1972) The growth of ctenophores and medusae. *Zoologicheskii Zhurnal*, **51**, 179–188. In Russian; English abstract.

Zaitsev, Y.P. and Polishchuk, L.N. (1984) An increase in the number of *Aurelia aurita* (L.) in the Black Sea. *Ekologiya Morya (Kiev)*, **17**, 35–46. In Russian; English abstract.

Zakutskiy, V.P., Kuropatkin, A.P. and Gargopa, Y.M. (1988) Long-range estimation of abundance and biomass of medusae in the Azov Sea based on the degree of salinization. *Hydrobiological Journal (English Translation of Gidrobiologicheskii Zhurnal)*, **24**, 17–21.

Zavodnik, D. (1987) Spatial aggregations of the swarming jellyfish *Pelagia noctiluca* (Scyphozoa). *Marine Biology (Berlin)*, **94**, 265–269.

Zavodnik, D. (1991) On the food and feeding in the northern Adriatic of *Pelagia noctiluca* (Scyphozoa), in *UNEP: Jellyfish Blooms in the Mediterranean. Proceedings of the II Workshop on Jellyfish in the Mediterranean Sea*, UNEP, Athens, pp. 212–216.

Zubkoff, P.L. and Linn, A.L. (1975) Isozymes of *Aurelia aurita* scyphistomae obtained from different geographical locations, in *Isozymes IV Genetics and Evolution*, (ed. C.L. Markert), Academic Press, New York, pp. 915–930.

Index

Page numbers appearing in **bold** refer to definitions and page numbers appearing in *italics* refer to illustrations.

www.ingramcontent.com/pod-product-compliance
Ingram Content Group UK Ltd.
Pitfield, Milton Keynes, MK11 3LW, UK
UKHW020135160926
475607UK00003B/51

* 9 7 8 0 4 1 2 4 5 1 1 0 2 *